ENERGY PROCESSING AND SMART GRID

IEEE Press
445 Hoes Lane
Piscataway, NJ 08854

ENERGY PROCESSING AND SMART GRID

JAMES A. MOMOH

IEEE PRESS SERIES ON POWER ENGINEERING

IEEE PRESS

WILEY

Published by John Wiley & Sons, Inc., Hoboken, New Jersey.
Published simultaneously in Canada.

Library of Congress Cataloging-in-Publication Data is available.

ISBN: 978-1-119-37614-9

Cover Design: Wiley
Cover Images: (Top image) © Sam Robinson/ Gerryimages; (Center image) © Chombosan/iStockphoto

Set in 10/12pt TimesLTStd by Aptara Inc., New Delhi, India

Printed in the United States of America.

10 9 8 7 6 5 4 3 2 1

CONTENTS

PREFACE

THE TRADITIONAL electrical engineering curriculum requires a basic course in electrical machinery, which includes three-phase networks, electromagnetics, transformers, synchronous generators, induction-machine and DC-machine modeling, construction, and performances analysis.

The knowledge of these concepts does not fully prepare students to understand, analyze, and participate in the design, development, and deployment of future grids with features of micro-smart-grid functions. The skills needed include knowledge of classical machine control, real-time measurements, renewable energy resources, storage technology, inverters and converters, and power electronics for processing different energy sources to serve different load models. This book introduces energy-processing concepts and topics needed by students in electrical engineering—and those outside of electrical engineering who will work in future grid development.

The intended audience includes undergraduates and first-year graduate students in a typical engineering program. It is assumed that these students have taken network analysis and electromagnetics courses and hence these topics are only summarized. Full treatment can be found in textbooks on standard machines and electromagnetics.

The idea of this book stems from a research grant from the Department of Energy/National Science Foundation future-grid initiative to the Power System Engineering Research Center/Industry University Collaboration Research Center Research and Workforce Award. As part of this grant, the Howard University team designed research and workforce education materials to contribute to the development of a new curriculum for teaching future energy conversion that will prepare students for handling research and education modules involving future-grid development.

The workforce-education module discussed in this text is a timely assignment since the electrical engineering program at Howard University—like at other universities—is reviewing its curriculum to accommodate new topics in future grid, since there appears to be no single book to present the balance of classical-machine and new trends in energy conversion and processing. This book serves as a reference for students as well as consultants who wish to have a quick overview of machines and energy processing and their relevance to microgrid with smart-grid functions.

Basic materials have been explained with illustrative examples and pictures. In the last few years, the materials in this book have been used to teach required undergraduate courses at Howard University, and some materials were used for teaching first-year graduate students who have a minimum background in energy processing for smart grid. We recommend that any materials too basic for the reader be skipped.

Throughout the book, we tried to minimize details on classical-machine concepts to allow for understanding and working knowledge of processing energy sources in the microgrid/smart-grid environment. We urge the reader to consult advanced books for further reading on these topics.

As noted, we have added some modern topics such as renewable energy, storage technologies, inverters and converters, power electronics, metering, and control for microgrid systems. From our experience thus far, this book will help kindle students' interest in old-machine analysis and also their pursuit of new topics in design and development of smart-grid and microgrid systems.

The book is organized into nine major sections:

(i) Basic concepts of network analysis applied to power and electromagnetic concepts as relevant to design and understanding of electromechanical systems, which form the backbone of energy processing

(ii) Fundamentals of machinery functions, construction, modeling, and performance for transformers, synchronous machines, induction machines, and DC machines

(iii) Renewable-energy and storage-technology options for sustainability of energy needs in future grids

(iv) Design of inverter and converter and power electronics for energy processing from different sources to serve different loads

(v) Microgrid applications design, technology, and operational management

(vi) Review and evaluation of metering communication and control for enabling different functions for operating and managing smart-grid and microgrid systems

(vii) Smart-grid design, architecture, security protocols, and real-time measurements

(viii) Different test beds and their features with useful energy-processing technology

(ix) Exercises for design, testing, and evaluation of energy processing as experimental case studies; this involves a 'power game' within an energy-processing platform with OPAL RT in the loop

'KNOWLEDGMENTS

WOULD LIKE to acknowledge the vision of the Power System Engineer-earch Center (PSERC)/Industry University Collaboration Research Center :), and Department of Energy (DOE) to encourage deployment of workforce- materials along with research in support of design, development, and nent of advanced tools needed for the future grid. The future grid—and its bility, security, efficiency, and affordability—is fundamental, and thus energy processing. Furthermore, electromagnetic machines, power electron-age, inverters/converters, tools for the design of the microgrid and smart grid ntial. Knowledge of communications, controls, and standards is essential for ; the structure of the future grid. Therefore, my hope is that this book will te to research materials that prepare undergraduate and graduate students to for research and careers in the evolving power grid and smart grid industries. he new course in energy processing and smart grid at Howard University neet the challenge of electrical engineering curriculum revision at the under- level to reduce credit hours for graduation from 130 to 120 as in other engi- schools. It has been taught as an undergraduate junior-level course and as a r graduate special topic course.

ecial thanks are due to several students who helped in the preparation, -solving, and exercises in the book. We also acknowledge the pioneering energy processing, machines, and smart grid and microgrid by several

)REWORD

INCREASINGLY, ELECTRICITY is being produced from renewable :es—primarily solar and wind—in an effort to mitigate the consequences of warming resulting from high concentrations of atmospheric carbon dioxide :ed by burning fossil fuels. This new direction in energy production and envi- ntal protection requires an extensive re-engineering of the existing electric grid computerized future grid—or a smart grid—that accommodates the special teristics of renewable power-generating technologies. To advance grid mod- :ion, the Power Systems Engineering Research Center (PSERC) has been con- g research with its industry partners on the technical challenges facing the : power industry.

Keeping up with the education side of its mission, PSERC's faculty has devel- oftware tools, courses, a virtual library, and training materials to ensure the f a workforce well versed in implementing the future grid. This book, *Energy sing and Smart Grid*, developed as a part of this mission, is divided into three sections: energy conversion including renewables, power electronics interfaces, rements, and controls of the microgrid and smart grid, and test beds for the grid.

The author, James A. Momoh, has extensive experience in power system mod- control, planning, and operations which make him qualified to write a book sing these topics. This book is a valuable addition to IEEE John Wiley & Sons which addresses the development of educational tools to meet the needs of rent and future engineers that will be managing these complex cyber-physical is as well as innovators who will bring future transformations.

The book provides a fundamental understanding of energy processing and tech- es needed for building the smart grid. It covers major topics on the fundamen- smart grid, energy conversion and power electronics, cyber security, real-time rement, and state estimation techniques. The book is easy to use because it a language that is understandable to both novices and experts. It will serve as reference as well as self-study guide for graduate students, research teams, stem operators.

Vijay Vittal
Ira A. Fulton Chair Professor
Director PSERC
School of Electrical, Computer and Energy Engineering
Arizona State University

CHAPTER 1

INTRODUCTION

1.1 INTRODUCTION

The process of electric energy generation, transmission, and distribution is conducted via a large-scale central generation (CG) to meet increasing demand at industrial, commercial, and residential loads. The generation of power through central generation is done via coal-fired, fossil-fuel, hydro, and steam turbines. These resources are converted to electrical energy then processed through different technologies to meet growing load demands. Similarly, the transmission of power from generating units requires high investment, including numerous energy-processing technologies to transfer power to load centers. This is done via transformers and solid-state switching devices at very high voltages. Distribution-system networks are also significant for dispatching power at low voltage via step-down transformers to domestic consumers. Given the different direct current (DC) and alternating current (AC) load models and categories, energy-processing devices such as converters and inverters are needed especially when issues of safety, quality of power, security, and processing are important to meet different loads from various sources.

In recent years, many new distributed generation (DG) sources have been introduced with the aim of reducing losses and increasing reliability, regardless of economy of scale of CG, which can be huge when accounting for cost of transmission and reliability during failure.

DG created from solar, mini-hydro, wind, stirling engines, and fuel cells is of common interest. The debate on how to determine to what extent one should drive CG or DG is a research topic under the future grid initiative. In this paper [1], we were able to show three criteria for selecting which energy sources and processing options should be selected. The criteria included economy of scale, resilience, sustainability, and reliability. Regardless of the choice, appropriate energy processing tools and concepts are a dominant concern moving forward. It is therefore important for a book to be dedicated to addressing appropriate fundamental concepts and technologies for designing, building, and validating the performance of the future grid. That is the goal of this book.

In this book, we propose to provide an integrated foundational knowledge that harnesses the role of energy conversion machines and devices with some new topics

Energy Processing and Smart Grid, First Edition. James A. Momoh

for energy generation, transmission, distribution, delivery, and consumption. First, it is relevant to provide the working definition of *microgrid*.

The microgrid involves a stand-alone or grid-connected system to:

- match generation with load
- serve as a reserve margin
- help stabilize the power system
- increase reliability and affordability
- reduce the impact of threats on the bulk power system

In contrast, the smart grid is a two-way digital system with active participation of customers within the grid that includes renewable energy resources (RER) with the aim of sustainability and allows reasonable penetration levels and interoperability [2]. It possesses self-healing capabilities; cyber security; and real-time, design-based functions such as reconfiguration, demand-side management (DSM), demand response, and power quality.

We present eight points in this book to support control communications and energy processing for the smart grid. They are the basis of the seventeen chapters, categorized as follows:

1. A basic review of network (circuit) analysis and electromagnetics is provided. This includes discussion on the fundamental concepts of three-phase analysis of AC sources to different load configurations. The text assumes the sources are AC that can be converted to DC or AC/DC or DC/AC using inverters and converters, as discussed in Chapter 12. A set of hand-calculation exercises is presented although software tools such as NEPLAN [3], Electrical Transient and Analysis Program (ETAP) [4], and Personal Computer Simulation Program with Integrated Circuit Emphasis (PSPICE) [5] can also be used. Balanced or unbalanced conditions can also be analyzed using symmetrical components.
2. Electromagnetic concepts are introduced by providing the unified theory of Maxwell's equations and their applications for understanding machinery concepts, including transformers, synchronous and induction machines, and DC machines. The unified theories of Faraday, Lenz, and Ampère are given to illustrate the concept of electromagnetic computation. Equivalent analogous forms of magnetic circuits in electric circuits are given for magnetic circuits with rectangular and toroidal shapes with and without air gaps for different ferromagnetic materials.
3. Fundamental understanding of machines is discussed. This includes the conversion process and the role of Maxwell's equations. The construction and model of the machines using electrical network equivalents are given using short-circuit and open-circuit analyses for determining the equivalent parameters [6]. The power flow in each machine—accounting for conversion, electrical, mechanical, and stray losses—is given. Following this, the text provides a guide for computing efficiency of the power input relative to the output power

received. In addition, voltage regulation and control strategies to achieve optimum energy processing are discussed.

4. Fundamental knowledge of storage and renewable resources is important for development of future grids where sustainability and mobile power are needed. Storage is safe, inexpensive, and the reason for interest in the design of future environmentally friendly grids. The text provides a working model, description, size, and metrics of different resources and storage technologies. Software packages are recommended for studying the impact of RER for reliability and cost–benefit analysis for stand-alone distribution system topologies for the future grid.
5. Efficient technology to handle processing energy from one state or form to another includes inverters and converters, which provide AC/DC, DC/DC, or DC/AC conversion from a given resource to given loads. To minimize poor power quality, different electronic devices and filters are used. The text provides a fundamental knowledge of power electronics to allow the reader appropriate choices of electronic devices for energy processing.
6. In designing the operation and management of the smart grid and microgrids, real-time processing of energy and information is essential. We present an overview of real-time data such as voltage, current, power, and frequency, which measure the status of the grid via smart meters and phasor measurement units (PMUs). The formulations and specifications of the devices and their use in communication and control schemes are given. Work in areas of real-time voltage-stability management, power quality, frequency control, voltage/volt-ampere reactive (var), reconfiguration, and several other grid functions are included in the proposed exercises.
7. We provide a description for understanding the smart grid and microgrid, including functionality, architecture, and the test bed. The chapters integrating the concept of microgrids in this text are an evolving process. Research work and laboratory exercises are important activities for general electrical installations.
8. Finally, the design of microgrid systems with smart grid functions is described.

Our research involving the design and construction of a microgrid test bed consisting of solar power, super capacitors, batteries, metering devices, converters and inverters, and a real-time simulator is ongoing at Howard University. The introduction of the OPAL-RT real-time digital simulator (RTDS) to the array of research and educational equipment at the Center for Energy Systems and Control (CESaC) at Howard University is allowing us to develop functions such as power quality, voltage stability, voltage/var, DSM, and restoration.

BIBLIOGRAPHY

[1] J. A. Momoh, S. Meliopoulos, and R. Saint, *Centralized and Distributed Generated Power System – A Comparative Approach*, PSERC publication, June 2012.
[2] J. Momoh, *Smart Grid – Fundamentals of Design and Analysis*, Wiley-IEEE Press, 2012.
[3] https://www.neplan.ch
[4] https://etap.com
[5] www.pspice.com
[6] Mohamed E. El-Hawary, *Electrical Power Systems – Design and Analysis*, IEEE Press, 1983.

CHAPTER 2

ELECTRIC NETWORK ANALYSIS IN ENERGY PROCESSING AND SMART GRID

2.1 INTRODUCTION

In modern AC electric power systems (Figure 2.1), power is generated, transmitted, and distributed as balanced, three-phase AC. The three-phase system was independently invented by Galileo Ferraris, Mikhail Dolivo-Dobrovolsky, Jonas Wenström, and Nikola Tesla in the 1880s and is the most widely used means of transferring power through power grids. The three-phase system has the advantage of economy over the single-phase system because more power can be transmitted with significant cost savings in conductors per unit line length. The three-phase system may be configured as three-wire star, four-wire star, or three-wire delta system.

Because of advances in electronics, the future electric power system is headed in the direction of microgrids and smart grids. In anticipation of demands, researchers and students need to be equipped with relevant knowledge on the emerging trends in this area. This chapter introduces the fundamentals of electric power systems and the basic computational tools needed for the design and analysis of the future-generation power system.

2.2 COMPLEX POWER CONCEPTS

In electrical power systems, we are mainly concerned with flow in the electrical circuit, such as Voltage (V), Frequency (f), Current (I), and Power (P). To treat sinusoidal, steady-state behavior of an electric current, some further definitions are necessary.

Let

$$v(t) = V_m Cos(\omega t + \theta) \tag{2.1}$$

and

$$i(t) = I_m Cos\ (\omega t + \theta - \varphi) \tag{2.2}$$

Energy Processing and Smart Grid, First Edition. James A. Momoh

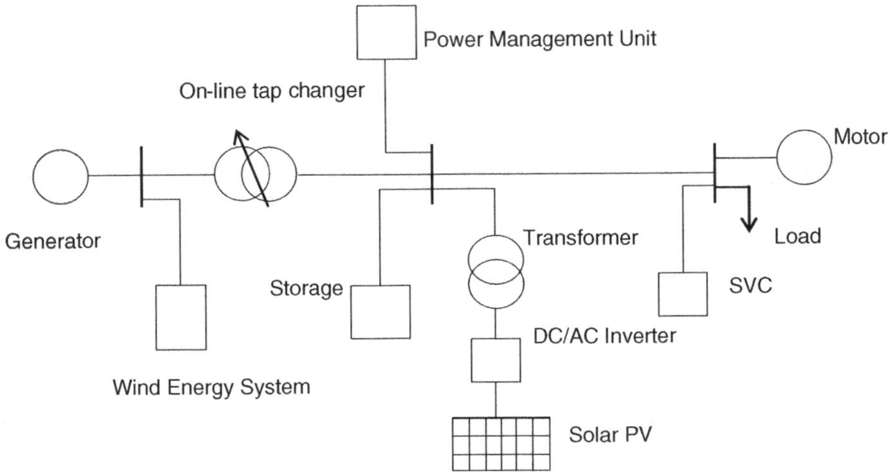

Figure 2.1 Simplified single-line diagram schematics of a modern electric power system.

and

$$Z = Z\angle\psi \tag{2.3}$$

where

$$I_m = \frac{V_m}{Z}. \tag{2.4}$$

In polar form,

$$V = V_m e^{j\varphi} = V_m \angle\varphi \tag{2.5}$$

or

$$V(t) = Re\left(Ve^{j\omega t}\right). \tag{2.6}$$

It is important to recall the following trigonometric identities:

$$Cos\alpha Cos\beta = \frac{1}{2}\left[Cos(\alpha-\beta) + Cos(\alpha+\beta)\right] \tag{2.7}$$

$$Sin(\omega t + \alpha) = Cos\left(\omega t - \frac{\pi}{2}\right). \tag{2.8}$$

For a voltage signal represented in terms of root mean square (rms) value:

$$V(t) = 100\sqrt{2}Cos\left(\omega t + 30°\right)\ Volts \tag{2.9}$$

$$V = 100\angle 30°$$

$$i(t) = 10\sqrt{2}Sin\left(\omega t + 30°\right) A \tag{2.10}$$

$$i(t) = 10\sqrt{2}Cos\left(\omega t + 30° - \frac{\pi}{2}\right) A \tag{2.11}$$

$$i(t) = 10\sqrt{2}Cos\left(\omega t - 60°\right) A. \tag{2.12}$$

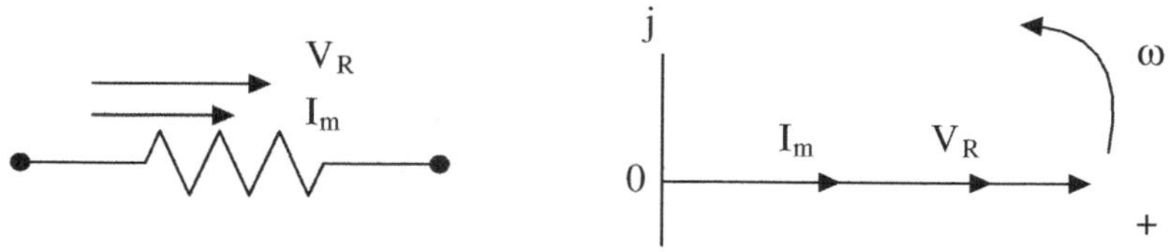

Figure 2.2 Phasor diagram of a purely resistive circuit.

For impedance, if the circuit is purely resistive, inductive, or capacitive, there will be difference in current and angle phase shift.

2.2.1 Purely Resistive Circuit

In a purely resistive circuit, Figure 2.2, the current is in phase with voltage:

$$V_R = RI_m. \tag{2.13}$$

2.2.2 Purely Inductive Circuit

Current lags behind V by 90° as seen in Figure 2.3:

$$V_L = j\omega LI_m. \tag{2.14}$$

2.2.3 Purely Capacitive Circuit

In a purely capacitive circuit Figure 2.4, current leads voltage by 90°:

$$V_c = \frac{I_m}{j\omega C} \tag{2.15}$$

2.2.4 Instantaneous Power

The instantaneous power is given by:

$$P(t) = \frac{V_m I_m}{2}\left[Cos\varphi + Cos\left(2\omega t - \varphi\right)\right] \tag{2.16}$$

$$P_m = \frac{V_m I_m}{2} Cos\varphi \tag{2.17}$$

$$V_m = \sqrt{2}V_{Rms} \tag{2.18}$$

$$P_{av} = V_{Rms}V_{Rms}Cos\varphi \tag{2.19}$$

where $Cos\varphi$ is the power factor.

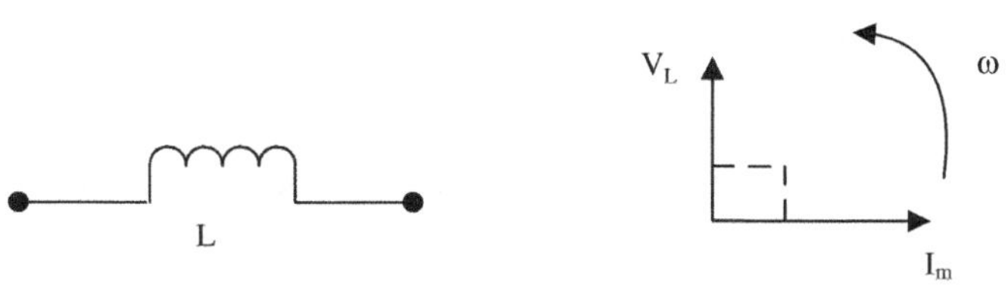

Figure 2.3 Phasor diagram of purely inductive circuit.

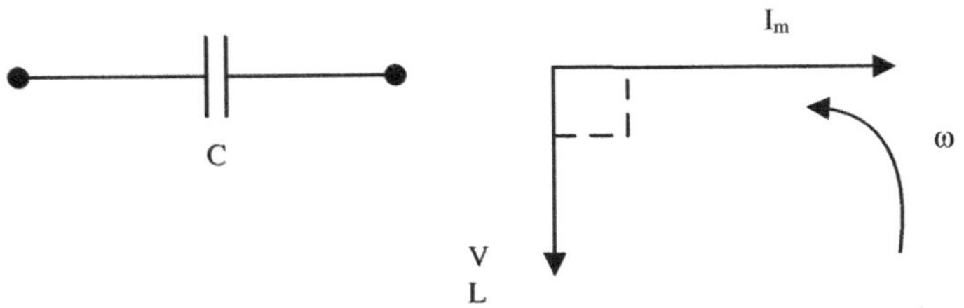

Figure 2.4 Phasor diagram of a purely capacitive circuit.

2.2.5 Power Factor

The power factor of a system is defined as:

$$Cos\varphi = \frac{Active\ Power}{Apparent\ Power} = \frac{P\ (KW)}{S\ (KVA)} \tag{2.20}$$

where $P = VICos\varphi$ and $Q = VISin\varphi$.

Q is the reactive power measured in kilo-var (kvar).

2.2.6 Complex Power

The apparent power S is then:

$$S = VI \tag{2.21}$$

$$I = I\angle\varphi, \tag{2.22}$$

and

$$I^* = I\angle - \varphi \tag{2.23}$$

$$S = VI^*. \tag{2.24}$$

If the load impedance is Z, then,

$$S = ZII^* = Z|I|^2. \tag{2.25}$$

In terms of load admittance,

$$S = VY^*I^* \tag{2.26}$$

$$= Y^*|V|^2. \tag{2.27}$$

2.3 REVIEW OF AC-CIRCUIT ANALYSIS USING PHASOR DIAGRAMS

Consider the AC circuit of Figure 2.5. The load, operating at a voltage V_L, draws a current I_L from the source whose voltage is V_s, through resistance R and inductive reactance jX.

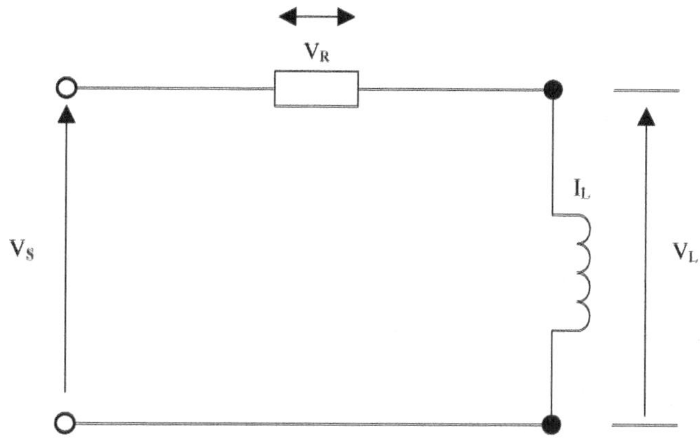

Figure 2.5 AC circuit analysis with phasor diagram.

By Kirchhoff's voltage rule:

$$V_s = V_L + V_R \tag{2.28}$$

$$V_s = V_L + IR \tag{2.29}$$

where

$$Z = (R + jX) \tag{2.30}$$

$$Z = |Z| \angle\theta. \tag{2.31}$$

If the load is operated at power factor $Cos\varphi$ and voltage V_L then

$$V_L = |V_L| \angle 0^\circ \tag{2.32}$$

$$I_L = |I| \angle\varphi. \tag{2.33}$$

Then Equation 2.29 may be rewritten as:

$$V_s = |V_L| \angle 0^\circ + (|I| \angle\varphi)(|Z| \angle\theta) \tag{2.34}$$

$$V_s = |V_L| \angle 0^\circ + |I|\,|Z| \angle\varphi + \theta. \tag{2.35}$$

A phasor diagram Figure 2.6 may be constructed based on Equation 2.35 from which the source voltage V_s may be determined.

2.4 POLYPHASE SYSTEMS

One of the methods of transmitting and distributing AC electric power is by means of polyphase systems. This is a system with three or more energized AC currents

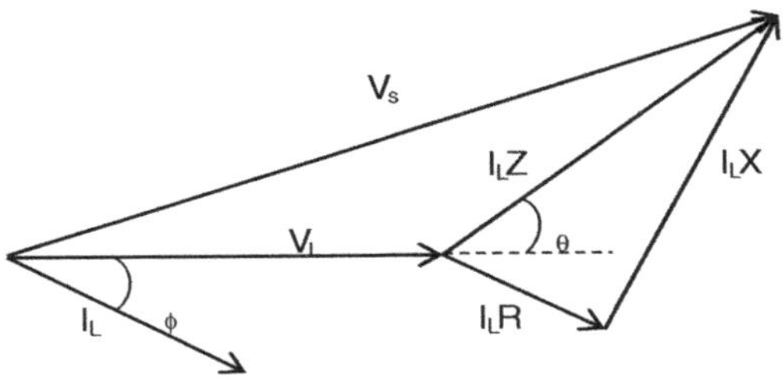

Figure 2.6 Phasor relationships of power system quantities.

carrying conductors with a phase deviation between them. For a balanced n-phase system, the phase difference is given by:

$$\emptyset_d = \frac{2\pi}{n} \ degrees. \tag{2.36}$$

For two-, three- ... six-phase systems, voltages will be out of phase by 90°, 120°, ... 60°, respectively.

Polyphase systems are particularly useful for transmitting power as more power can be transmitted than when a single phase is used.

2.4.1 Three-Phase Circuits

Power generation, transmission, and distribution are usually connected in a type of polyphase system for heavy utilization of AC electric power. These types of connections provide economic advantages as well as system stability and capacity control. Both voltage and currents are sinusoidal waveforms equal in magnitude, but are displaced from one another by 120° in time phase. Stator windings are connected in three-phase through a ground wire, leading to four wires. The center of the four wires (Figure 2.7) leads to a Y-connected system, where each is referred to as a phase and the fourth conductor is called the *neutral wire*, which has four-wire balanced connection.

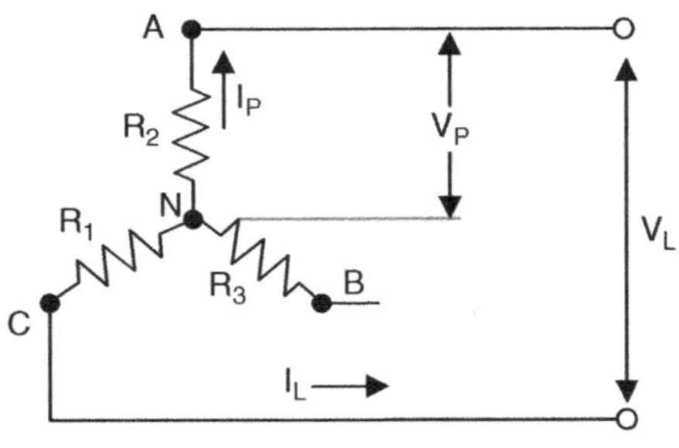

Figure 2.7 Equivalent Y diagram.

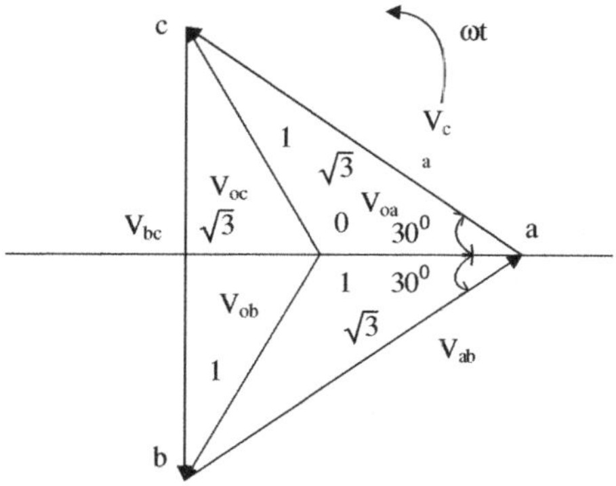

Figure 2.8 Phase and line voltage representation.

2.4.2 Balanced Y-Connected Three-Phase Source

In vector (phasor) form

$$\vec{V}_{AN} = \frac{V_L}{\sqrt{3}}\angle 90^\circ \tag{2.37}$$

$$\vec{V}_{BN} = \frac{V_L}{\sqrt{3}}\angle - 30^\circ \tag{2.38}$$

$$\vec{V}_{CN} = \frac{V_L}{\sqrt{3}}\angle - 150^\circ \tag{2.39}$$

$$\vec{V}_{AB} = \vec{V}_{AN} - \vec{V}_{BN} \tag{2.40}$$

$$\vec{V}_{AN} + \vec{V}_{BN} + \vec{V}_{CN} = 0 \tag{2.41}$$

$$\vec{V}_{AB} + \vec{V}_{BC} + \vec{V}_{CA} = 0. \tag{2.42}$$

Similarly, for Δ-connected system

$$\vec{V}_{BA} = \vec{V}_{AC} = \vec{V}_{CB} = \vec{V}_L. \tag{2.43}$$

2.4.3 Phase and Line Voltages: Delta Connected

The line-neutral and line–line voltages with proper relations are shown in Figure 2.8.

$$\vec{V}_{CB} = \sqrt{3}\vec{V}_{NB}. \tag{2.44}$$

$\sqrt{3} \times$ Line-Neutral, V_{BA} leads V_{NA} by 30°, V_{CB} leads V_{NB} by 30°.

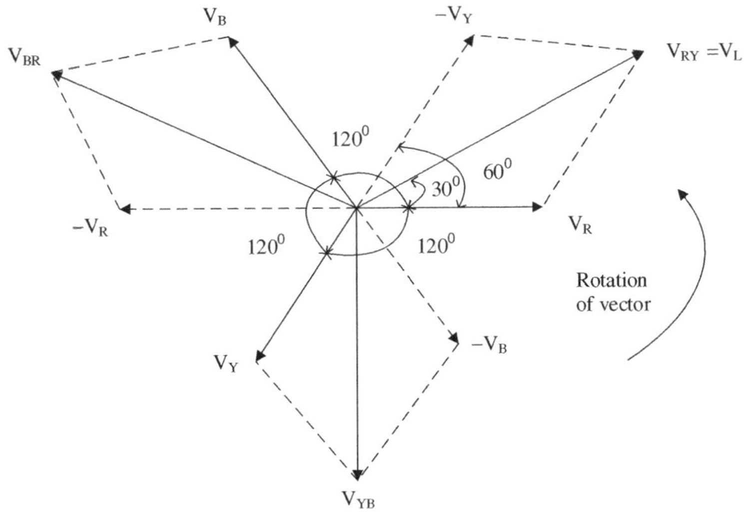

Figure 2.9 Equivalent Y-connected voltage phasor representation.

2.4.4 Equivalent Y-Connected Voltage Phasor Representation

Taking V_{NA} as a reference from Figure 2.9,

$$V_{RY} = \sqrt{3}V_{ph}\angle 30^\circ = V_L\angle 30^\circ \tag{2.45}$$

$$V_{YB} = \sqrt{3}V_{ph}\angle -90^\circ = V_L\angle -90^\circ \tag{2.46}$$

$$V_{BR} = \sqrt{3}V_{ph}\angle 150^\circ = V_L\angle 150^\circ. \tag{2.47}$$

2.4.5 Mesh or Delta Connection

$$V_{AB} = V_{AN} = V_{ph} \tag{2.48}$$

$$V_{BC} = V_{BN} = V_{ph} \tag{2.49}$$

$$V_{CA} = V_{CN} = V_{ph} \tag{2.50}$$

These connections are not properly balanced to maintain balanced voltage across each load. A delta-connected generator Figure 2.10 is possible but not desirable for two main reasons:

(a) Grounding is not possible with a delta-connected generator. For safety, a generator-neutral point typical of Y is the logical point of connection to ground.

(b) A delta connection of the coils of the generator provides a short-circuited path in which current can flow. Third harmonics in the coil voltages cause a disturbance, which produces power loss and lowers the efficiency of the generator.

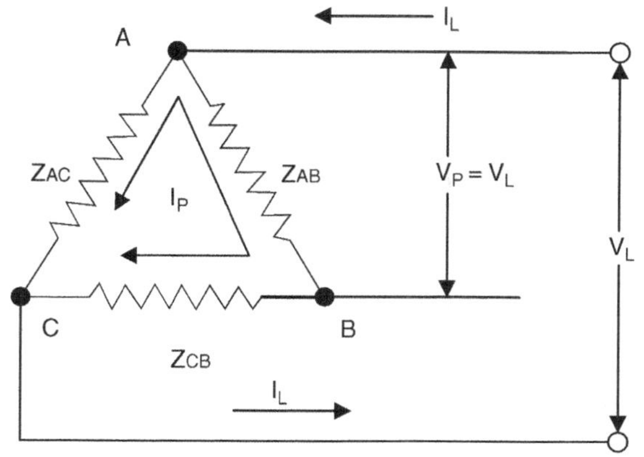

Figure 2.10 Mesh or delta connection.

2.5 THREE-PHASE IMPEDENCE LOADS

A delta-connected load or Y-connected load (Figure 2.11) uses the same configuration as discussed in Section 2.8.

$$V_{CB} = V\angle 0^\circ \tag{2.51}$$

$$V_{AC} = V\angle - 120^\circ \tag{2.52}$$

$$V_{AB} = V\angle - 240^\circ \tag{2.53}$$

From which the individual phase currents may be computed as

$$I_{CB} = \frac{V_{CB}}{Z_{CB}} \tag{2.54}$$

$$I_{AC} = \frac{V_{AC}}{Z_{AC}} \tag{2.55}$$

$$I_{BA} = \frac{V_{BA}}{Z_{BA}}. \tag{2.56}$$

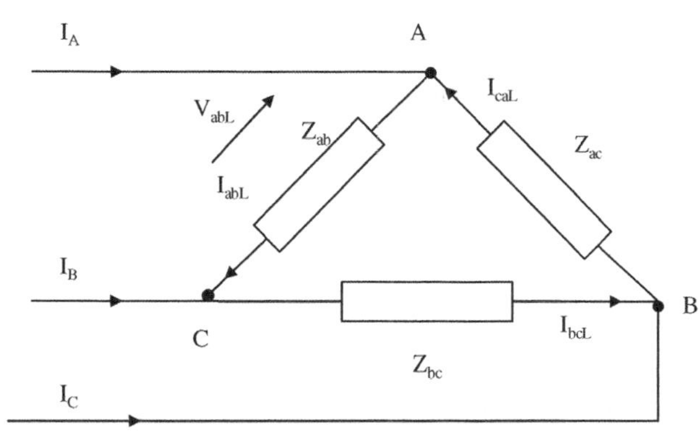

Figure 2.11 Delta-connected three-phase loads.

The line currents are

$$I_A = I_{AC} - I_{BA} \tag{2.57}$$
$$I_B = I_{BA} - I_{CB} \tag{2.58}$$
$$I_C = I_{AC} - I_{BA}. \tag{2.59}$$

Also, under a balanced condition,

$$I_A + I_B + I_C = 0. \tag{2.60}$$

In general, the sum of the three line currents in a three-phase, three-wire system is zero. This implies that power is also balanced.
Total power

$$P = P_{CB} + P_{AC} + P_{BA}. \tag{2.61}$$

These are also given as real power for each phase, i.e.,

$$P_{CB} = V_{CB} I_{CB} Cos\left(\theta_{CB}^{V} - \theta_{CB}^{I}\right) \tag{2.62}$$
$$P_{AC} = V_{AC} I_{AC} Cos\left(\theta_{AC}^{V} - \theta_{AC}^{I}\right) \tag{2.63}$$
$$P_{BA} = V_{BA} I_{BA} Cos\left(\theta_{BA}^{V} - \theta_{AC}^{I}\right). \tag{2.64}$$

Accordingly, if

$$P_{AC} = P_{BA} = P_{CB} = V_{ph} I_{ph} Cos\theta, \text{ where } \theta = \left(\theta_{BA}^{V} - \theta_{AC}^{I}\right) \tag{2.65}$$

then $|P_p|$ are equal and out of phase 120° apart and the line currents are $\sqrt{3} I_{phase} = I_L$. Similarly, voltages are equal and out of phase by 120°.

Therefore, we write: $V_{CB} = V_{Ph} = V_L$ for delta connection

$$I_{CB} = I_{Ph} = \frac{I_L}{\sqrt{3}}.$$

Total power

$$P_{Total} = 3 V_{ph} I_{ph} \; Cos\left(\theta_v - \theta_i\right).$$

Therefore,

$$\begin{aligned} P_{Total} &= 3 V_L \frac{I_L}{\sqrt{3}} \; Cos\;\left(\theta_v - \theta_i\right) \\ &= \sqrt{3} V_L I_L \text{Cos}\theta. \end{aligned} \tag{2.66}$$

Cosθ is referred to as the power factor.

The same applies to a Y-connected system (Figure 2.12)

$$P_{Total} = \sqrt{3} V_L I_L Cos\left(\theta_v - \theta_i\right). \tag{2.67}$$

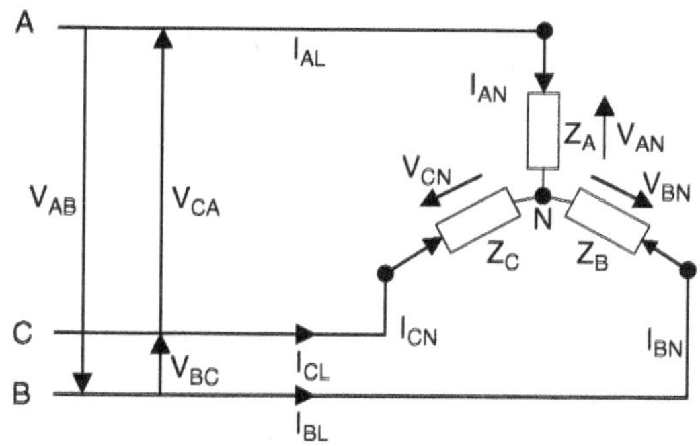

Figure 2.12 Y-connected load.

A Y-connected load can have four wires, as shown:

$$I_N = I_A + I_B + I_C \tag{2.68}$$

$$V_{CB} = V_L\angle 0°, V_{AC} = V_L\angle -120°, V_{AB} = V_L\angle -240° \tag{2.69}$$

$$V_{AN} = \frac{V_L}{\sqrt{3}}\angle -90°, V_{BN} = \frac{V_L}{\sqrt{3}}\angle 150°, V_{CN} = \frac{V_L}{\sqrt{3}}\angle 30° \tag{2.70}$$

The phase and line currents have the same value

$$I_{AN} = I_A = \frac{V_{AN}}{Z_A} \tag{2.71}$$

$$I_{BN} = I_B = \frac{V_{BN}}{Z_B} \tag{2.72}$$

$$I_{CN} = I_C = \frac{V_{CN}}{Z_C}. \tag{2.73}$$

This leads to:

$$P_{AN} = V_{AN}I_{AN}\,Cos\left(\theta_v - \theta_i\right)$$
$$P_{BN} = V_{BN}I_{BN}Cos\left(\theta_v - \theta_i\right)$$
$$P_{CN} = V_{CN}I_{CN}Cos\left(\theta_v - \theta_i\right).$$

So that total power

$$P = P_{AN} = P_{BN} = P_{CN} = 3P_{AN} \tag{2.74}$$
$$P = 3V_{AN}I_{AN}Cos\left(\theta_v - \theta_i\right)$$

$$V_{AN} = V_{ph} = \frac{V_L}{\sqrt{3}} \tag{2.75}$$

$$I_{AN} = I_A = I_L. \tag{2.76}$$

Total power

$$= 3V_{ph}I_{ph}\,Cos\left(\theta_v - \theta_i\right)$$
$$= \sqrt{3}V_LI_LCos\theta. \tag{2.77}$$

Example 1

Find the positive-sequence, line-to-line voltages in the balanced Y-connected generator if the line-to-neutral voltages have a magnitude of 12 kV and phase *a* is the reference.

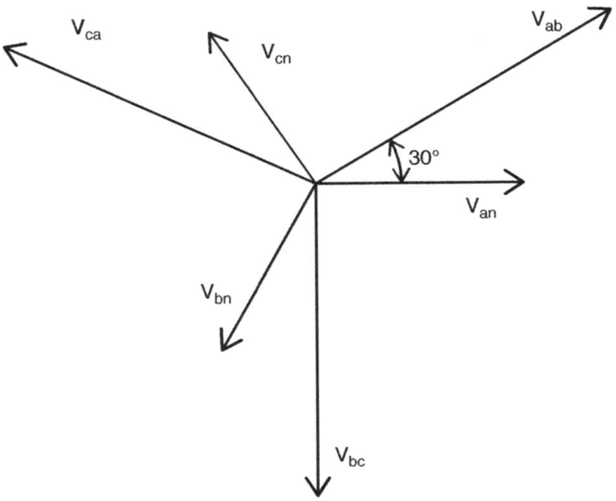

$$V_{L-L} = \sqrt{3}V_{L-N}\angle +30^\circ$$
$$20.8\angle +30^\circ \text{ KV}$$

Hence,

$$V_{ab} = 20.8\angle 30^\circ \text{ KV}$$
$$V_{bc} = 20.8\angle -90^\circ \text{ KV}$$
$$V_{ca} = 20.8\angle 150^\circ \text{ KV.}$$

Example 2

Given a Y-connected load of impedance $Z_Y = 10\angle 30^\circ\ \Omega$ and a balanced positive sequence source with $V_{ab} = 14.4\angle 30^\circ$ kV, find the load currents.

Solution:

$$I_Y = \frac{V_{L-N}}{Z_Y}$$

$$I_{AB} = \frac{V_{AB}}{Z_Y} = \frac{14.4\angle 0^\circ}{10\angle 30^\circ} = 1.44\angle -30^\circ$$

$$I_{BC} = \frac{V_{BC}}{Z_Y} = \frac{14.4\angle -120^\circ}{10\angle 30^\circ} = 1.44\angle -150^\circ$$

$$I_{CA} = \frac{V_{CA}}{Z_Y} = \frac{14.4\angle 120^\circ}{10\angle 30^\circ} = 1.44\angle 90^\circ$$

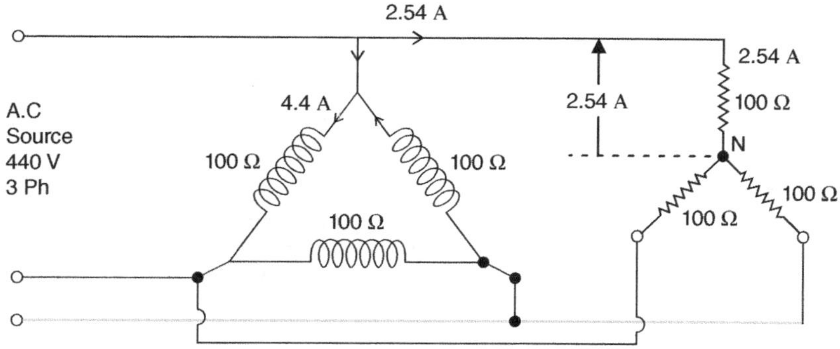

Example 3
Three noninductive resistances, each of 100 ohms, are connected in star to a three-phase 440 V supply. Three inductive coils, each of reactance 100 ohms connected in delta are also connected to the supply. Calculate the (a) line current and (b) power factor of the system.

Solution:
Voltage across each phase resistance $= V_\phi = \frac{V_L}{\sqrt{3}} = \frac{V_L}{\sqrt{3}} = 254V$

Current flowing through each resistor $= \frac{254}{100} = 2.54\, ohm$
For three inductive coils connected in delta, $V_\phi = V_L = 440V$

$$\text{phase current} = \frac{440}{100} = 4.40A$$

$$\text{line current} = \sqrt{3}I_\phi = \sqrt{3} \times 4.40 = 7.62A$$

Total line current drawn from the source = (2.53 − j7.62) = 8.032 A

$$\text{power factor angle } \phi = Tan^{-1}\left(\frac{7.62}{2.54}\right) = 71.57°$$

$$\text{hence, power factor} = Cos\phi = 0.3162.$$

2.6 TRANSFORMATION OF Y TO DELTA AND DELTA TO Y NETWORKS

In power-network computations, it is often required to convert a Y network to a delta (Δ) and vice versa. Consider the Y connection of impedances shown in Figure 2.13. Given Z_A, Z_B, *and* Z_C,

$$Z_{\Delta 31} = \frac{Z_A Z_B + Z_B Z_C + Z_C Z_A}{Z_B} \tag{2.78}$$

$$Z_{\Delta 12} = \frac{Z_B Z_C + Z_B Z_A + Z_A Z_C}{Z_C} \tag{2.79}$$

$$Z_{\Delta 23} = \frac{Z_A Z_B + Z_B Z_C + Z_C Z_A}{Z_A}. \tag{2.80}$$

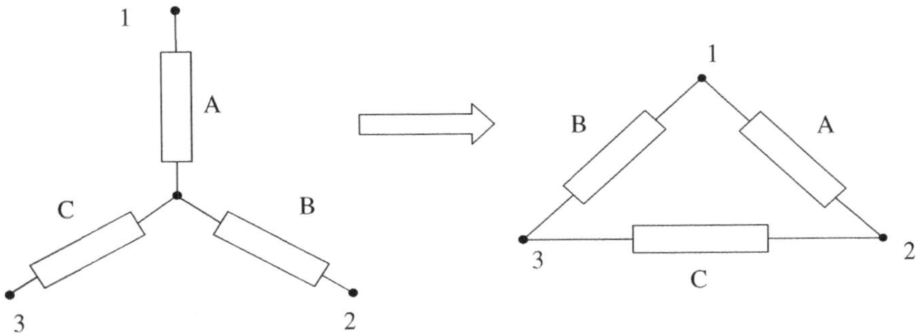

Figure 2.13 Star-delta conversion of impedances.

For balance, if

$$Z_A = Z_B = Z_C = Z_{YA}$$

then

$$Z_{\Delta A} = Z_{\Delta B} = Z_{\Delta C} = 3Z_{YA}. \tag{2.81}$$

For conversion of delta to Y, the diagram and process is as shown in Figure 2.14:

$$Z_{YA} = \frac{Z_{31}Z_{12}}{Z_{31} + Z_{12} + Z_{23}}$$
$$Z_{YB} = \frac{Z_{12}Z_{23}}{Z_{31} + Z_{12} + Z_{23}}$$
$$Z_{YC} = \frac{Z_{31}Z_{23}}{Z_{31} + Z_{12} + Z_{23}}$$

If

$$Z_A = Z_B = Z_C = Z_{\Delta \text{A}}$$

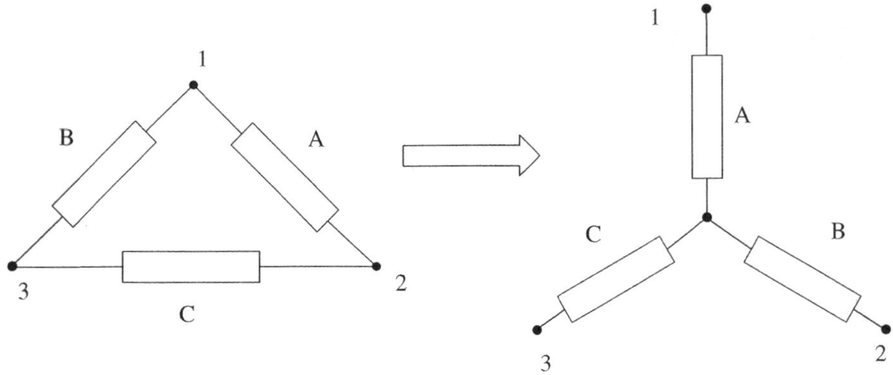

Figure 2.14 Delta-star impedance conversion.

then

$$Z_{YA} = Z_{YB} = Z_{YC} = \frac{Z_{\Delta A}}{3}.$$

2.7 SUMMARY OF PHASE AND LINE VOLTAGES/CURRENTS FOR BALANCED THREE-PHASE SYSTEMS

Table 2.1 shows a summary of phase and line voltages and currents for balanced three phase systems.

Example 1
A balanced, delta-connected load having an impedance 20 – j15Ω is connected to a delta-connected, positive-sequence generator having $V_{ab} = 330\angle 0°$V. Calculate the phase currents of the load and the line currents.

Solution:
The load impedance per phase is: $Z_\Delta = 20 - j15 = 25\angle -36.87°\Omega$.
Since $V_{AB} = V_{ab}$, the phase currents are:

$$I_{AB} = \frac{V_{AB}}{Z_\Delta} = \frac{330\angle 0°}{25\angle -36.87°} = 13.2\angle 36.87°$$

$$I_{BC} = I_{AB}\angle -120° = 13.2\angle -83.13°A$$

$$I_{CA} = I_{AB}\angle +120° = 13.2\angle 156.87°A$$

For delta load, the line current always lags the corresponding phase current by 30° and has a magnitude $\sqrt{3}$ times that of the phase current. Hence, the line currents are:

$$I_a = I_{AB}\sqrt{3}\angle -30° = (13.2\angle 36.87°)\left(\sqrt{3}\angle -30°\right)$$

$$= 22.86\angle 6.87°A$$

$$I_b = I_a\angle -120° = 22.86\angle -113.13°A$$

$$I_c = I_a\angle +120° = 22.86\angle 126.87°A.$$

Example 2
A balanced *abc sequence*, Y-connected source with $V_{an} = 100\angle 10°$ is connected to a Δ-connected, balanced load 8 + j4Ω per phase. Calculate the phase and line currents.

Option 1

$$Z_\Delta = 8 + j4 = 8.944\angle 26.57°\Omega$$

If the phase voltage $V_{an} = 100\angle 10°$, then the line voltage is

$$V_{ab} = V_{an}\sqrt{3}\angle 30° = 100\sqrt{3}\angle(10 + 30)° = V_{AB}$$

TABLE 2.1 Summary of Phase and Line Voltages/Currents for Balanced, Three-Phase Systems

Connection		Phase Voltages/Currents	Line	
Y–Y	Voltage	$V_{YAN} = V_p\angle 0°$ $V_{YBN} = V_p\angle -120°$ $V_{YCN} = V_p\angle 120°$	$V_{YAB} = \sqrt{3}V_p\angle 30°$ $V_{YBC} = V_{YAB}\angle -120° = \sqrt{3}V_p\angle -90°$ $V_{YCA} = V_{YAB}\angle 120° = \sqrt{3}V_p\angle 150°$	1.1
	Current	**Phase currents same as line currents**	$I_A = \frac{V_{YAN}}{Z_Y} = \frac{V_p<0°}{Z_Y} = I_A\angle \pm\theta$ where $+\theta$ is for leading pf and $-\theta$ is for lagging pf $I_B = I_A\angle -120° \pm\theta$ $I_C = I_A\angle 120° \pm\theta$	
Y − Δ	Voltage	$V_{YAN} = V_p\angle 0°$ $V_{YBN} = V_p\angle -120°$ $V_{YCN} = V_p\angle 120°$	$V_{YAB} = V_{\Delta 12} = \sqrt{3}V_p\angle 30°$ $V_{YBC} = V_{\Delta 23} = \sqrt{3}V_p\angle -90°$ $V_{YCA} = V_{\Delta 31} = \sqrt{3}V_p\angle 150°$	1.2
	Current	$I_{\Delta a} = \frac{V_{\Delta 31}}{Z_\Delta} = \frac{I_{\Delta 1}}{\sqrt{3}}$ $I_{\Delta b} = \frac{V_{\Delta 12}}{Z_\Delta} = \frac{I_{\Delta 2}}{\sqrt{3}}$ $I_{\Delta c} = \frac{V_{\Delta 23}}{Z_\Delta} = \frac{I_{\Delta 3}}{\sqrt{3}}$	$I_{\Delta 1} = I_A = I_A\angle \pm\theta$ $I_{\Delta 2} = I_B = I_A\angle -120° \pm\theta$ $I_{\Delta 3} = I_A\angle 120° \pm\theta$	1.3
Δ − Δ	Voltage	$V_{\Delta 12} = \sqrt{3}V_p\angle 30°$ $V_{\Delta 23} = \sqrt{3}V_p\angle -90°$ $V_{\Delta 31} = \sqrt{3}V_p\angle 150°$	**Same as phase voltage**	
	Current	$I_{\Delta a} = \frac{V_{\Delta 31}}{Z_\Delta} = \frac{I_{\Delta 1}}{\sqrt{3}}$ $I_{\Delta b} = \frac{V_{\Delta 12}}{Z_\Delta} = \frac{I_{\Delta 2}}{\sqrt{3}}$ $I_{\Delta c} = \frac{V_{\Delta 23}}{Z_\Delta} = \frac{I_{\Delta 3}}{\sqrt{3}}$	$I_{\Delta 1} = I_A\angle \pm\theta$ $I_{\Delta 2} = I_A\angle -120° \pm\theta$ $I_{\Delta 3} = I_A\angle 120° \pm\theta$	1.4
Δ − Y	Voltage	$V_{YAN} = V_p\angle 0°$ $V_{YBN} = V_p\angle -120°$ $V_{YCN} = V_p\angle 120°$	$V_{YAB} = V_{\Delta 12} = \sqrt{3}V_p\angle 30°$ $V_{YBC} = V_{\Delta 23} = \sqrt{3}V_p\angle -90°$ $V_{YCA} = V_{\Delta 31} = \sqrt{3}V_p\angle 150°$	1.5
	Current	$I_{\Delta a} = \frac{V_{\Delta 31}}{Z_\Delta} = \frac{I_{\Delta 1}}{\sqrt{3}}$ $I_{\Delta b} = \frac{V_{\Delta 12}}{Z_\Delta} = \frac{I_{\Delta 2}}{\sqrt{3}}$ $I_{\Delta c} = \frac{V_{\Delta 23}}{Z_\Delta} = \frac{I_{\Delta 3}}{\sqrt{3}}$	$I_{\Delta 1} = I_A = I_A\angle \pm\theta$ $I_{\Delta 2} = I_B = I_A\angle -120° \pm\theta$ $I_{\Delta 3} = I_A\angle 120° \pm\theta$	

or

$$V_{AB} = 173.2\angle 40°\text{V}.$$

The phase currents are

$$I_{AB} = \frac{V_{AB}}{Z_\Delta} = \frac{173.2\angle 40°}{8.944\angle 26.57°} = 19.36\angle 13.43°A$$
$$I_{BC} = I_{AB}\angle -120° = 19.36\angle -106.57°A$$
$$I_{CA} = I_{AB}\angle 120° = 19.36\angle 133.43°A.$$

The line currents are

$$I_a = I_{AB}\sqrt{3}\angle - 30^\circ = \sqrt{3}\left(19.36)\angle(13.43 - 30)^\circ\right)$$
$$= 33.53\angle - 16.57^\circ A$$
$$I_b = I_a\angle - 120^\circ = 33.53\angle - 136.57^\circ A$$
$$I_c = I_a\angle 120^\circ = 33.53\angle 103.43^\circ A.$$

Option 2

Use a single-phase analysis approach.

Solution:

Convert the delta-connected load to the star using the delta–star conversion procedure where

$$Z_Y = {}^{Z_\Delta}\!/_3$$

hence

$$I_a = \frac{V_{an}}{Z_\Delta/3} = \frac{100\angle 10^\circ}{2.981\angle 26.57^\circ} = 33.54\angle - 16.57^\circ.$$

Recall that line current and phase current are the same in a star-connected system.

Hence, using the *abc sequence*, line currents are obtained as:

$$I_b = I_a\angle - 120^\circ = 33.53\angle - 16.57^\circ - 120^\circ A$$
$$= 33.53\angle - 136.57^\circ A$$

Also

$$I_c = I_a\angle 120^\circ = 33.53\angle - 16.57^\circ + 120^\circ A$$
$$= 33.53\angle 103.43^\circ A.$$

Example 3

A balanced, delta-connected load with an impedance $20 - j5\Omega$ is connected to a delta-positive sequence generator with $V_{ab} = 330\angle 0^\circ$. Calculate the phase currents of the load and the line currents.

Solution:

The load impedance per phase is

$$Z_\Delta = 20 - j5\Omega = 25\angle - 36.87^\circ\Omega.$$

For a delta-connected system, phase voltage is equal to line voltage, i.e., $V_{AB} = V_{ab}$, hence the phase currents are

$$I_{AB} = \frac{V_{AB}}{Z_\Delta} = \frac{330\angle 0^\circ}{25\angle - 36.87^\circ} = 13.2\angle 36.87^\circ A$$
$$I_{BC} = I_{AB}\angle - 120^\circ A = 13.2\angle - 83.13^\circ A$$
$$I_{CA} = I_{AB}\angle 120^\circ A = 13.2\angle 156.87^\circ A.$$

For a delta load, the line current always lags the corresponding phase current by 30° and has a magnitude of $\sqrt{3}$ times that of the phase current. Hence, the line currents are

$$\begin{aligned} I_a = I_{AB}\sqrt{3}\angle 30° &= (13.2\angle 36.87°)\left(\sqrt{3}\angle -30°\right) \\ &= 22.86\angle 6.87° \\ I_b = I_a\angle -120° &= 22.86\angle -113.13°A \\ I_c = I_a\angle 120° &= 22.86\angle 126.87°A. \end{aligned}$$

Example 4

A balanced, Y-connected load with a phase impedance of $40+j25\Omega$ is supplied by a balanced, positive-sequence delta-connected source with a line voltage of 210 V. Calculate the phase currents. Use I_{ab} as a reference.

Solution:

The load impedance is

$$Z = 40 + j25\Omega = 47.17\angle 32°\Omega$$

and the source voltage is $V_{ab} = 210\angle 0°$.

When the delta-connected source is transformed into a Y-connected source,

$$V_{an} = \frac{V_{ab}}{\sqrt{3}}\angle -30° = 121.2\angle -30°V.$$

The line currents are

$$\begin{aligned} I_a &= \frac{V_{ab}}{Z_Y} = \frac{121.2\angle -30°}{47.12\angle -32°} = 2.57\angle -62°A \\ I_b &= I_a\angle -120° = 2.57\angle -178°A \\ I_c &= I_a\angle 120° = 2.57\angle 58°A \end{aligned}$$

which are the same as the phase currents.

2.8 PER-UNIT SYSTEMS

The power-system quantities such as voltage, current, and power are normalized for ease of computation. In power-system analysis, per unit or percent of specified base values typically are expressed as measurements. The advantages of using per unit measurements are:

(a) ease of computation and ease of comparison of results for various power systems that may have different base quantities,

(b) early detection of calculation errors, especially when device parameters fall in small ranges, and

(c) elimination of ideal transformers as windings, such that voltages, currents, and external impedances and admittances, expressed in per unit, do not change when they are referred from one side of the transformer to another. This leads to computational savings in a power system with hundreds of transformers and helps to eliminate errors in calculation.

The per unit definition allows an actual quantity to be normalized to unity and facilitates the comparison of all other measured values to that base-unit value. By definition, the per-unit value of a quantity X is given by

$$X_{per\ unit} = \frac{X_{Actual}}{X_{base}}. \tag{2.82}$$

The actual quantity is the value of the quantity in actual units such as watts, vars, or hertz. The base values are determined by the mass quantity allowed for a device in terms of V, I, P, or f by the operation.

The base value has the same unit as the actual quantity, hence per unit is dimensionless. For electrical laws to be valid in the per unit system, the following relations must be used for other bases:
given that

$$S = VI^*,$$

where

$$S\angle\phi = V\angle\alpha I\angle - \beta \tag{2.83}$$

then

$$S_{base/\phi} = Q_{base/\phi} = P_{base/\phi} \tag{2.84}$$

such that

$$\frac{S\angle\phi}{S_{base}} = \frac{V\angle\alpha I\angle - \beta}{S_b} \tag{2.85}$$

$$S_{base} = V_{base} I_{base} \tag{2.86}$$

$$\frac{S\angle\phi}{S_{base}} = \frac{V\angle\alpha I\angle - \beta}{V_{base} I_{base}} = \frac{V}{V_{base}} \cdot \frac{I}{I_{base}} = V_{p.u.} I^*_{p.u.} \tag{2.87}$$

$$I_{base} = \frac{S_{base}}{V_{baseLN}} \tag{2.88}$$

$$Z_{base} = R_{base} = X_{base} = \frac{V_{baseLN}}{I_{base}} = \frac{V^2_{baseLN}}{S_{base/\phi}} \tag{2.89}$$

$$Y_{base} = G_{base} = B_{base} = \frac{1}{Z_{base}}. \tag{2.90}$$

Note that:

(a) The value of $P_\phi = Q_\phi = S_\phi$ applies to the entire system and is the same for the entire circuit.

(b) The ratio of voltage bases on either side of a transformer is selected to be the same as the ratio of the transformer voltage ratings. With these two rules, it is evident that per unit impedance is the same for transformers when referred from one side to the other.

Example 1

Consider a 100 V sinusoidal source in series with a 3Ω resistor, an 8Ω inductor, and a 4Ω capacitor. Taking base voltage as 100 V and base VA as 500VA and working in per unit:

(a) Draw the circuit diagram representing each component by its per unit value.

(b) Compute the series impedance and the current flowing in the circuit.

Solution:

$$I_{base} = \frac{S_{base}}{V_{base}}$$

$$= \frac{500}{100} = 5.0\,\text{A}$$

$$Z_{base} = \frac{V_{base}}{I_{base}}$$

$$Z_{base} = \frac{100}{5} = 20\Omega$$

or

$$Z_{base} = \frac{V_{base}^2}{S_{base}}$$

$$Z_{base} = \frac{(100)^2}{500} = 20\Omega$$

Converting the circuit component values to per unit:

$$I_{p.u.} = \frac{V}{V_{base}} = \frac{100\angle 0^\circ}{100} = 1.0\angle 0^\circ$$

$$R_{p.u.} = \frac{R}{Z_{base}} = \frac{3}{20} = 0.15$$

$$X_{Lp.u.} = \frac{X_L}{Z_{base}} = \frac{8}{20} = 0.40$$

$$X_{c\ p.u.} = \frac{X_c}{Z_{base}} = \frac{4}{20} = 0.20$$

The circuit is:

(a)

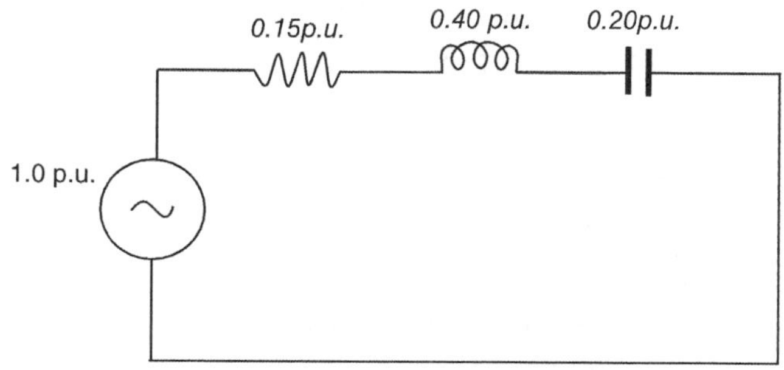

(b) $Z_{p.u.} = 0.15 + j(0.4 - 0.2)$
$= 0.15 + j0.2$
$0.25\angle 53.1°$

(c) $I_{p.u.} = \dfrac{V_{p.u.}}{Z_{p.u.}} = \dfrac{1\angle 0°}{0.25\angle 53.1°}$
$4.0\angle - 53.1°$

Illustrative Problems and Examples

1. Three impedances of $4 + j3\Omega$ are Δ connected and tied to a three-phase 208 V power line. Find:
   (a) I_{ph}, I_L, P, S, Q
   (b) power factor for this load

Solution:

$$V_L = V_{ph}\ (\Delta \text{ connection})$$

$$Z_1 = Z_2 = Z_3 = Z = 4 + j3\ \Omega$$

$$Z = \sqrt{4^2 + 3^2} = 5$$

$$Z = 5\angle 36.87°\ \Omega$$

$$V_{ph} = 208\ \text{V}$$

$$I_{ph} = V_{ph}/Z$$

$$I_{ph} = \frac{208}{4 + j3}$$

$$I_{ph} = 41.6\angle - 36.87°\ \text{A}$$

For Δ connection,

$$
\begin{aligned}
I_L &= \sqrt{3} I_{ph} \\
&= \sqrt{3} \times 41.6\angle - 36.87^\circ \text{ A} \\
&= 72\angle - 36.87^\circ \text{ A} \\
\text{S} &= \sqrt{3} V_L I_L \\
&= \sqrt{3} \times 208 \times 72 \\
&= 25.9 \text{ KW} \\
Q &= \sqrt{3} V_L I_L \sin\theta \\
&= \sqrt{3} \times 208 \times 72 \text{ Sin } (36.87^\circ) \\
&= 15.6 \text{ kW} \\
\text{P} &= \sqrt{3} V_L I_L \cos\theta \\
&= \sqrt{3} \times 208 \times 72 \times 0.8 \\
&= 20.75 \text{ KW} \\
\text{power factor} &= \text{Cos}(\theta) \\
\theta &= 36.87^\circ \\
\cos^{-1}(36.87^\circ) &= 0.8 \\
\text{power factor} &= 0.8
\end{aligned}
$$

2. If v is $141.4\cos(\omega t + 60^\circ)$ V and i is $14.14\,sin(\omega t + 120^\circ)$ A, find both V and I:

(a) maximum value

(b) rms value

(c) phasor expression using voltage as the reference

(d) phasor expression using current as the reference

(e) whether circuit in which the voltage and current exist is capacitive or inductive

Solution:

(a)

$$
\begin{aligned}
\text{V} &= 141.4\,\text{Cos}(\omega t + 60^\circ) \text{ Volts} \\
\text{I} &= 14.14\,\text{Sin}(\omega t + 120^\circ) \text{ A} \\
V_{max} &= 141.4 \text{ V} \\
I_{max} &= 14.14 \text{ A}
\end{aligned}
$$

(b)

$$
\begin{aligned}
V_{rms} &= V_{max}/\sqrt{2} = 141.4/\sqrt{2} = 99.98 \text{ V} \\
I_{rms} &= I_{max}/\sqrt{2} = 14.14/\sqrt{2} = 99.9 \text{ A}
\end{aligned}
$$

(c)

$$
\begin{aligned}
\text{V} &= 141.4\,\text{Cos}(\omega t + 60^\circ) \text{ Volts} \\
\text{I} &= 14.14\,\text{Cos}(\omega t + 120^\circ - 90^\circ) \text{ A} \\
\text{I} &= 14.14\,\text{Cos}(\omega t + 30^\circ) \text{ A}
\end{aligned}
$$

The voltage leads the current by 30°.

(d) The current lags the voltage by 30°.

(e) Hence the circuit is inductive.

3. Express each of these currents as a phasor

(a) $800\cos\left(\omega t + 130°\right)$ A

(b) $480\cos\omega t - 640\sin\omega t$ A

(c) $550\,sin\left(\omega t + 40°\right) + 420\,sin\left(\omega t + 110°\right)$ A

Solution:

(a)
$$\begin{aligned} I &= 800\cos(\omega t + 130°)\,A \\ I &= I_m\angle\Phi\,(\text{phasor}) \\ I_m &= 800\,A, \Phi = 130° \\ I &= 800\angle 130°\,A \end{aligned}$$

(b)
$$\begin{aligned} &480\cos(\omega t) - 640\sin(\omega t) \\ &= 480\cos(\omega t) - 640\cos(\omega t - 90°) \\ &= 480\angle 0° - 640\angle - 90° \end{aligned}$$

(c)
$$\begin{aligned} &550\sin(\omega t + 40°) + 420\sin(\omega t + 110°)\,A \\ &= 550\cos(\omega t + 40° - 90°) + 420\cos(\omega t + 110° - 90°)\,A \\ &= 550\cos(\omega t + -50°) + 420\cos(\omega t + 20°)\,A \\ &= 550\angle - 50° + 420\angle 20°\,A \end{aligned}$$

2.9 CHAPTER SUMMARY

In this chapter we presented the fundamentals of electric power network analysis concepts and principles, which are essential in analysis and computations in any electric power system. The nature and behavior of electric currents and voltage signals, and the effects of their interactions with discrete circuit components, are important base knowledge for students, as is awareness of the three-phase system as a better, more reliable carrier of electric power. Circuit analysis at this level is equally beneficial to dealing with more advanced electric power-system issues. Overall, fundamental computational skills needed for any analysis of the power system was presented here. This information is particularly applicable for competitive work in the micogrid.

EXERCISES

1. Determine the line current drawn from a three-phase, 440 V line by a three-phase, 10HP motor operating at rated load. The motor has an efficiency of 91.6 percent and a power factor of 0.85 lagging. Also calculate the real and reactive power drawn by the motor.

2. In Problem 4, if the line connecting the motor has an impedance of $0.5 + j1.0\Omega$, find the line–line voltage at the supply bus, assuming the motor bus voltage is 440 V.

3. Determine the phase and line currents of the delta-connected, positive-sequence load shown in Figure Q3.

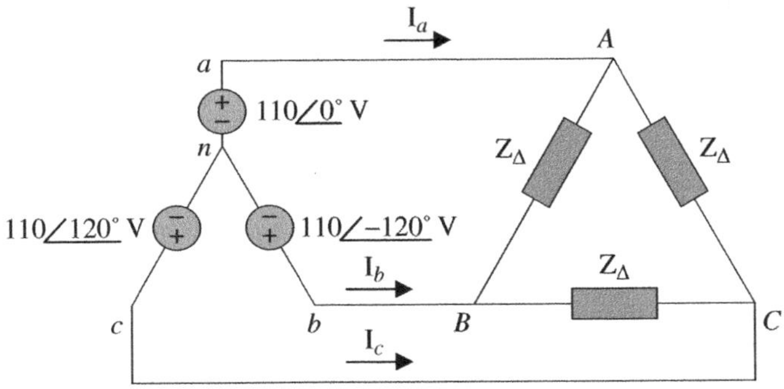

Figure Q3 A Wye-Delta Circuit.

given that $Z_{\Delta} = 60 < 45°\ \Omega$.

4. For the three-phase Y–Δ circuit shown in Figure Q4, calculate the line and phase currents of the load. Assume that $Z_L = 12 + j2\ \Omega$.

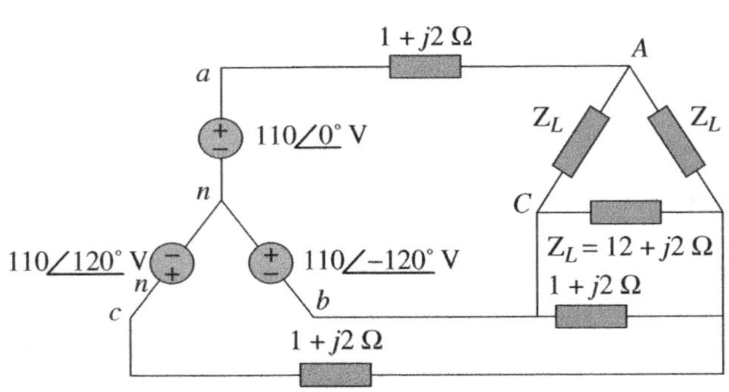

Figure Q4 A Wye-Delta Circuit.

5. Determine the current drawn from a three-phase, 440 V line by a three-phase 15HP motor operating at full load, 90 percent efficiency, and 80 percent power-factor lagging. Find the values of P and Q drawn from the line.

6. If the impedance of each of the three lines connecting the motor of Problem 5 to a bus is $0.3 + j0.85$, find the line-to-line voltage at the bus that supplies 440 V at the motor.

7. A balanced delta load consisting of pure resistance 15ohm per phase is in parallel with a balanced Y load having phase impedances of $(8 + j6)$ ohm. Identical impedances of $(2 + j5)$ ohm are in each of the three lines connecting the combined loads to a 110 V, three-phase supply. Find the current drawn from the supply and line voltage at the combined loads.

8. A balanced, delta-connected load having an impedance $12 - j10\Omega$ is connected to a delta connected, positive-sequence generator having $V_{ab} = 210\angle 0°$ V. Calculate the phase currents of the load and line currents.

9. A balanced, Y-connected load with a phase impedance of $60 + j25\Omega$ is supplied by a balanced, positive-sequence, delta-connected source with a line voltage of 220 V. Using I_{ab} as a reference, calculate the phase currents.

BIBLIOGRAPHY

[1] B. L. Theraja and A. K. Theraja, *A Textbook of Electrical Technology*, 1st multicolour ed., S. Chand, New Delhi, 2005.

[2] J. Casazza and F. Delea, *Understanding Electric Power Systems: An Overview of the Technology, the Marketplace, and Government Regulations*, John Wiley & Sons, Hoboken, NJ, 2003.

[3] C. A. Gross, *Power System Analysis*, John Wiley & Sons, New York, 1979.

[4] S. J. Chapman, *Electric Machinery Fundamentals*, 5th ed., McGraw-Hill, New York, 2012.

[5] T. Gönen, *Electrical Machines with MATLAB*, 2nd ed., Taylor & Francis Group, Boca Raton, FL, 2011.

[6] W. L. Matsch and J. D. Morgan, *Electromagnetics and Electromechanical Machines*, 3rd ed., John Wiley & Sons, Hoboken, NJ, 1986.

[7] M. S. Sarma, *Electric Machines: Steady-State Theory and Dynamic Performance*, WMC Brown Publishers, Dubuque, IA, 1985.

[8] C. K. Alexander and M. N. O. Sadiku, *Fundamentals of Electric Circuits*, 5th ed., McGraw-Hill, New York, 2013.

[9] J. A. Momoh, *Electric Power Systems Applications of Optimization*, 2nd ed., CRC Press, Boca Raton, FL, 2008.

CHAPTER 3

MAGNETIC SYSTEMS FOR ENERGY PROCESSING

3.1 INTRODUCTION

This chapter deals with a wide range of electromagnetic devices used for conversion of electrical energy to mechanical energy, or simply a process of transforming electrical energy to a useful form. The electric machine systems for energy processing are:

- Generators, connected to prime movers (steam-, gas-, or hydro-turbine systems), convert mechanical energy to electrical energy for production of electrical power.
- Transformers process (transform) electric energy from the most convenient generation voltage to a voltage level suitable for transmission to a load point and into voltage appropriate for distribution.
- Electric motors are used to convert electrical energy to mechanical energy for drive systems and further energy processing.
- Electronic converters process electric power from AC to DC or DC to AC with controllable voltage, current, or frequency.
- Drive-system controls of motors and converters are capable of controlling the speed or position of mechanical loads.

3.2 MAGNETIC FIELDS

The conversion of power or energy from one machine to another is achieved using magnetic fields that depend on a variety of materials, structures, locations, and intensities. To understand how the magnetic field works or changes as a function of time, we employ Maxwell's equations, and its application to different machine types.

Energy Processing and Smart Grid, First Edition. James A. Momoh

Maxwell's equations may be expressed in differential and integral forms, summarized in the following four laws.

Ampère's Law

$$\nabla \times \boldsymbol{H} = \boldsymbol{J}_c + \frac{\partial \boldsymbol{D}}{\partial t}$$

$$\oint H.dl = \int \left(\boldsymbol{J}_c + \frac{\partial \boldsymbol{D}}{\partial t} \right) .dS \tag{3.1}$$

The line integral of the magnetic field strength H on a closed geometric-integration loop *l* (*magnetic circulation voltage*) is equal to the total electric current flowing through the area A limited by this loop (*magneto-motive force* and *ampere-turns*), inclusive of the displacement current, which is neglected. This relates the intensity of the magnetic field to the current producing it; it is the principle behind the generator/motor action, whereas Lenz's law states that the induced emf gives rise to a current whose magnetic field opposes the change in the original magnetic flux.

Faraday's Law

E, the induced voltage = electromotive force (emf). Faraday's law is given as

$$\nabla \times \boldsymbol{E} = -\frac{\partial \boldsymbol{B}}{\partial t}$$

$$\oint E.dl = \int \left(-\frac{\partial \boldsymbol{B}}{\partial t} \right) .dS \tag{3.2}$$

The line integral of the electric field strength E on a closed geometric integration loop (*electric circulation voltage*) is equal to the negative, time-dependent variation of the total magnetic flux that penetrates the area A limited by this loop, i.e., induced voltage in a coil of conductor $e = \frac{d(\phi)}{dt}$ (enclosed flux ϕ).
For N coils,

$$e = \oint E.dl = \frac{d(N\phi)}{dt} \tag{3.3}$$

$$\lambda = N\phi \tag{3.4}$$

$$e = \frac{d\lambda}{dt}. \tag{3.5}$$

Faraday's law leads to transformer action for energy processing.

Gauss's Law of Electricity

$$\nabla.\mathbf{D} = \rho$$

$$\oint \mathbf{E}.d\mathbf{A} = \frac{q}{\varepsilon_o} \tag{3.6}$$

The implication of Gauss's law of electricity is that the total electric field penetrating a closed surface of any volume is the electric charges inside this volume.

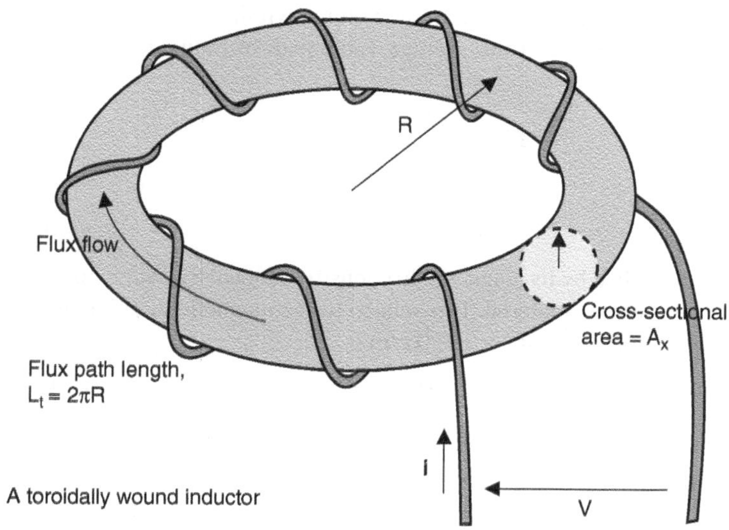

Figure 3.1 Toroidal coil.

Gauss's Law of Magnetism

$$\nabla.\mathbf{D} = 0$$

$$\oint \boldsymbol{B}.d\boldsymbol{A} = 0 \tag{3.7}$$

This implies that the total magnetic flux penetrating a closed surface of any volume is zero.

Faraday's law and Ampère's law are very important in electrical machine design. At their simplest, an equation can be employed to determine the voltages induced in the windings of an electrical machine. It is also necessary—for instance, in the determination of losses caused by eddy currents in a magnetic circuit, and when determining the skin effect in copper. With respect to flux linkage in an electric machine's windings, if there are N turns of winding, the flux does not link all these turns ideally, but with a ratio of less than unity. Hence, we may denote the effective turns of winding by k_w N, ($k_w < 1$). In electrical machines theory, the factor kw is known as the winding factor,

This formulation is essential to electrical machines and is written as

$$e = -k_w N \frac{d}{dt} \int \boldsymbol{B}.d\boldsymbol{S} = -k_w N \frac{d\phi}{dt} = -\frac{d\psi}{dt}. \tag{3.8}$$

For Ampère's law, consider a toroidal non-ferromagnetic material, shown in Figure 3.1.

Magnetic field is effectively confined to the volume within the toroid (cores). If H_r is the value of flux intensity at the average value path of radius r, it may be expressed as

$$H_r = \frac{NI}{2\pi r} \tag{3.9}$$

or

$$2\pi H_r r = NI. \tag{3.10}$$

The magnetic field is described by its magnetic flux density, denoted by a vector quantity $\vec{B}$. For different non-ferrous material, B is related to the magnetic field intensity H by the constant μ_0 given by $\mu_0 = 4\pi \times 10^{-7}$H/m

$$\vec{B} = \mu_0 H \tag{3.11}$$

$$\phi = \vec{B}A \tag{3.12}$$

$$A = \pi r^2 \tag{3.13}$$

3.3 EQUIVALENT MAGNETIC AND ELECTRIC CIRCUITS

To analyze the concept of magnetic force:

$$mmf = Ni = \mathcal{F} = \int H.dl = \int J.dA \tag{3.14}$$

$$\mathcal{F} = Hl = \frac{Bl}{\mu_0} = \frac{\phi l}{\mu_0 A} \tag{3.15}$$

where

$$J = {}^{I}/_{A}.$$

The constant of proportionality between $\mathcal{F}$ and ϕ is related by reluctance $\Re$, given by

$$\text{reluctance} \quad \Re = \frac{\mathcal{F}}{\phi} = \frac{l}{\mu_0 A}. \tag{3.16}$$

The relation between the variables may be modeled by the equivalent magnetic circuit seen in Figure 3.2.

The flux leakage λ of the toroidal core

$$\lambda = N\phi = NBA \tag{3.17}$$

$$\lambda = N\mu_o \text{HA} \quad (\text{where B} = \mu_o H) \tag{3.18}$$

$$\lambda = \frac{N^2 \mu_o Ai}{l}, \quad \left(\text{where H} = \frac{Ni}{l}\right) \tag{3.19}$$

$$L = \frac{\lambda}{i} = \frac{N^2 \mu_o A}{\ell} \tag{3.20}$$

$$e = \frac{d\lambda}{dt} = L\frac{di}{dt}. \tag{3.21}$$

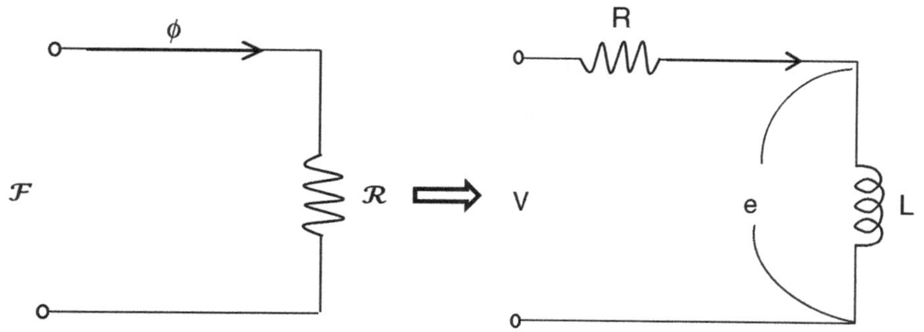

Figure 3.2 Magnetic and electric circuit.

So, using equivalent electric circuit

$$V = Ri + e \tag{3.22}$$

$$V = Ri + L\frac{di}{dt} \tag{3.23}$$

3.3.1 Differences between Magnetic and Electric Circuits

(a) Electric current flows around an electric circuit, but flux does not flow.

(b) Under a constant temperature, the resistance of an electric circuit is constant and is independent of current density; however, the reluctance of a magnetic circuit depends on flux, hence flux density.

(c) The flow of an electric current in an electric circuit involves the expenditure of energy. In a magnetic circuit, energy is needed to create flux but not to maintain it.

3.4 OVERVIEW OF MAGNETIC MATERIALS

Examples of magnetic materials with high values of relative permeability μ_r are cobalt, aluminum, tungsten, and nickel. These are called *ferromagnetic materials* because they are relatively easy to magnetize. These ferromagnetic materials are classified as hard or soft materials. Soft materials include: soft steel, iron (Figure 3.3), nickel, cobalt, some rare-earth elements, and many alloys of elements. Hard ferromagnetic materials such as copper–nickel alloys and chromium–steel alloys are alloyed with aluminum, nickel, and cobalt.

3.5 HYSTERESIS LOOPS AND HYSTERESIS LOSSES IN FERROMAGNETIC MATERIALS

For ferromagnetic materials, the magnetizing curve shown in Figure 3.4 is not reversible; that is, if H is increased until the material is saturated, when H is reduced

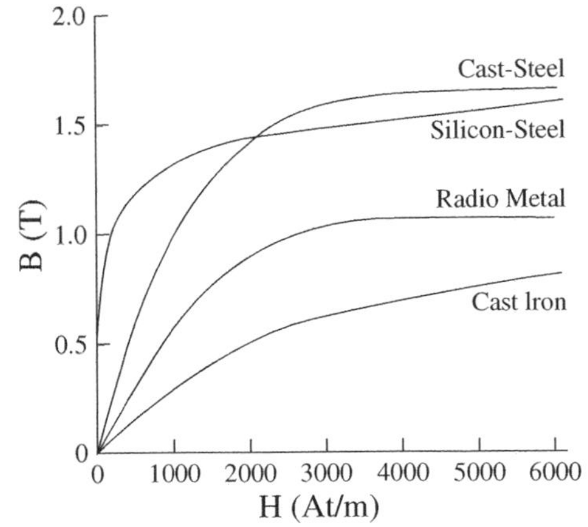

Figure 3.3 BH curve for some ferromagnetic materials.

again the value of B does not reduce to zero along the same line. Figure 3.5 illustrates a hysteresis loop.

In Figure 3.5 [11], the point 0, where the axes cross, represents a ferromagnetic material that is unmagnetized. As current in the coil increases, H also increases and between points 0 and 1, flux density, B increases following the magnetization curve. At point 1, the material has reached saturation and B no longer increases. To demagnetize the ferromagnetic material, we gradually reduce the current in the coil. As stated earlier, the graph will not follow the same path it did when the current increased; instead it goes from point 1 through point 2 then down to point 3.

At point 2, $H = 0$, therefore the current has reached zero, but there is still some *remnant flux density* so that the material is still partially magnetized. The current

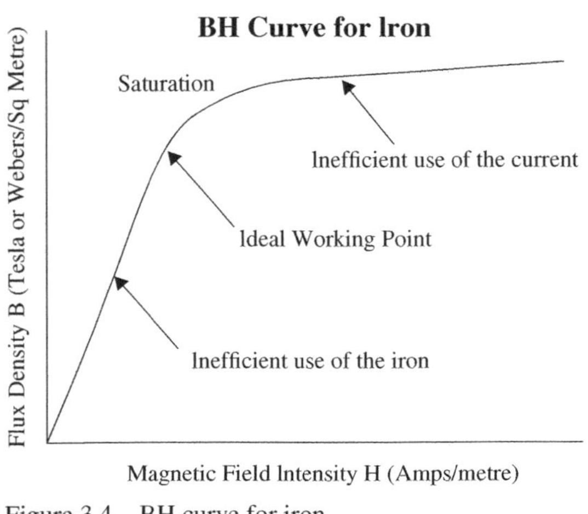

Figure 3.4 BH curve for iron.

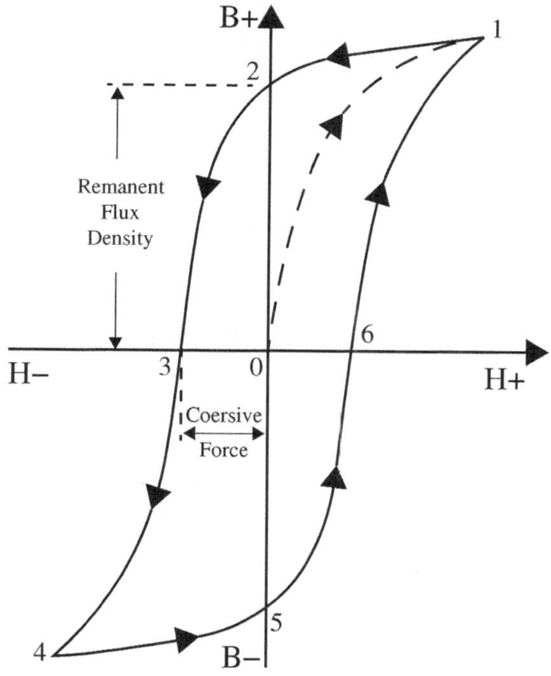

Figure 3.5 A hysteresis loop for a solenoid's ferromagnetic core.

is now reversed so that H is in the opposite direction as it was before; during this process, H has a negative value, therefore when the current is increased, the value of H reduces. At point 3, the material is finally demagnetized and the value of H at this point is called the *coercive force*.

If the reversed current increases further, we reach point 4, where the material saturates so that the magnetic poles of the domains face in the opposite direction to those at point 1. The reversed current is now reduced and reaches zero at point 5; however, once again, some flux remains. If the current is now increased in the original direction, all the flux would have gone at point 6 and saturation is reached once more at point 1.

3.5.1 Hysteresis Loss

Since the coercive force must be applied to overcome the remanent magnetism, work is done in completing the hysteresis loop and the energy concerned appears as heat in the magnetic material. This heat represents a form of energy loss known as *hysteresis loss*. The amount of loss depends on the material's value of coercive force. For example, by adding silicon to iron, a material with a very small coercive force can be made; such materials typically contain 5 percent silicon and have a very narrow hysteresis loop. Materials with narrow hysteresis loops are easily magnetized and demagnetized, and known as *soft magnetic materials*.

Hysteresis loss is always a problem in AC transformers where the current is constantly changing direction. The magnetic poles in the core cause losses because they constantly reverse direction. Rotating coils in DC machines also incur hysteresis losses.

3.6 DEFINITIONS

- **Residual magnetic flux density (B_{res}):** When H is equal to zero, there is a residual value of magnetic flux density B_{res} in the core. The magnitude of B depends on the material. Thus the residual flux-density effect creates a permanent magnet.
- **Coercive force or coercivity:** This is the negative value of magnetic field intensity (H_c) required to reduce the flux density B to zero.
- **Saturation:** Any further increase in H in the negative direction causes the magnetic core to be magnetized with opposite polarity. Increasing the current in a negative direction further results in saturation.
- **Hysteresis loop:** As the direction of the current in the core reverses back and forth, a loop is formed; this is referred to as a *hysteresis loop*.

3.7 MAGNETIC CIRCUIT LOSSES

Magnetic materials are subject to time-varying flux, which normally leads to energy loss in the material in the form of magnetic losses. Such magnetic losses are called iron or core losses. There is no core loss if flux does not vary with time.

Core losses are defined as the sum of hysteresis and eddy-current losses, given by

$$W_c = K_e f^2 + K_n f \quad \text{(Watts)} \tag{3.24}$$

where K_e = proportionate constant depending on the core materials, f = supply frequency, and K_n = property dependency on the magnetic material.

(a) Hysteresis Losses

According to Charles Steinmetz, *hysteresis loss* caused by energy needed to cause the reversal in the hysteresis loop is determined empirically as

$$P_n = VfK_nB_m^n \tag{3.25}$$

where V is the volume of ferromagnetic material, f is the frequency (Hz), K_n is the property called dependency on the magnetic material. K_n for soft, 1 mm, silicon, sheet steel are 0.025, 0.001, 0.0001, respectively. B_m is maximum flux density and n is the *Steinmetz exponent*, which it varies from 1.5 to 2.5 or (2.0).

(b) Eddy-Current Losses

Because iron is a conductor, time-varying magnetic fluxes induce opposite voltage and currents called *eddy currents*, which circulate within the core. The undesirable circulating current flow around the flux core is so large that it can demagnetize the magnets. Thus, eddy currents establish flux that opposes the original change imposed by the core.

The equation for the eddy-currents loss calculation is

$$P_e = K_e \mathrm{V}(\mathrm{f}\ t_i B_m)^2 \tag{3.26}$$

where K_e is the proportionate constant depending upon the core materials, V is the volume of ferromagnetic material (m^3), t_i is the lamination thickness, and B_m is the maximum flux density (T).

(c) I^2R Losses

I^2R losses are found in windings of the machine. By convention, these losses are computed on the basis of the DC resistance of the winding at 75°C. I^2R losses depend on the effective resistance of the winding under the operating frequency and flux conditions.

(d) Mechanical Losses

Mechanical losses consist of brush and bearing friction and windage losses caused by running a machine at high speed while unloaded and unexcited. Frequently these losses are lumped with core losses and determined at the same time.

(e) Stray Losses

Stray losses arise from nonuniform current distribution in the copper winding. There is no formula for its calculation. The usual approach is to take 1 percent of the output of the DC machine. For synchronous and induction machines, we can find it through testing.

In general, the total losses contribute to determining a machine's efficiency, and this is used to compare different electromagnetic-based machines. The contribution of these losses for generators, motors, and transformers is usually determined to calculate their efficiency.

(f) Magnetic Fringing

In practice, not all the magnetic flux produced in a magnetic circuit will be concentrated within the core. Apart from the leakage flux, which will appear in the surrounding free space, if a gap exists within the magnetic circuit, the flux tends to spread out, as shown in Figure 3.6 [11]; this effect is known as *fringing*.

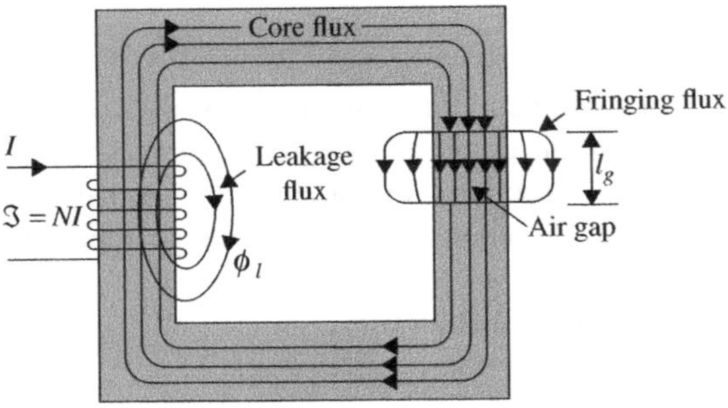

Figure 3.6 Magnetic flux leakage and fringing.

3.8 PRODUCING MAGNETIC FLUX IN AIR GAP

The production of magnetic flux is confined to a curved magnetic path in a toroid, as seen in Figure 3.7 [6]. The creation of flux in an air gap is also necessary for electrifying.

With an iron gap

$$\phi_{iron} = \phi_{air}(\mathrm{W_b}) \tag{3.27}$$

$$B_{iron} = B_{air}\,(\mathrm{T}) \tag{3.28}$$

$$H_{iron} = \frac{B_{air}}{\mu_o}\,(\mathrm{A/m}) \tag{3.29}$$

If ϕ is the flux in the core of the toroid, $\mathcal{R}_{core}$ and $\mathcal{R}_g$ are the reluctances of the core and the air gap, respectively then

$$\mathcal{F} = NI = \phi\left(\mathcal{R}_{core} + \mathcal{R}_g\right)$$

$$\phi = \frac{\mathcal{F}}{\left(\mathcal{R}_{core} + \mathcal{R}_g\right)} = \frac{NI}{\left(\mathcal{R}_{core} + \mathcal{R}_g\right)}. \tag{3.30}$$

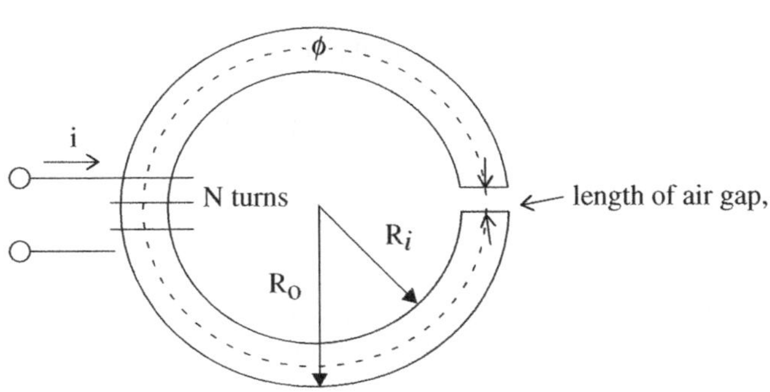

Figure 3.7 Toroid with air gap.

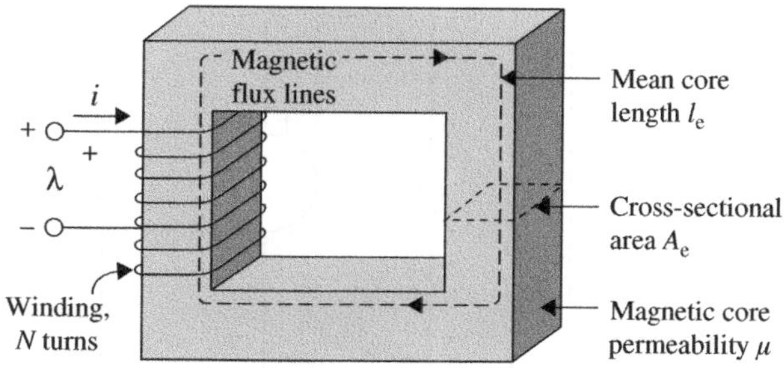

Figure 3.8 Rectangular magnetic circuit.

3.9 RECTANGULAR-SHAPED MAGNETIC CIRCUITS

Let us consider a simple magnetic circuit, as shown in Figure 3.8 [12], with a single core material having uniform cross-sectional area A and mean length of flux path l. Reluctance offered to the flow of flux is $\mathfrak{R}$. The corresponding electrical representation is rather simple. The equivalent electrical circuit is also drawn beside the magnetic circuit, as shown in Figure 3.9 for comparison.

Although in the actual magnetic circuit, there is no physical connection of the winding and the core, in the magnetic-circuit representation, magnetomotive force (mmf) and reluctance are shown to be connected.

$$\mathcal{F} = Ni = \oint H.dl \tag{3.31}$$

$$\mathcal{F} = H_{iron}l_{iron} + H_{air}l_{air} \tag{3.32}$$

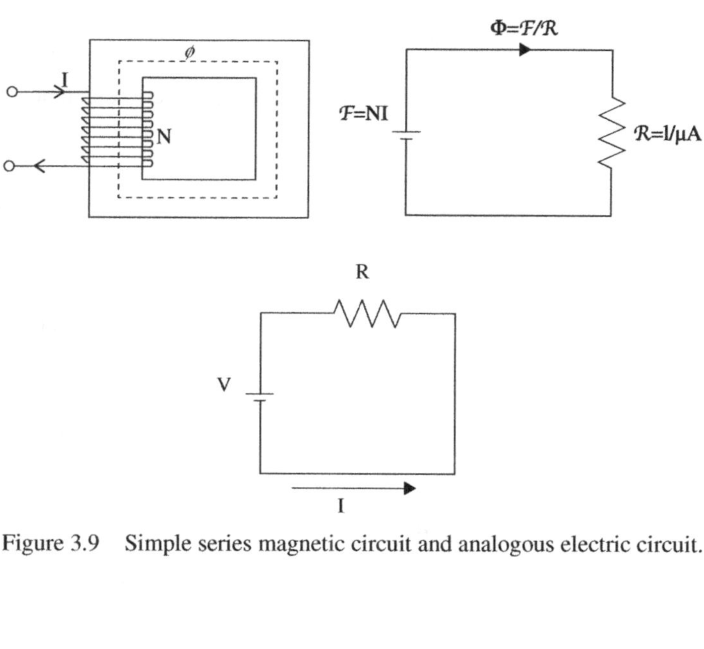

Figure 3.9 Simple series magnetic circuit and analogous electric circuit.

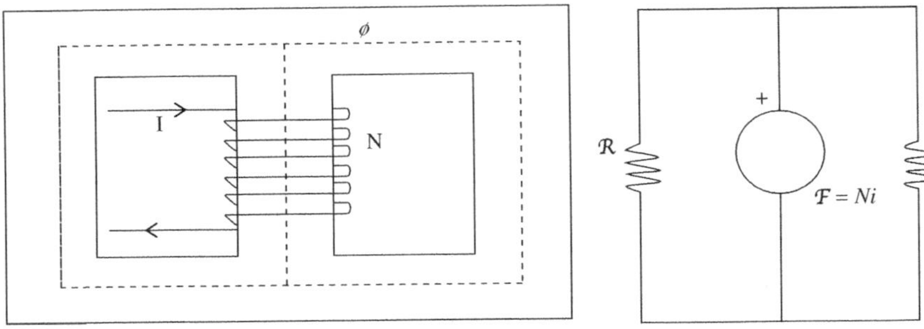

Figure 3.10 Parallel magnetic circuit.

$$I = \frac{V}{R} \tag{3.33}$$

$$\phi = \frac{\mathcal{F}}{R} \tag{3.34}$$

$$\phi = \frac{\mu NIA}{l} = \frac{NI}{l/\mu A} \tag{3.35}$$

$$\mathfrak{R} = \frac{l}{\mu A} \tag{3.36}$$

where $\mathcal{F} = \phi\mathfrak{R}$ and $\mathfrak{R} = \frac{\mathcal{F}}{\phi}$, NI is mmf in ampere-turns, ϕ is flux in weber, and $\mathfrak{R}$ is reluctance.

Let permanence of the magnetic circuit

$$\mathcal{P} = \frac{1}{\mathfrak{R}} \tag{3.37}$$

therefore

$$\phi = \mathcal{F}\mathfrak{R}. \tag{3.38}$$

R can be connected in series or parallel (Figure 3.10) so that the equivalent reluctance is $\mathfrak{R}_{eq}$.

For a system of n series reluctances,

$$\mathfrak{R}_{eq} = \mathfrak{R}_1 + \mathfrak{R}_2 + \cdots \mathfrak{R}_n. \tag{3.39}$$

For reluctances in parallel

$$\frac{1}{\mathfrak{R}_{eq}} = \frac{1}{\mathfrak{R}_1} + \frac{1}{\mathfrak{R}_2} + \cdots \frac{1}{\mathfrak{R}_n} \tag{3.40}$$

or permanence

$$\mathcal{P}_{eq} = \mathcal{P}_1 + \mathcal{P}_2 + \cdots \mathcal{P}_n \tag{3.41}$$

if

$$\phi = \text{BA} = \frac{\mu NIA}{l} \tag{3.42}$$

and $\lambda = N\phi$

$$\lambda = \frac{\mu N^2 IA}{l} = \frac{N^2 \mu IA}{l}.$$

If

$$L = \frac{\lambda}{I} = \frac{N^2 \mu IA}{Il}$$

but

$$\frac{1}{\Re} = \frac{\mu A}{l}$$

therefore

$$L = \frac{N^2}{\Re}\ (\mathrm{H}) \tag{3.43}$$

The inductance L is defined as the flux linkage per ampere of current flow in the coil. It is measured in henrys (H).

Illustrative Problems and Examples

(1) What conditions are necessary for a magnetic field to produce a force on a wire?

Solution:
$F = BIL \sin\theta$

- **(a)** There must be a relative angle between the current and the magnetic field.
- **(b)** The electric current must be flowing in the wire path.
- **(c)** The wire must be placed within the magnetic flux of the field.

(2) Distinguish between hysteresis and eddy-current loss and describe how to minimize eddy-current loss in a core.

Solution:

Hysteresis loss is due to the heating of the core as a result of the material molecular-structure reversal, which occurs as the magnetic flux alternates. Example: The power required for the continuous reversal of the elementary magnets of which iron is composed can be minimized using a core with small areas of hysteresis loop.

Eddy-current loss is due to continuous induced current path in the coil resulting in excessive heating of the coil. It can be minimized by using a core material of high resistivity. It can also be minimized by using some thin sheets of lamination for the core.

(3) What is the equivalent expression for an induced voltage?

Solution:
The expression for induced voltage can be written as:

$E = -N\,(d\Phi/dt)$

where E = induced voltage

N = number of turns

$(d\Phi/dt)$ = rate of change of flux

(4) Write the expression for reluctance. What is its unit?

Solution:
The expression for reluctance can be written as:

reluctance $(R) = l/(\mu \times A)$

where μ = permeability of the material

A = cross-sectional area

l = length (mean length).

The unit is ampere-turn/weber.

(5) A magnetic circuit has a continuous core of a ferromagnetic material. The coil is supplied from a battery and draws a certain amount of exciting current, producing flux in the core. If an air gap is then introduced in the core, the exciting current will:

(a) increase, (b) remain the same, (c) decrease, or (d) become zero.

Solution:
The air gap will not change the exciting current. So, the exciting current remains the same. This is because the air gap is introduced to solve the problem of excessive flux produced by the high level of current.

(6) Explain hysteresis in terms of magnetic domain theory.

Solution:
Hysteresis is loss due to the heating of the core as a result of the material molecular structure reverse, which occurs as the magnetic flux alternates.

$$P_H = K_H f B^n_{max}$$

(7) What are eddy-current losses? What can be done to minimize these losses?

Solution:
Eddy-current loss is due to continuous induced current in the core. It results in excessive heating of the core. It can be minimized by using a core material of high resistivity or by using a thin sheet of lamination for the core.

$$P_E = K_E f^2 B^2_{max}$$

(8) What conditions are necessary for a magnetic field to produce a voltage on a wire?

Solution:
The conditions for a magnetic field to produce voltage on a wire is that the conductor must be at an angle with the magnetic field, i.e.,

$$\theta \neq 0°$$
$$F = BIL \sin\theta$$

(9) What is Ampère's and Lenz's law?

Solution:
Ampere's law states that the line integral of the magnetic field intensity (H) around a closed path is equal to the total current enclosed by the path.

$$\oint Hdl = I_{enclosed}$$

For a conductor in circular form

$$H2\pi r = 2$$

Lenz's law states that the direction-induced emf in the coil is such that it tends to oppose the motion of the flux producing it.

3.10 CHAPTER SUMMARY

The magnetic effect of electric current and its applications were the central objectives of this chapter. The centrality of Maxwell's equations to the development of various electromagnetic devices cannot be overemphasized. Working examples and applications of Maxwell's equations to a power energy system were provided. This is important because from generation through transmission down to distribution, critical components of the system are built and operated to provide quality and reliable power based on the principle of magnetic effect of electric current. This fundamental knowledge is essential to students who want to understand next-generation electric power systems as many of the components for power generation, protection, measurement, regulation, and security in a smart/microgrid environment are based on this essential principle. It is therefore essential foundational knowledge the student must embrace. The concepts described in this chapter are used in modeling electromechanical energy processing systems and components.

EXERCISES

1. Calculate the self-inductance L of a toroid with a square cross-section with inner radius a, outer radius b = a + h, (height h), and N square windings.

2. Consider a copper core as seen in Figure Q2 with susceptibility $X_m = -9.7 \times 10^{-6}$, length of core L = 1 m, gap length g = .01 m, cross-sectional area A = .1 m, current I = 10A, N = 5 turns. Find B_g.

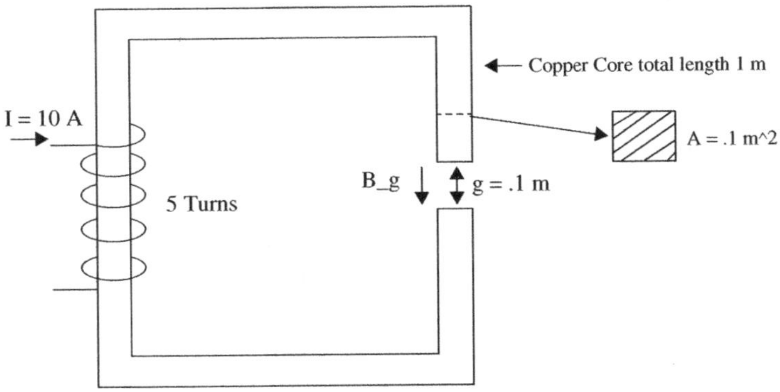

Figure Q2 Copper Core Magnetic Circuit.

3. 2mWb is to be produced in the air gap of the magnetic circuit shown in Figure Q3. How many ampere-turns must the coil provide to achieve this? Relative permeability μ_r of the core material may be assumed to be constant and equal to 5,000. All the dimensions shown are in cm and the sectional area is 25cm^2 throughout.

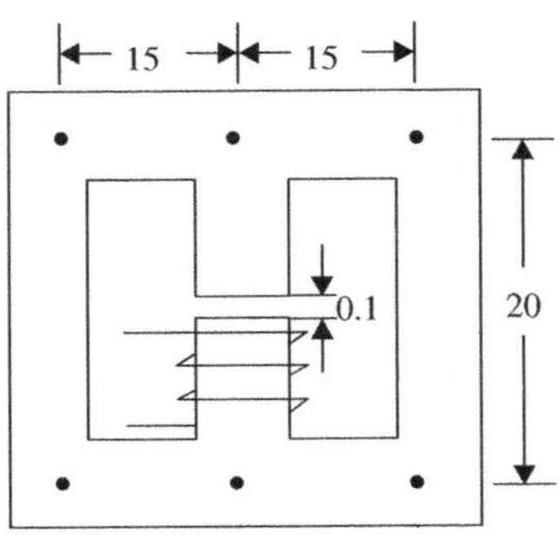

Figure Q3 Magnetic Circuit.

4. The core loss in an iron-core reactor is 600 W, of which 400 W is hysteresis loss when the applied voltage is 220 V and the frequency is 60 Hz. Neglecting winding resistance:
   (a) What is the core loss if the frequency is 10,000Hz?
   (b) What is the core loss if the frequency is 60 Hz and the voltage is 440V?
   (c) What can we conclude from (a) and (b)?

BIBLIOGRAPHY

[1] C. A. Gross, *Power System Analysis*, John Wiley & Sons, Hoboken, New Jersey, 1979.
[2] S. J. Chapman, *Electric Machinery Fundamentals*, 5th ed., McGraw-Hill, New York, 2012.
[3] T. Gönen, *Electrical Machines with MATLAB*, 2nd ed., Taylor & Francis Group, Boca Raton, FL, 2011.
[4] W. L. Matsch and J. D. Morgan, *Electromagnetics and Electromechanical Machines*, 3rd ed., John Wiley & Sons, Hoboken, NJ, 1986.
[5] M. S. Sarma, *Electric Machines: Steady-State Theory and Dynamic Performance*, WMC Brown Publishers, Dubuque, IA, 1985.
[6] R. Ramshaw and R. G. van Heeswijk, *Energy Conversion: Electric Motors and Generators*, Saunders College Publishing, Philadelphia, 1990.
[7] G. R. Slemon, *Electric Machines and Drives*, Addison-Wesley Publishing, New York, 1992.
[8] S. Falcones, R. Ayyanar, and X. Mao, "A DC–DC Multiport-Converter-Based Solid-State Transformer Integrating Distributed Generation and Storage," *IEEE T. Power Electr.* 28, 5 (May 2013).
[9] B. L. Theraja and A. K. Theraja, *A Textbook of Electrical Technology*, 1st multicolour ed., S. Chand, New Delhi, 2005.
[10] D. Gerling, *Electrical Machines: Mathematical Fundamentals of Machine Topologies*, e-book edition, Springer, New York, 2015.
[11] S. A. Nasar, *Electric Energy Conversion and Transmission*, Macmillan Publishing Company, New York, 1985.
[12] A. E. Fitzgerald, C. Kingsley, and S. D. Umans, *Fitzgerald & Kingsley's Electric Machinery*, 4th ed., McGraw-Hill, New York, 1983.

CHAPTER 4

TRANSFORMERS

4.1 INTRODUCTION

Electric energy processing involves conversion, transfer, transmission, control, and storage of energy. Sources of energy can be converted into many forms (such as chemical, mechanical, electrical, sunlight, and wind, and audio, visual, or otherwise) including data processing by computer. This energy when transformed can be transferred and distributed. Potential energy resulting from rainfall in lakes and rivers can be converted to kinetic energy of seawater and tides, which could supply 50 percent of total world energy demand. Other forms of energy exist, such as kinetic energy of wind and solar. Other forms of energy include geothermal energy and fossil fuels such as oil, coal, and gas. Mechanical to electrical energy as simplified energy processing supply chain is shown in Figure 4.1.

Energy sources that are transformed and transferred are sent from the source to the consumer in different forms as needed. They are easily measurable, observable, and controllable to meet industrial, commercial, and residential loads. The creative and associative voltage and current achieve the production of power transfer. For mechanical power force and velocity are associated with mechanical power. But what is the link that enables one form of power to be converted to another?

Lorentz's force equation provides the answer; it is the force experienced by an electric charge externally by applied field, electric field, and magnetic field. The vector equation is given as

$$F = q\ [E + (V \times B)] \tag{4.1}$$

where F is the electromagnetic force (Newton), q is the charge (Coulombs), E is the external electric field (V/m), V is the velocity of the charge (m/s), and B is the external magnetic field (Tesla).

Since motion of charge in a conductor constitutes electric current, Lorentz's equation can be written as

$$F = q\ [E + l\ (I \times B)] \tag{4.2}$$

where I is the current in the conductor (amperes), and l is the effective length in meters of the conductor subjected to the magnitude of the magnetic field.

Energy Processing and Smart Grid, First Edition. James A. Momoh

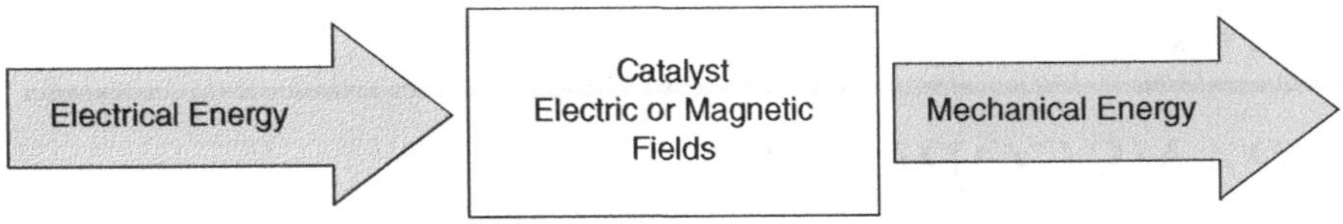

Figure 4.1 Electric or magnetic fields as catalysts for electro-mechanical energy conversion.

Whenever there is field acting on a charge, it is expected that energy conversion is possible. Therefore, electric and magnetic fields serve as catalysts that enable electromechanical energy conversion to take place. Without the presence of electric and magnetic fields, no electromechanical energy conversion is possible. Examples of electromechanical energy conversion devices include electric rotating machines (generator/motors), pickups, microphones, voltmeters, ammeters, wattmeters, relays, and transformers. Some of them are presented here for illustration; they all satisfy Lorentz's equation.

Finally, machine design or simply magnetic field machines (which allow conversion) is governed by physical laws of energy-conversion processing. The design must be optimized according to some criteria to enable efficiency and low cost of operation or production.

4.2 FIRST TWO MAXWELL'S LAWS

The conversion of electro-mechanical energy involves two basic empirical Maxwell's laws.

(a) Faraday's law states that electromotive force induced in a coil of wire is proportional to the rate of change of magnetic flux linking the coil of wire.

$$e = -\frac{d\lambda}{dt} \tag{4.3}$$

$$e = BLV \tag{4.4}$$

Or by vector notation

$$e = \ell.(B \times V) \tag{4.5}$$

(b) Lorentz's law states that the current-carrying conductor will experience a mechanical force when placed in a magnetic field. This force is proportional to the magnitude of the current, the magnetic field, and the effective length of the conductor subjected to the magnetic field.

$$F = qE + L\ (I \times B) \tag{4.6}$$

Note that Lorentz's force has direction when

$$F = BIL \sin \alpha\ (Newton)$$
$$F = i(I \times B) \tag{4.7}$$

α gives the direction of the magnetic field.

Using the right-hand rule, if ℓ conductors carry a current i and are subject to a uniform magnetic field of density B (Tesla), when the conductor is arranged in perpendicular position to the direction of magnetic field,

$$\alpha = 90^\circ$$
$$\sin\alpha = 1.$$

Then, a force is produced given by

$$F_{max} = Bil \tag{4.8}$$

This direction leads to generator actions. However, reversing the direction of α using the left-hand corkscrew rule then

$$F_{max} = -BIL. \tag{4.9}$$

The Faraday Case
Assume

$$e = -\frac{d\lambda}{dt}.$$

This leads to

$$-d\varphi = B \times d\,(area) = BlVdt.$$

Therefore

$$e = \frac{BlV\,dt}{dt} = BlV$$
$$e = BlV \sin\alpha \sin\beta$$

at maximum

$$e_{max} = BlV$$
$$e = \ell.(\mathbf{B} \times \mathbf{V})\,. \tag{4.10}$$

4.3 TRANSFORMERS

A transformer is a static device that has no moving parts but allows transfer of AC power at different voltage levels. The two electric windings wrapped in a common core can be air but are usually of ferromagnetic material. The transfer is done by electromagnetic induction (Faraday's law). Transformers are grouped in different forms based on their application and use.

4.3.1 Power and Distribution Transformers

Power and distribution transformers are used primarily for transforming voltages from one level to another. Power transformers in particular are used at transmission

Figure 4.2 Power and distribution transformer. *Courtesy of ABB Inc.*

levels to boost voltage at generation (600V–13.2 kV) to transmission voltages ranging from 132 kV to 1,000 kV and above. At the point of distribution, they are also used to step down the voltage to distribution level voltages, as seen Figure 4.2 [20].

Distribution transformers are most common at load centers. They operate at a minimum of 120/240 V (single phase/three phase) for domestic applications, while distribution voltages for commercial and industrial areas can be much higher. Distribution transformers may be mounted on plinths a little above ground level or pole mounted (see Figure 4.3 (a) and (b)) [5], [20].

4.3.2 Isolating Transformers

Isolating transformers, as shown in Figure 4.4, [22] are used to electrically isolate electric circuits from each other or to block DC signal while maintaining AC continuity between circuits. They can also eliminate electromagnetic noise in many circuits.

4.3.3 Instrument Transformers

Instrument transformers are used to measure high voltages and large currents with standard small-range voltmeters, ammeters, and wattmeters, then transform voltage or current to relatively small/safe values. They are classified as potential (voltage) transformers and current transformers, and are single-phase transformers used to step

(a) (b)

Figure 4.3 Distribution transformers. (a) Pole-mounted distribution transformer (b) Ground-mounted distribution transformer. *Courtesy of ABB Inc.*

down voltage or currents to safe values. Figures 4.5 and 4.6 [19] reveal the schematic diagrams and the real diagrams of instrument transformers.

4.3.4 Communication Transformers

Communication transformers are used in electrical devices for impedance matching for improved power transfer, for isolation of DC current, and as input transformers, output transformers, and insulation apparatuses between circuits. They can be

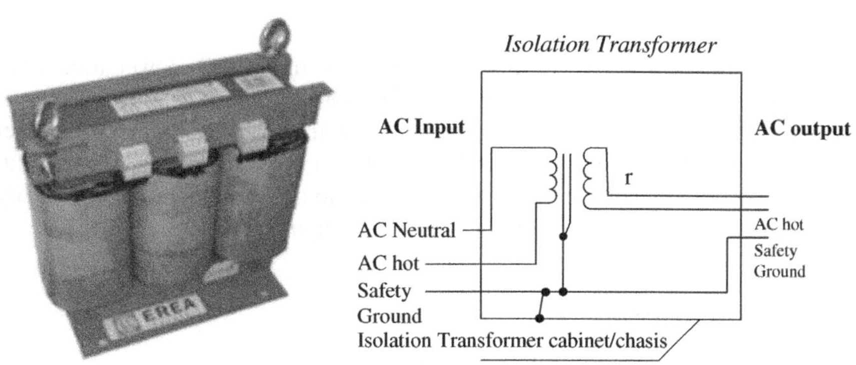

(a) (b)

Figure 4.4 (a) Physical isolating transformer. (b) Isolating system schematic. *Courtesy of EREA Energy Engineering BVBV.*

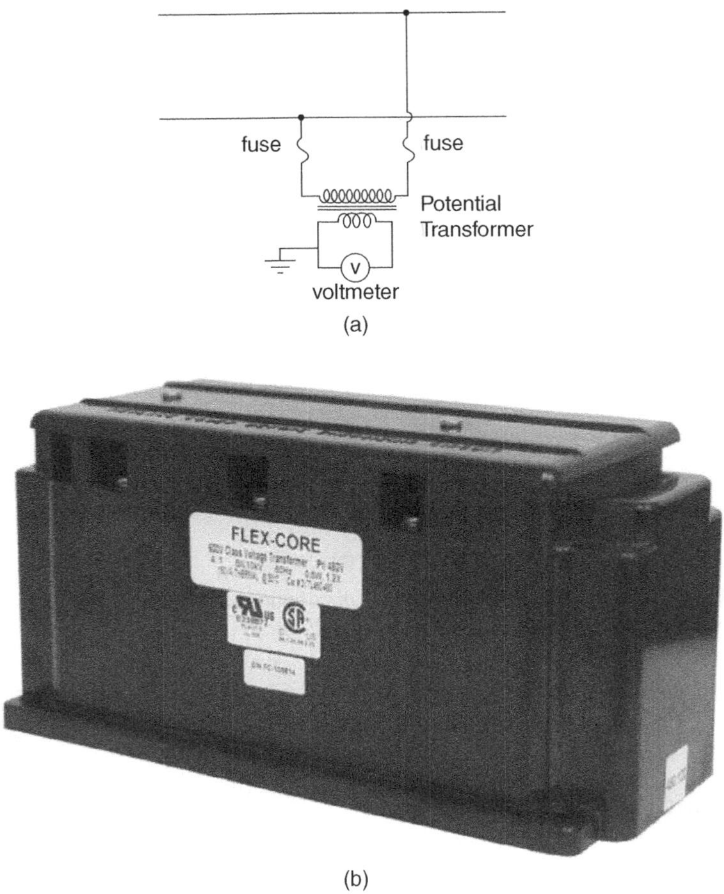

Figure 4.5 (a) Schematic of a potential transformer. (b) Three-phase physical voltage transformer. *Courtesy of Flex-Core.*

used in frequencies from 60 Hz to 400 Hz or more. In general, power transformers are designed to work at 50 Hz or 60 Hz frequencies. Transformers designed at 50 Hz or 60 Hz can be used at higher frequencies, but transformers designed at higher frequencies cannot operate at lower (50 Hz or 60 Hz) because the core would saturate and the secondary voltage would not be similar to or proportional to the primary voltage. A communication transformer is shown in Figure 4.7 [21].

4.3.5 Autotransformers

The two windings of an autotransformer are not connected to each other. Hence power is transferred from one side to the other, and only by induction. There is no electrical isolation between input side and output side—so power from primary to secondary is by induction. Figure 4.8 [22] shows an autotransformer.

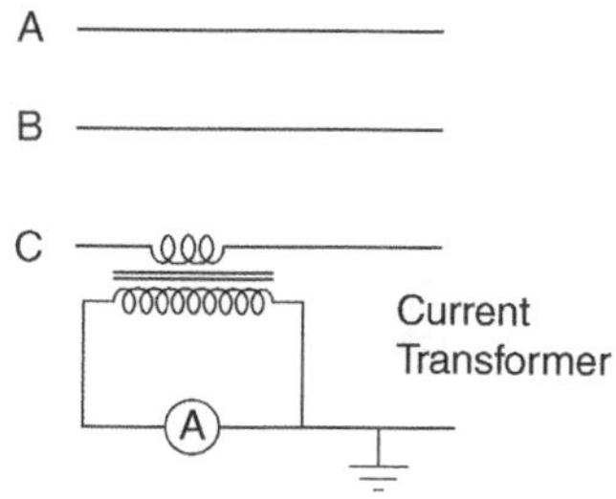

(a)

(b)

Figure 4.6 (a) Schematic of a current transformer. (b) Solid-core current transformer. *Courtesy of Flex-Core.*

4.3.6 Transformer Construction and Sizes

Magnetic cores of transformers in power systems are built in core type or shell type. In either case, they are made of laminated silicon steel sheets to minimize eddy-current loss/core loss. Silicon contents decrease magnetic losses by 3 percent and 97 percent.

Transformers are constructed in different shapes and sizes for different purposes.

- **(a) Distribution transformers** include single- or three-phase transformers with ratings of 500kVA at voltage levels of 69 kV.
- **(b) Power transformers** are rated in MVA for high voltage and typically used for interstate power transfer.

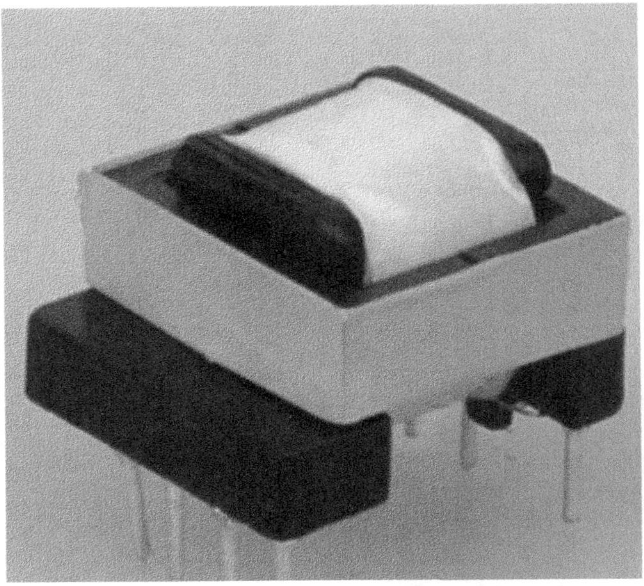

Figure 4.7 Communication transformer. *Courtesy of Hangzhou Smart Technology.*

Most transformer windings are immersed in a tank of oil for better insulation and cooling. When no ferromagnetic materials but only air is present, such transformers are called *air-core transformers*, have poor magnetic coupling, and are only used in electronic circuits. In this chapter, the focus is on the *iron-core transformer*.

4.4 IDEAL SINGLE-PHASE TRANSFORMER MODELS

The transformer winding is modeled as the ideal case, where primary winding is referred to as N_P and secondary winding as N_S. From Figure 4.8, we assume:

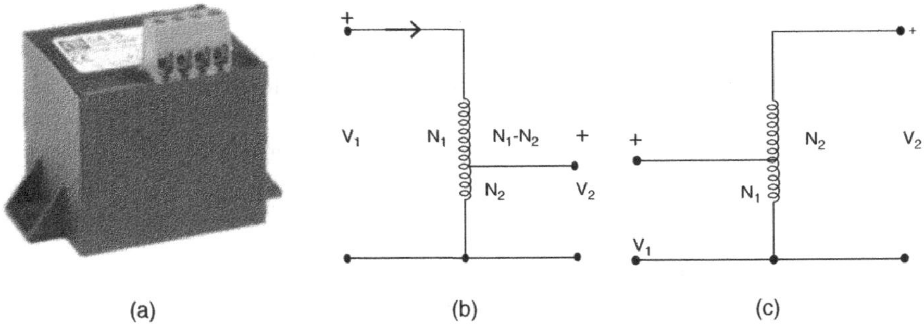

Figure 4.8 Autotransformer and its models. (a) Physical autotransformer (b) Step-down model (c) Step-up model. *Courtesy of EREA Energy Engineering BVBV.*

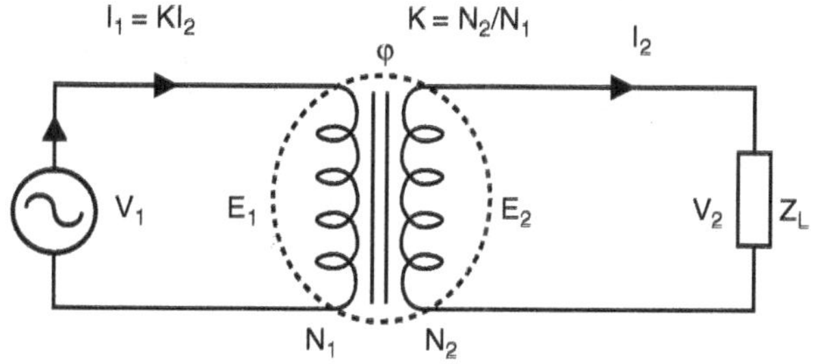

Figure 4.9 Ideal transformer model.

(a) The winding resistance is negligible.

(b) Magnetic field intensity required to magnetize the core is negligible.

(c) The core losses are negligible.

(d) The magnetic flux is confined to the ferromagnetic core.

The flux generated in the core is time varying; emf e_1 will be induced in a coil linking the flux. This leads to induced emf as in the primary winding N_1 and induced emf e_2 in the coil of secondary (N_2) turns winding.

The input power in the transfer without losses is absorbed power by the losses, hence the property of the ideal transformer is a loss-free machine. If a voltage V_1 is applied to the terminals of N_1 turns winding, as shown in Figure 4.9, a magnetizing current will flow, producing a flux φ and flux linkage λ in the winding.

The rate of change of flux with respect to time given e_1 induced voltage is written in mathematical form as

$$e_1 = V_1(t) = \frac{d\lambda_1}{dt} = N_1 \frac{d\varphi}{dt}. \tag{4.11}$$

The flux φ links N_2 turns winding, producing flux linkage λ_2, gives

$$e_2 = V_2(t) = \frac{d\lambda_2}{dt} = N_2 \frac{d\varphi}{dt}. \tag{4.12}$$

Rewriting to account for N turns

$$e_1 = -\frac{d\lambda_1(t)}{dt} = -N_1 \frac{d\varphi}{dt} \tag{4.13}$$

$$e_2 = -\frac{d\lambda_2(t)}{dt} = -N_2 \frac{d\varphi}{dt}. \tag{4.14}$$

Combining both equations

$$\frac{V_1}{V_2} = \frac{e_1(t)}{e_2(t)} = \frac{N_1}{N_2} = a \tag{4.15}$$

where a is called the *transformer ratio*.

This equation provides a ratio of instantaneous values of induced voltages as a function of the turn ratio is a constant.

Suppose the load circuit is now connected to secondary N_2 winding, causing current i_2 to flow to a sink. Also, consider the flux path around the core. Flux around the closed path must be zero, therefore we can write

$$N_1 i_1 = N_2 i_2, N_1 i_1 - N_2 i_2 = 0. \tag{4.16}$$

Thus, the current ratio is

$$\frac{i_1}{i_2} = \frac{N_2}{N_1} = \frac{1}{a}. \tag{4.17}$$

The last equation for the current ratio is written

$$\frac{i_1}{i_2} = \frac{N_2}{N_1} = \frac{1}{a}$$

with the product

$$VI = \frac{V_1(t)}{V_2(t)} \times \frac{i_1}{i_2} = 1. \tag{4.18}$$

This confirms the loss-free ideal transformer, where the power input equals the power output at all instances of time. This is called the *power invariance principle* that means VA is conserved.

Option 1

Given the turn ratio

$$\frac{V_1}{V_2} = a \tag{4.19}$$

and

$$\frac{i_1}{i_2} = \frac{1}{a}. \tag{4.20}$$

Then

$$\frac{V_1/V_2}{i_1/i_2} = a\left(\frac{1}{a}\right)^{-1} = a^2. \tag{4.21}$$

Similarly,

$$\frac{V_1}{V_2} \times \frac{i_2}{i_1} = a^2$$

$$\frac{V_1}{i_1} \times \frac{i_2}{V_2} a^2. \tag{4.22}$$

If V_2 is load voltage, I_2 is load current.

So, the ideal transformer connected to the terminals of windings of 2 to the load impedance Z_{in} of the source will see a^2Z_{load}. That means we have transferred the load impedance from the secondary-winding side of the transformer to the primary winding side.

Example

A power of 10 kW is to be supplied to an electric furnace at a voltage 50/230 V, 60 Hz.

(1) Compute the turn ratio.

(2) Determine the primary and secondary currents.

(3) Determine the impedance by 230 V supply.

Turn ratio

$$\frac{V_1}{V_2} = \frac{N_1}{N_2} = \frac{230}{50} = 4.6 = a$$

Secondary current

$$\frac{1}{a} = \frac{i_1}{i_2} = \frac{1}{4.6}$$

$$4.6\, i_1 = i_2$$

$$I_2 = \frac{P_2}{V_2} = \frac{10,000}{50} = 200A$$

$$i_i = \frac{2,000}{4.6} = 43.5$$

load

$$z_2 = \frac{V_2}{i_2} = \frac{50}{200} = 0.25 + j0$$

equating

$$Z_1' = \left(\frac{N_1}{N_2}\right)^2 Z_2$$

$$(4.6)^2.(0.25 + j0)$$

$$= (5.29 + j0)\,\Omega$$

4.5 MODELING A TRANSFORMER INTO EQUIVALENT CIRCUITS

An ideal transformer does not account for the actual practical behavior of a transformer. To include the effects of imperfection, a more accurate model of the equivalent circuit is needed. The circuit equivalent must include:

(a) Winding resistance. For both the primary and secondary resistance, typically the winding resistance is maintained at about 0.1 percent of the terminal voltage of large transformers.

(b) Magnetizing inductance. To account for core material used in designing the transformer, the permeability μ_r of the material must be determined. The value allows us to determine the flux such that $N_1\varphi = \lambda$, which are the flux linkages in the transformer's magnetic property. Given the cross-sectional area of the core, the core density is given by

$$B = \varphi/A. \tag{4.23}$$

The corresponding field intensity H around the core path length is given as

$$B = \mu_0\mu_1 H. \tag{4.24}$$

So, the model of magnetizing current i_{im} must flow in N_1 and then by Ampère's circuital law

$$N_1 i_{im} = Hl. \tag{4.25}$$

Therefore, for mostly

$$\lambda = N_1\varphi = N_1 BA = \frac{{N_1}^2\mu_1\mu_o A}{l} i_{im} = L_{im} i_{im} \tag{4.26}$$

for secondary side

$$L_{2m} = \left(\frac{N_2}{N_1}\right)^2 L_{im},$$

hence,

$$i_{2m} = \left(\frac{N_1}{N_2}\right) i_{1m} \tag{4.27}$$

Again, magnetizing current is small in many transformers and may be specified as determined from a series of tests.

4.5.1 Modeling of Transformer Losses

The transformer core is made from ferromagnetic materials so it suffers the same problem of magnetic losses, such as eddy and hysteresis losses. They are expressed based on an empirical formula for core losses as

$$P_c = K_c\left(\frac{dB}{dt}\right)^2 \text{ W/m}^3. \tag{4.28}$$

In terms of induced voltage

$$e_1 = \frac{d\lambda}{dt} = N_1 A \frac{dB}{dt} \tag{4.29}$$

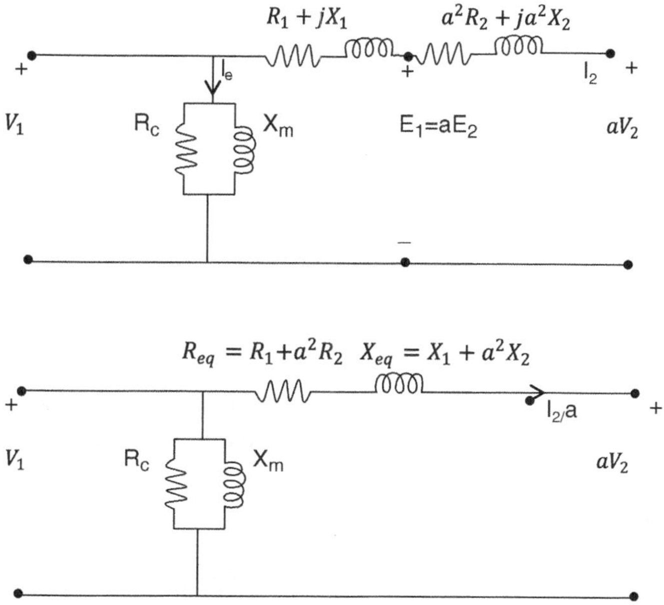

Figure 4.10 Approximate equivalent circuit of real transformer (parameters referred to primary side).

which gives:

$$P_c = Al\rho_c \tag{4.30}$$

where A is the area, l is the length, ρ_c is the core loss density.

$$P_c = AlK_c\left(\frac{e_1}{N_1A}\right)^2 = \frac{e_1^{\,2}}{R_{ic}} \tag{4.31}$$

where $R_{ic} = \frac{N_1^{\,2}A}{K_c l}$, which can also be determined experimentally.

Since core loss is small for most transformers it is negligible. But to compute the actual transformer efficiency and temperature we need to include all the losses.

4.5.2 Transformer-Equivalent Circuits

We can take advantage of the turn ratios to refer the parameters of a transformer from the secondary side to the primary side and vice versa without showing the ideal transformer.

Alternative 1: Referred to Primary Side

Figure 4.12 shows the final approximate equivalent circuit of a transformer referred to primary side neglecting shunt branch and winding resistance. Firstly, on Figure 4.10, the approximate equivalent circuit of a real transformer referred to primary side is given. Secondly Figure 4.11 shows the approximation when shunt branch is neglected.

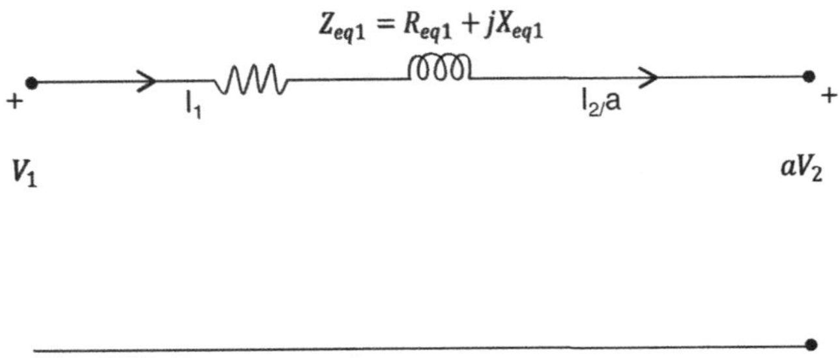

Figure 4.11 Final approximate equivalent circuit referred to the primary neglecting shunt branch.

The final approximate equivalent circuit neglects the transformer resistance and the shunt arm parameters since their effect in real transformer operation is negligible.

Alternative 2: Referred to Secondary Side

Alternatively, we may refer to parameters of transformer to the secondary side, as seen on Figure 4.13. The approximate equivalent circuit of the transformer is shown in Figure 4.14, which uses an equivalent impedance. Combining the secondary resistance and inductance with the primary resistance and inductance referred to the secondary and writing the resistances and reactances in terms of overall equivalent quantities, we have the equivalent impedance approximation.

Further, neglecting shunt branch will give Figure 4.15. Finally, approximate equivalent circuit when resistance is neglected will result, Figure 4.16.

Writing the summary of the equivalent circuit, if the transformation ratio is

$$a = \frac{N_2}{N_1} = \frac{V_2}{V_1}$$

R_1, R_2, X_1, and X_2 are the primary and secondary resistance and reactance, respectively.

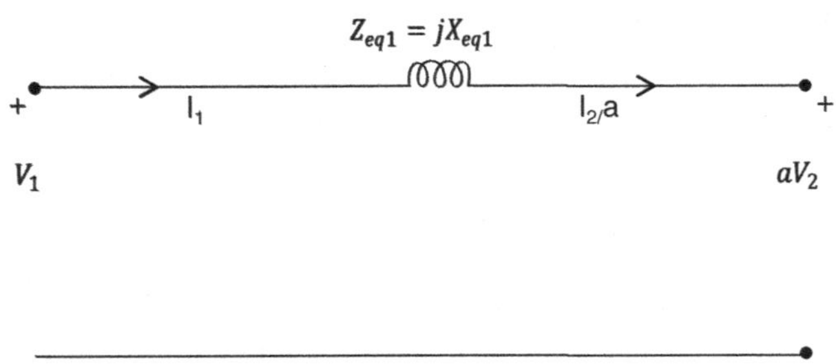

Figure 4.12 Final approximate equivalent circuit referred to the primary neglecting shunt branch and winding resistance.

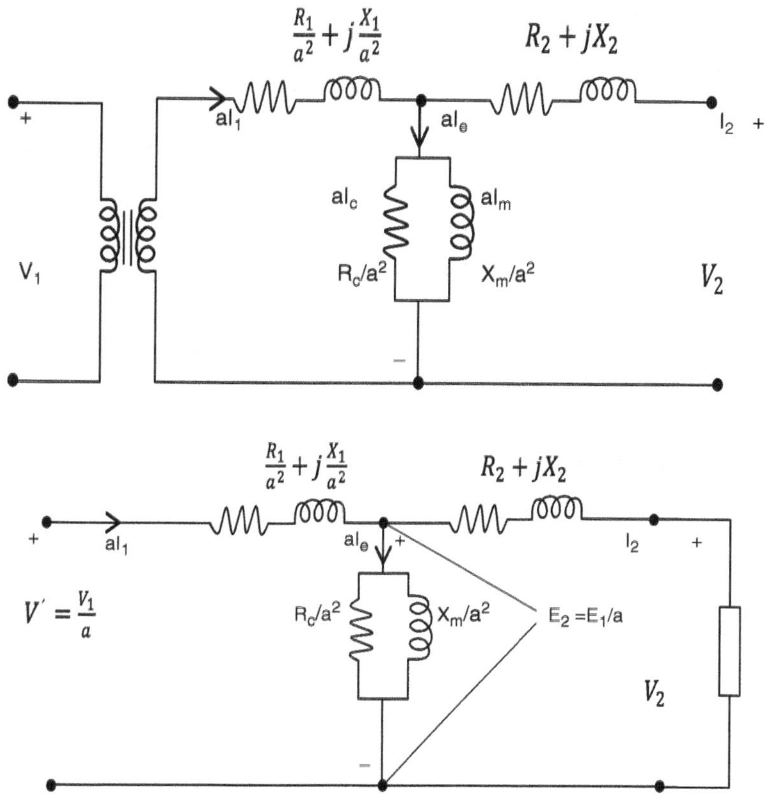

Figure 4.13 Approximate equivalent circuit of real transformer with parameters referred to the secondary side.

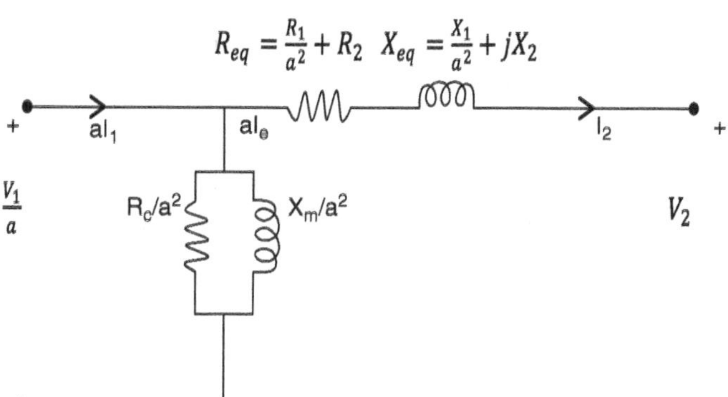

Figure 4.14 Approximate equivalent circuit of real transformer (parameters referred to the secondary side).

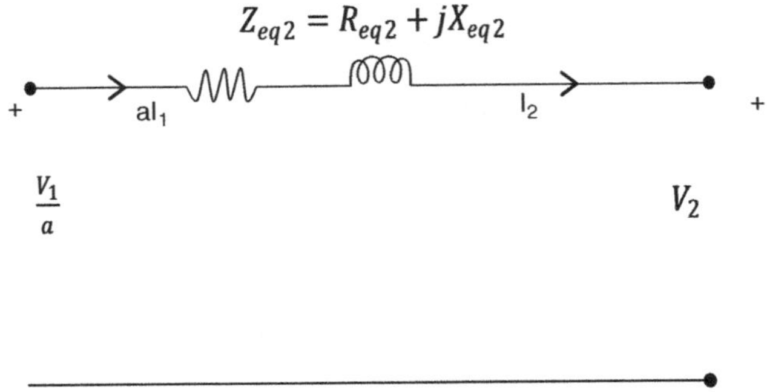

Figure 4.15 Final approximate equivalent circuit referred to the secondary, neglecting shunt branch.

Primary resistance referred to the secondary is

$$R_1' = \frac{R_1}{a^2}$$

$$X_1' = \frac{X_1}{a^2}. \tag{4.32}$$

Overall resistance and reactance on the secondary side is

$$R_1' + R_2 = \frac{R_1}{a^2} + R_2 = R_{eq2}$$

$$X_1' + X_2 = \frac{X_1}{a^2} + X_2 = X_{eq2} \tag{4.33}$$

$$Z_{eq2} = \left(\frac{R_1}{a^2} + R_2\right) + \left(\frac{X_1}{a^2} + X_2\right)$$

$$= R_{eq2} + \mathrm{j}X_{eq2}. \tag{4.34}$$

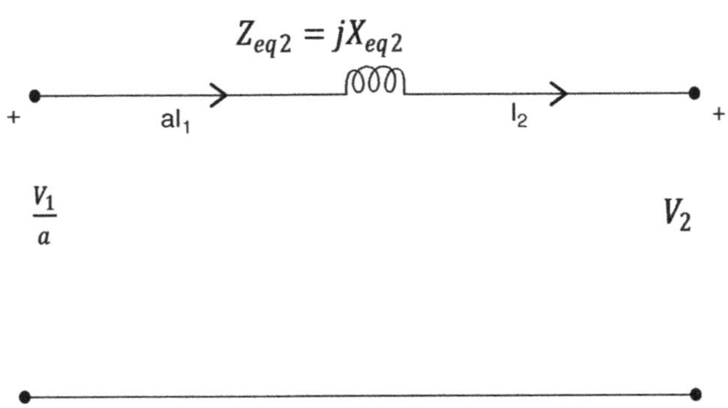

Figure 4.16 Final approximate equivalent circuit referred to the primary neglecting shunt branch and winding resistance.

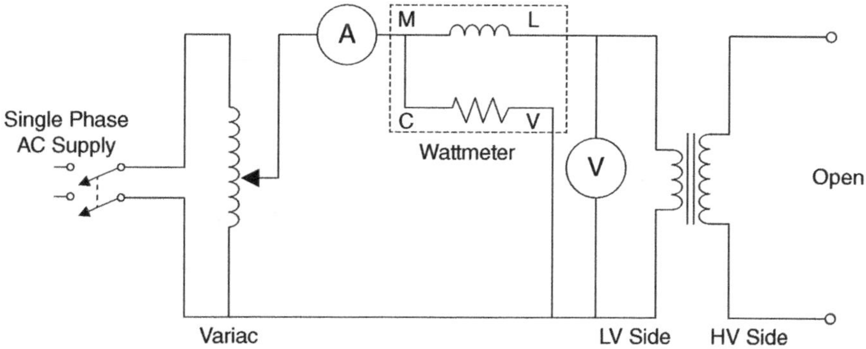

Figure 4.17 Circuit for open-circuit test.

Referring secondary parameters to primary side is

$$R_2' = a^2 R_2$$

$$X_2' = a^2 R_2. \tag{4.35}$$

Overall resistance and reactance on the primary side is

$$R_1 + R_2' + = R_1 + a^2 R_2 = R_{eq1}$$

$$X_1 + X_2' + = X_1 + a^2 X_2 = X_{eq1} \tag{4.36}$$

$$Z_{eq2} = (R_1 + a^2 R_2) + (X_1 + a^2 X_2)$$

$$= R_{eq1} + jX_{eq1}. \tag{4.37}$$

4.5.3 Determination of Equivalent-Circuit Parameters

The parameters of transformers that make up the model in Section 4.4.2 consist of R_1, R_2, X_1, X_2, G_m, and B_m. X_1 and X_2 are the leakage reactance, which accounts for magnetic flux that does not couple all turns. X_m is associated with mutual flux of the two windings and the resistance R_c accounts for losses due to hysteresis and eddy current. By performing two tests (open circuit and short circuit), given the transformer voltage and power rating, we can determine the values of theses parameters under loading conditions.

4.6 TRANSFORMER TESTING

Transformer testing comprises the open-circuit and short-circuit tests.

4.6.1 Open-Circuit Test

The simple test setup is shown in Figure 4.17.

The yields for the equivalent circuit during open circuit is shown in Figure 4.18.

At $I_2 = 0$, R_1, X_1, X_2, and R_2, do not contribute to the equivalent circuit, but X_m and R_m do. We determine them as follows:

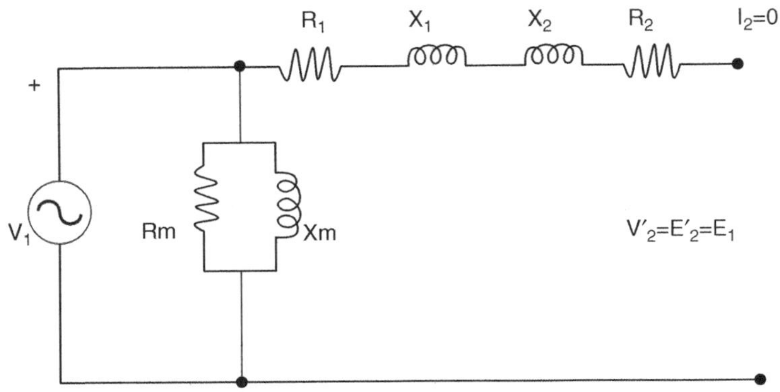

Figure 4.18 The equivalent circuit during open circuit.

(a) voltmeter reading is the open-circuit voltage $= V_{oc}$
(b) ammeter reading is the open-circuit current $= I_{oc}$
(c) wattmeter reading, which accounts for iron loss $= P_{oc}$

A tabular record of the results and other computations derived therefrom may be put up as shown in Table 4.1.

Using the results of the open-circuit test, we compute

$$Y_{eq} = \frac{1}{R_m} + \frac{1}{jX_m} = G_m - jB_m \tag{4.38}$$

$$Y_{eq} = Y_{open\ cct} = \frac{I_{oc}}{V_{oc}} \angle \theta_{oc} \tag{4.39}$$

where θ_{oc} is the angle of Y_{eq} (admittance) from the open circuit, then cos θ_{oc} is the power factor, which is found as:

$$\frac{P_{oc}}{V_{oc} I_{oc}}. \tag{4.40}$$

$$\theta_{oc} = Cos^{-1} Pf_{oc} \text{ for a given } P_{oc}$$

For a given transformer θ_{oc} is always lagging behind the power factor as shown

$$Y_{eq} = Y_{oc} = \frac{I_{oc}}{V_{oc}} \angle - \theta_{oc}. \tag{4.41}$$

TABLE 4.1 Open-Circuit Test Results

S/N	Open-Circuit Voltage V_{oc} (V)	Open-Circuit Current I_{oc} (A)	Core Loss P_{oc} (W)	Power-Factor Angle θ_{oc} (degrees)	G_c (Mho)	B_c (Siemens)	Y_c (Siemens)

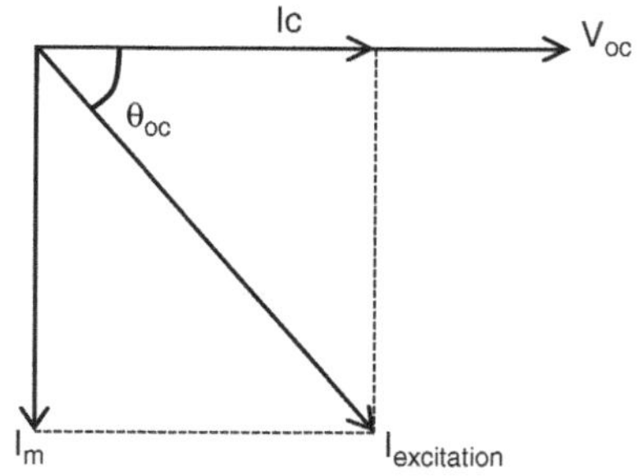

Figure 4.19 No-load phasor diagram.

Therefore

$$Y_{eq} = Y_{oc} = G_c - jB_m, \tag{4.42}$$

where $G_c \geq G_m$ for core losses, so we have $\frac{1}{R_c} = G_c$ and $\frac{1}{X_m} = B_m$, respectively, computing

$$G_c = G_{oc} = \frac{P_{oc}}{V_{oc}^2} \tag{4.43}$$

and

$$B_m = B_{oc} = \sqrt{\left(Y_{oc}^2 - G_{oc}^2\right)}. \tag{4.44}$$

The value of R_c and X_m are now known from Equations 4.43 and 4.44.

Procedure for Open-Circuit Test Computation

(a) Using V_{oc} and I_{oc}, compute Y_{oc}, which is a complex quantity with magnitude Y_{oc} and argument θ_{oc}, where θ_{oc} is the power-factor angle on open circuit.
(b) Note that $Y_{oc} = G_c + jB_m$.
(c) Compute $G_c = G_{oc} = \frac{P_{oc}}{V_{oc}^2}$.
(d) Compute $B_m = B_{oc} = \sqrt{\left(Y_{oc}^2 - G_{oc}^2\right)}$.
(e) Obtain R_c and X_m from the inverses of G_c and B_m from the open-circuit test

$$R_c = \frac{1}{G_{oc}} \; and \; X_m = \frac{1}{B_m}.$$

This completes the shunt-admittance pointer identification. Its no-load phasor diagram is determined and presented in Figure 4.19.

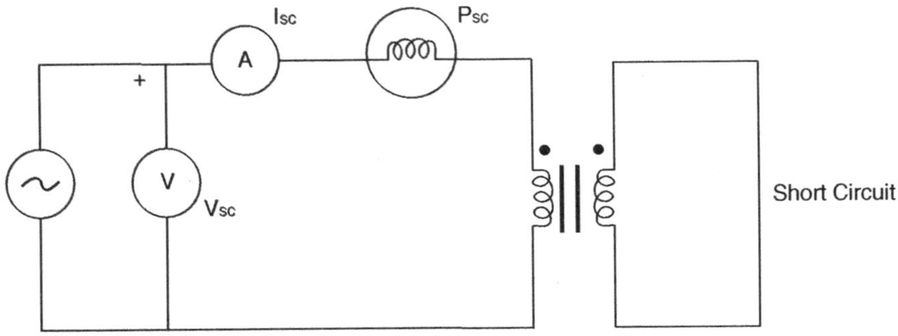

Figure 4.20 Set-up for short-circuit test.

4.6.2 Short-Circuit Test

The short-circuit test is called the *copper loss test* and helps determine the series parameters of the primary and secondary winding model. Short circuiting one winding performs the test; typically, the primary winding is energized while the secondary winding is short circuited. Assuming the core losses are ignored and $P_{sc} = P$ at short circuit, the power measured by the wattmeter is the copper loss $P_{cu.}$ The set-up for conducting the test is shown in Figure 4.20.

Upon short circuiting the system above, we redraw the equivalent circuit diagram as shown in Figure 4.21.

From the power measured, P_{sc}, V_{sc}, and I_{sc} are known experimentally. Hence, a table of values can be set up as shown in Table 4.2.

Using Table 4.2, compute

$$\theta_{sc} = Cos^{-1}\frac{P_{sc}}{V_{sc}I_{sc}}. \tag{4.45}$$

This is identically equal to

$$\theta_{sc} = Cos^{-1} PF_{sc}. \tag{4.46}$$

Again, the power factor is always lagging, hence the negative θ_{sc}.

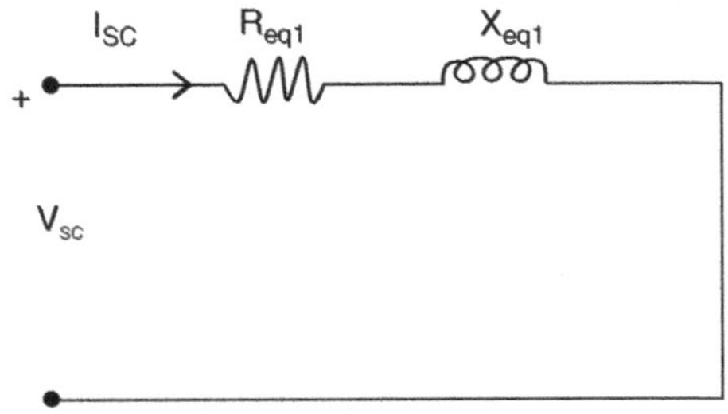

Figure 4.21 Equivalent circuit diagram.

TABLE 4.2 Short-Circuit Test Results

S/N	Short-Circuit Voltage V_{sc} (V)	Short-Circuit Current I_{sc} (A)	Copper Loss P_{oc} (W)	Power-Factor Angle θ_{sc} (degrees)	Z_{eq} (Ohm)	R_{eq} (Ohm)	X_{eq} (Ohm)

We now compute

$$Z_{eq1} = \frac{V_{sc}}{I_{sc}} \angle -\theta_{sc} \tag{4.47}$$

$$Z_{eq1} = R_{eq1} + jX_{eq1} \tag{4.48}$$

with respect to secondary side added

$$R_{eq1} = R_1 + a^2 R_2 \tag{4.49}$$

$$X_{eq1} = X_1 + a^2 X_2. \tag{4.50}$$

Alternatively, using the wattmeter, voltmeter, or ammeter readings, we can find

$$R_{eq1} = R_{sc} + \frac{P_{sc}}{I_{sc}^2} \tag{4.51}$$

$$X_{eq1} = X_{sc} = \sqrt{Z_{eq}^2 - R_{eq}^2} \tag{4.52}$$

where

$$Z_{eq1} = Z_{sc} = \frac{V_{sc}}{I_{sc}}. \tag{4.53}$$

Under this condition, the phasor equivalent of the transformer is given as shown in Figure 4.22.

Procedure for Short-Circuit Test Calculation

(a) Ammeter reads the short circuit current I_{sc}.

(b) Voltmeter reads the short circuit voltage V_{sc}.

(c) Wattmeter reads the transformer copper loss $P_{sc} = V_{sc} I_{sc} \cos\theta_{sc}$.

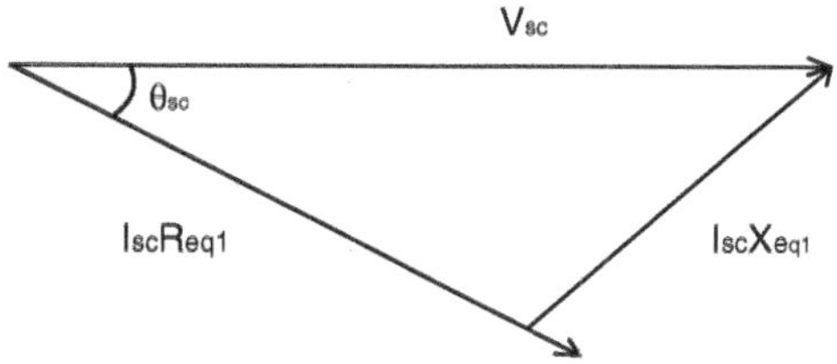

Figure 4.22 Phasor equivalent of the transformer.

(d) Compute power factor θ_{sc} using P_{sc}, V_{sc} and I_{sc}

$$\theta_{sc} = Cos^{-1}\frac{P_{sc}}{V_{sc}I_{sc}}$$

Note that θ_{sc} is always lagging, so always append the negative sign.

(e) Compute the equivalent impedance of the transformer referred to the primary side Z_{eq1} using, V_{sc}, I_{sc}, and θ_{sc} (note that Z_{eq1} is a complex quantity whose real and imaginary parts are R_{eq1} and X_{eq1}, respectively).

(f) Recall that $R_{eq1} = R_1 + a^2R_2$, where R_1 and R_2 are the primary and secondary resistances, respectively.

(g) Also, $X_{eq1} = X_1 + a^2X_2$, where X_1 and X_2 are the primary and secondary reactances, respectively
Note that if R_2' is secondary resistance referred to the primary side,

$$R_2' = a^2R_2.$$

In the same vein,

$$X_2' = a^2X_2.$$

Note that a is the transformation ratio.

Example

A 200/1,000 V, 50 Hz, single-phase transformer gave the following test results:

- open-circuit test (LV/primary side): 200 V, 1.2A, 90 W
- short-circuit test (HV/secondary side): 50 V, 5A, 110 W

Calculate the parameters of the equivalent circuit referred to the LV side.

Solution:

$$\text{Transformation ratio} = \frac{V_2}{V_1} = \frac{200}{1,000} = \frac{1}{5}$$

Shunt branch parameters from the open-circuit test: We need

$$R_c = \frac{V^2}{P_i} = 200^2/90 = 44\,\text{ohm},$$

$$I_c = 200/444 = 0.45\text{A}$$

$$I_m = \left(I_e^2 - I_c^2\right)^{0.5}$$

$$I_m = \left(1.2^2 - 0.45^2\right)^{0.5} = 1.11\text{A}$$

$$X_m = 200/1.11$$
$$= 180.2\,\text{ohm}.$$

Since the short-circuit test has been conducted from the HV side, the parameters will be computed on the secondary/HV side and then referred to primary/LV side by appropriately transforming them.

From the short-circuit test readings,

$$Z_{eq2} = \frac{V_{sc}}{I_{sc}} = 50/5 = 10\,\text{ohm}$$

$$R_{eq2} = \frac{P_{sc}}{I_{sc}^2} = 110/(5)^2 = 4.40\,\text{ohm},$$

$$X_{eq2} = \sqrt{Z_{eq2}^2 - R_{eq2}^2}$$

$$= \left(10^2 - 4.4^2\right)^{0.5} = 8.9\,\text{ohm}.$$

Referring these to the LV/primary side using the turns ratio

$$R_{eq1} = a^2 R_{eq2} = (1/5)^2 \times 4.4$$

$$= 0.176\,\text{ohm}$$

$$X_{eq1} = a^2 X_{eq2} = (1/5)^2 \times 8.98$$

$$= 0.36\,\text{ohm}.$$

Equivalent circuits can be drawn with R_c and X_m calculated from above and R_{eq1} and X_{eq1} as above.

4.7 TRANSFORMER SPECIFICATIONS

The information on a transformer's nameplate (as seen Figure 4.23) includes:

- kVA/MVA rating
- voltage rating
- impedance
- turn ratio
- input kVA (slightly less than output kVA)
- possibly a figure of merit in percent, which provides the transformer impedance
- sometimes an excitation current
- manufacturer and serial number

Other types of transformers include:

- autotransformer
- phase-shifter transformer
- on-load tap changer (OLTC)

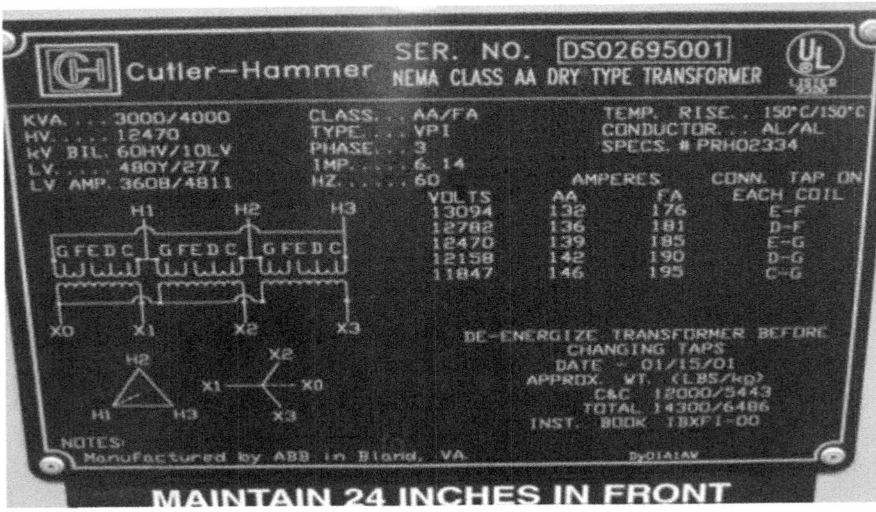

Figure 4.23 Typical transformer nameplate showing its parameters.

4.8 THREE-PHASE POWER TRANSFORMERS

Since the generated power is mostly three phase and transmitted also as three phases, the transfers are also so used. A three-phase transformer connection is necessary to step up or step down voltage levels at different source locations. The three-phase transformer is built for each phase of the primary and secondary winding associated with it. The following configurations are possible:

- Y: Y
- Δ: Δ
- Y: Δ
- Δ: Y

Each of these is described in Table 4.3, including some of the challenges or regulations required.

4.9 NEW ADVANCES IN TRANSFORMER TECHNOLOGY: SOLID-STATE TRANSFORMERS — AN INTRODUCTION

Transformers are indispensable components of electric power systems. They are rugged and have an average efficiency of about 97.5 percent. However, they possess some undesirable properties, including sensitivity to harmonics, voltage drop under load, the need for protection from system disruptions and overload, protection of the system from problems arising at or beyond the transformer, environmental concerns regarding mineral oil, and performance under DC-offset-load unbalances.

TABLE 4.3 Three-Phase Power Transformer Configuration

Schematic Diagram	Wiring Diagram	Descriptions
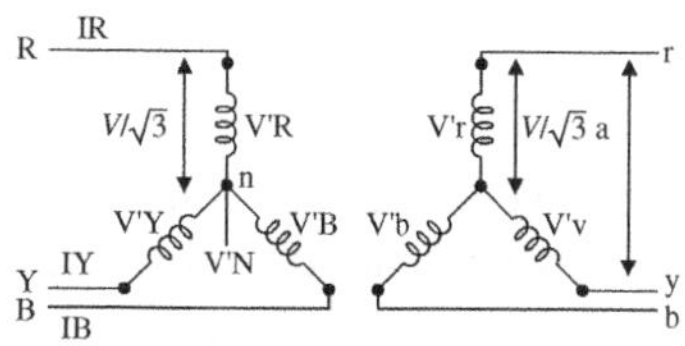	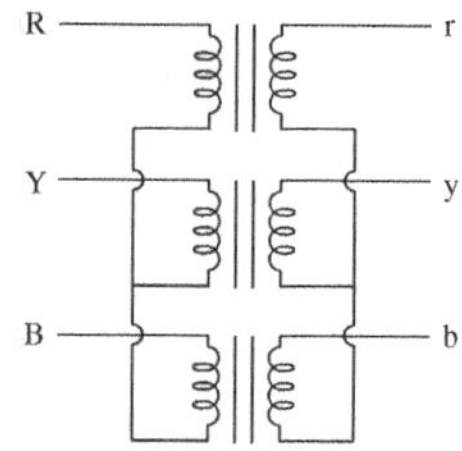	Y: Y There is no phase difference. It creates no problem as long as it is solidly grounded.
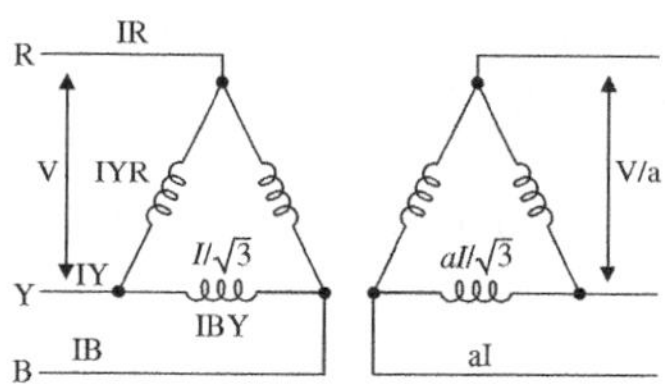	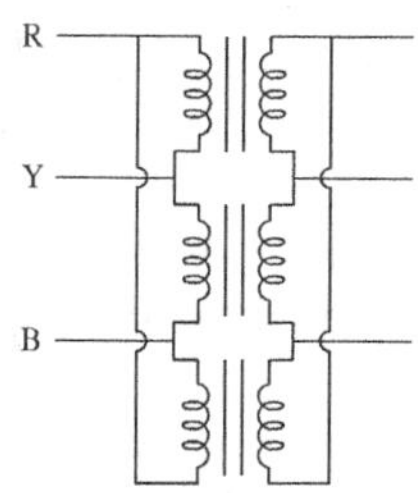	Δ: Δ Under unbalanced conditions, the line currents are many times greater than winding currents when the third harmonics in the excitation current are ignored. No phase shift, no problem with unbalanced load or harmonics.
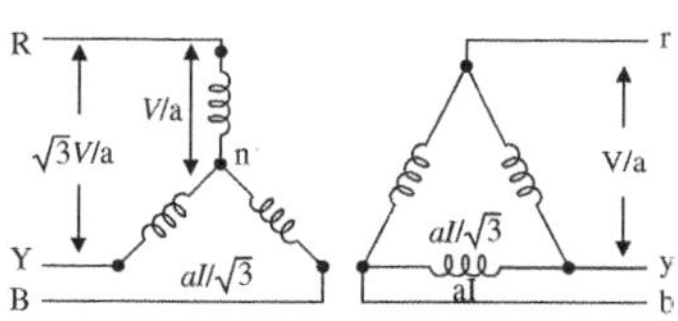	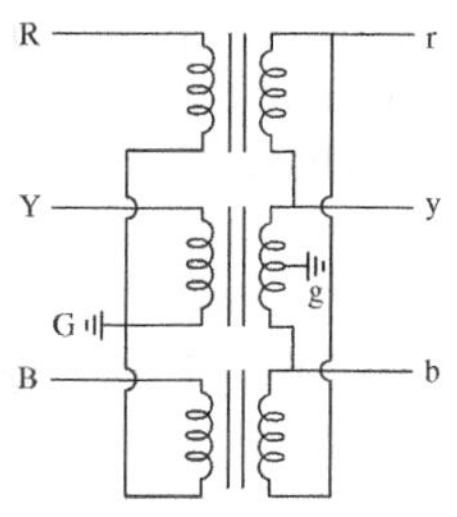	Y: Δ There is no problem with third-harmonic components in the voltages, since they are absorbed in the circulating current on the delta side. This connection can be used with unbalanced loads but Y must be grounded.
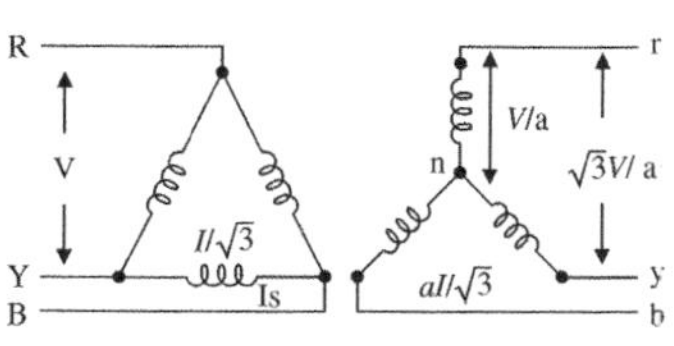	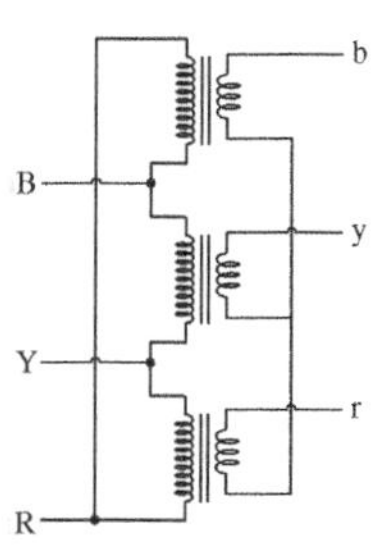	Δ: Y There is no third-harmonic component in voltage. The secondary voltage lags the primary voltage by 30°.

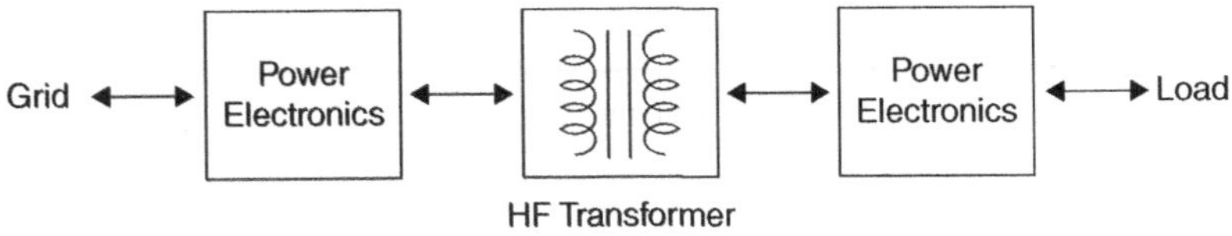

Figure 4.24 Basic solid-state transformer structure.

Consequently, research has been focused recently on ways overcome such concerns. With the advancement of power-electronic circuits and devices, the all-solid-state transformer is a viable option to replace conventional copper- and iron-based transformers for better power quality.

Solid-state transformers (SSTs) are a collection of high-powered semiconductor components, conventional high-frequency transformers, and control circuitry used to provide a high level of flexible control to power-distribution networks, thereby facilitating the smooth conversion of AC to DC and DC to AC as required. Add some communication capability and the entire package is often referred to as a *smart transformer*.

The basic structure of an SST is depicted in Figure 4.24. The isolation is achieved through a high-frequency (HF) transformer. The grid voltage is converted into HF AC voltage through power-electronics-based converters before being applied to the primary side of the HF transformer. The opposite process is performed on the HF transformer's secondary side to obtain an AC and/or DC voltage for the load.

Several SST configurations have been proposed. They are (Figure 4.25):

- single-stage with no DC link
- two-stage with low-voltage DC (LVDC) link

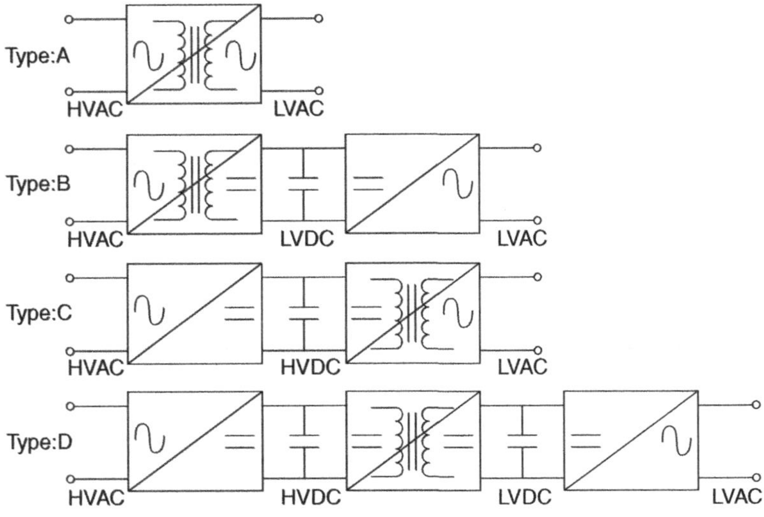

Figure 4.25 Different SST configurations.

- two-stage with high-voltage DC (HVDC) link
- three-stage with both HVDC and LVDC links

The DC link of the third configuration is not appropriate for distributed energy resources (DER) integration since it is high voltage and has no isolation from the grid; therefore, topologies under that classification are not practical for SST implementation.

Advantages of the SST

Apart from the fact that it can step up or step down AC voltage levels, the SST has the following advantageous functionalities:

- allows two-way power flow
- inputs or outputs AC or DC power
- actively changes power characteristics such as voltage and frequency levels
- improves power quality (reactive power compensation and harmonic filtering)
- provides efficient routing of electricity based on communication between utility provider, end-user site, and other transformers in the network
- greatly reduces the physical size and weight of individual transformer packages with equivalent power ratings

Illustrative Problems and Examples

Problem 1

(a) List the types of losses that occur in a transformer.

Solution:

The losses that occur in transformer are:

- core losses: hysteresis loss and eddy-current loss
- copper or ohmic losses
- stray losses

(b) What problems are associated with the Y–Y phase transformer connection?

Solution:

Problems associated with the Y–Y phase transformer connection include:

- third harmonic component of each phase will be in phase with each other, adding up and becoming large
- voltage drop at unbalanced load
- over excitation in fault condition
- neutral shifting
- over voltage at light loads

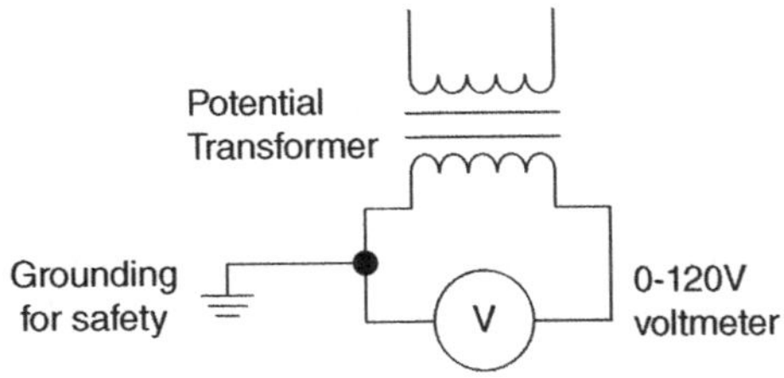

Figure P1 A potential transformer.

(c) Can a 60 Hz transformer be operated on a 50 Hz system? What actions are necessary to enable this operation?

Solution:

A 60 GHz transformer cannot be operated on a 50 Hz system. To enable this operation, a current limiter device (inductor or resistor) should be installed in series with the windings so that impedance can be increased to reduce the extra current.

(d) What is a potential transformer? How is it used?

Solution:

A potential transformer (as seen in Figure P1) is an instrument transformer used to measure high voltages with a standard, small-range voltmeter, which transforms the voltage to relatively small or safe values for metering. They are single-phase transformers used to step down voltages to safe (metering) values.

Problem 2

A 60 Hz, 120/24 V, single-phase ideal transformer has 300 turns in its primary winding. Determine the following:

(a) value of its primary and secondary currents

(b) number of turns in its primary windings

(c) maximum flux Φ_m in the core

Solution:

(a) We need more parameters to determine particular values of the currents. In terms of N_S and N_P,

$$N_P/N_S = V_P/V_S$$
$$300/N_S = 120/24 \Rightarrow N_S = 60$$

Therefore,

$$I_P = I_S \times N_S/N_P = I_S \times 60/300 = I_S/5$$

(b) $N_P = 300$, $N_S = 60$

(c) $V_P = 4.44\ f\Phi N_P$
So,

$$\Phi = V_P/\left(4.44\ f N_P\right)$$
$$= 120/(4.44 \times 60 \times 300)$$
$$= 1.5 \times 10^{-3}\ \text{Wb}$$

Problem 3

A single-phase, 2,500/250 V, ideal transformer has a load of 10 ∠40° connected to its secondary. If the primary of the transformer is connected to a 2,400 V line, determine the following:

(a) secondary current

(b) primary current

(c) input impedance as seen from the line

(d) output power of the transformer in kVA and kW

(e) The input power of the transformer in kVA and kW

$$N_S/N_P = 250/2,500$$
$$V_S = V_P \times N_S/N_P = 2,400 \times 250/2,500 = 240V$$
$$\text{load } Z_L = 10\angle 40°$$

Solution:

(a) Secondary current

$$I_S = 250V/(10\angle 40) = 25\angle -40°\ A$$

(b) Primary current

$$I_P = N_S/N_P \times I_S = 250/2,500 \times 24\angle -40° = 2.4\angle -40°\,A$$

(c) Input impedance

$$Z = E_P/I_P = 2,400/(2.4\angle -40°)$$
$$= 1,000\angle 40\ \text{ohm}$$

(d) $S_{out} = I_s V_s = 24\angle -40 \times 240 = 5,760\angle -40°$
Hence,

$$S_{out} = 5,760\text{VA} \quad \text{or} \quad 5.76\text{kVA}$$
$$S_{out} = 4,412.4 - j3,702.5$$
$$P_{out} = 4,412.4\text{W} = 4.412\text{kW}$$

(e) $S_{in} = I_p V_p = 2,400 \times 2.4 \times < -40°$
$= 5,760\angle -40°$
$|S_{in}| = 5,760\text{VA}$

Since,

$$S_{in} = 5,760\angle -40^\circ = 4,412.4 - j3,702.5$$

Therefore,

$$P_{in} = 4,412.4\text{W}$$

4.10 CHAPTER SUMMARY

The chapter provided concepts of transformer modeling and control, and the general use of transformers for power-energy systems. Transformers are critical components of any electric power system for step up or step down of voltage, instrumentation, and communication. Built on the principle of electromagnetic induction—an outflow of Maxwell's equations—transformers are powerful tools in the design of controls for electric power systems, whether operating in normal or fault modes. Advances in solid-state-transformer technology are needed for smart/microgrid systems where solid-state inverters are used. Development of commercial versions are bound to enhance their application in microgrids.

Essential concepts for transformers of different applications—including large-power transformers and distributed transformers needed for relay-protection control schemes—have been presented. Principles of modeling parameters and performance measures, such as efficiency and regulation, were also presented, and several exercises were given as illustrative examples of transformer operation and design.

EXERCISES

1. **(a)** List the types of losses that occur in a transformer.
   (b) Why does the power factor of a load affect the voltage regulation of a transformer?
   (c) What are the problems associated with the Y–Y phase transformer connection?
   (d) Can a 60 Hz transformer be operated on a 50 Hz system? What actions are necessary to enable this operation?
   (e) What is a potential transformer? How is it used?

2. A 100,000kVA, 230/115 kV, Δ–Δ, three-phase power transformer has a per unit resistance of 0.01 and a per unit reactance of 0.055. The excitation branch elements are $R_C = 1.00$ per unit and $X_m = 2.0$ per unit.
   (a) If this transformer supplies a load of 100MVA at 0.75 power factor lagging, draw the phasor diagram of one of the transformers.
   (b) What is the voltage regulation of the transformer bank under these conditions?
   (c) Sketch the equivalent circuit referred to the low-voltage side of one phase of this transformer.
   (d) Calculate the entire transformer impedances referred to the low-voltage side.

3. The results of tests on a single-phase, 10kVA, 480/120 V, 60 Hz distribution transformer are as shown in Table Q3.

TABLE Q3 Transformer Test Results

Open-Circuit Test (on Secondary Side)	Short-Circuit Test (on Primary Side)
$V_{oc} = 480V$	$V_{sc} = 1,140V$
$I_{oc} = 5.1A$	$I_{sc} = 1.40A$
$P_{oc} = 640W$	$P_{sc} = 560W$

(a) Find the equivalent circuit of this transformer referred to the low-voltage side of the transformer.

(b) Find the transformer's voltage regulation at rated conditions, with 0.8 power factor lagging and 0.8 power factor leading.

(c) Determine the transformer's efficiency at rated conditions and 0.8 power factor lagging.

(d) How does the efficiency of the transformer at rated conditions and 60 Hz compare to the same physical device running at 50Hz?

4. A single-phase 2,500/250 V, ideal transformer has a load of $10\angle 40°$ connected to its secondary. If the primary of the transformer is connected to a 2,400 V line, determine the following:

(a) secondary current

(b) primary current

(c) input impedance as seen from the line

(d) output power of the transformer in kVA and kW

(e) input power of the transformer in kVA and kW

5. Assume that an ideal transformer is used to step down 13.8–2.4 kV and that it is fully loaded when it delivers 100kVA. Determine the following:

(a) its turns ratio

(b) rated current for each winding

(c) load impedance referred to the high-voltage side, corresponding to full load

(d) load impedance referred to the low-voltage side, corresponding to full load

6. If a 60 Hz, 120/24 V, single-phase ideal transformer has 300 turns in its primary winding, determine the following:

(a) value of its primary and secondary currents

(b) number of turns in its primary windings

(c) maximum flux Φ_m in the core

7. (a) Distinguish between a current and potential transformer.

(b) Distinguish between a power and autotransformer.

(c) Will a transformer designed to work in the United States at 60 Hz work in Europe/Africa, where the frequency is 50Hz?

(d) Consider a core loss $P_{core} = 16$ W, volt-amperes for a given core $(VI)_{rms} = 20$VA, and induced voltage $V = 274/\sqrt{2} = 194V_{rms}$ when the winding has 200 turns. Find the power factor, core-loss current I_c, and the magnetizing current I_m.

8. A 15kVA, single-phase transformer with primary voltage of 2,100 V has a primary resistance of 1.0 ohm and a secondary resistance of 0.01 ohm. Find the equivalent secondary resistance and the full-load efficiency at 0.8 power factor if the iron loss of the transformer is 80 percent of full-load copper loss. Take the transformation ratio to be $\frac{1}{10}$.

9. Figure Q9 shows an equivalent circuit of an ideal transformer with an impedance R2 + J X2 = 1 + j4, connected in series with the secondary. The turns ratio N1/N2 = 5:1.

(a) Draw an equivalent circuit with the series impedance referred to the primary side.

(b) For a primary voltage of 120 V rms and a short connected across the secondary terminals (V2 = 0), calculate the primary current and the current flowing in the short.

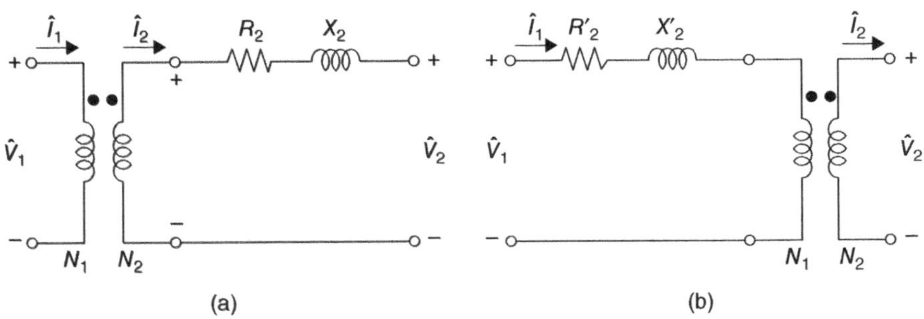

Figure Q9 Equivalent circuit diagram of the transformer.

10. The secondary winding of an ideal transformer has a terminal voltage of $v_s(t) = 282.8 \sin 377t$ V. The turns ratio of the transformer is 0.5. If the secondary current of the transformer is $i_s = 7.07 \sin(377t - 36.87°)$ A, what is the primary current of this transformer? What are the voltage regulation and efficiency? The impedances of the transformer referred to the primary side are:

$$R_{eq} = 0.20\,\Omega, \quad X_{eq} = 0.80\,\Omega, \quad R_c = 300\,\Omega, \quad X_m = 100\,\Omega$$

11. A 1,000VA, 230/115 V transformer was tested to determine its equivalent circuit. The results of the tests are shown in Table Q11.

TABLE Q11 Transformer Test Results

Metering	Open-Circuit Test (on Secondary Side)	Short-Circuit Test (on Primary Side)
Voltmeter	$V_{oc} = 115$V	$V_{sc} = 17.1$V
Ammeter	$I_{oc} = 0.11$A	$I_{sc} = 8.7$A
Wattmeter	$P_{oc} = 3.9$W	$P_{sc} = 38.1$W

(a) Find the equivalent circuit of this transformer referred to the low-voltage side of the transformer.

(b) Find the transformer's voltage regulation at rated conditions and (i) 0.8 power factor lagging, (ii) 1.0 power, and (iii) 0.8 power factor leading.

(c) Determine the transformer's efficiency at rated conditions and 0.8 power factor lagging.

12. A 10kVA, 480/120 V conventional transformer is to be used to supply power from a 600 V source to a 480 V load. Consider the transformer to be ideal, and assume that all insulation can handle 600 V.

(a) Sketch the transformer connection that will do the required job.

(b) Find the kVA rating of the transformer in the configuration.

(c) Find the maximum primary and secondary currents under these conditions.

13. Consider a 25kVA, 2,400/240 V, 60 Hz transformer with $Z_1 = 2.533 + j2.995\Omega$ and $Z_2 = (2.5333 + j2.995) \times 10^{-2}\ \Omega$ referred to the primary and secondary sides, respectively. Find V_1, aV_2, I_1, I_2, R_{eq1}, and X_{eq1}.

14. The transformer describe in Problem 10 is connected at the receiving end of a feeder that has an impedance of $0.3 + j1.8\Omega$. Let the sending-end voltage magnitude of the feeder be 2,400 V. Also, there is a load connected to the secondary side of the transformer that draws rated current from the transformer at 0.85 lagging power factor.

(a) Neglecting the excitation current of the transformer, determine the secondary-side voltage of the transformer under such conditions.

(b) Draw the associated phasor diagram.

15. Consider a 75kVA 2,400/240 V, 60 Hz transformer subjected to open-circuit and short-circuit tests. The results of the tests are shown in Table Q15.

TABLE Q15 Transformer Test Results

Metering	Open-Circuit Test (on Secondary Side)	Short-Circuit Test (on Primary Side)
Voltmeter	$V_{oc} = 2,400V$	$V_{sc} = 200V$
Ammeter	$I_{oc} = 0.50A$	$I_{sc} = 31.25A$
Wattmeter	$P_{oc} = 75W$	$P_{sc} = 800W$

Assuming that the transformer is operated at 0.92 lagging power factor, determine the following:

(a) equivalent impedance, resistance, and reactance of the transformer all referred to the primary side

(b) total loss, including the copper loss and core loss at full load

(c) efficiency of the transformer

(d) percentage voltage regulation of the transformer

BIBLIOGRAPHY

[1] J. Casazza and F. Delea, *Understanding Electric Power Systems: An Overview of the Technology, the Marketplace, and Government Regulations*, John Wiley & Sons, Hoboken, NJ, 2003.

[2] H. Saadat, *Power System Analysis*, 3rd ed., PSA Publishing, 2010.

[3] A. E. Fitzgerald, C. Kingsley, and S. D. Umans, *Fitzgerald & Kingsley's Electric Machinery*, 4th ed., McGraw-Hill, New York, 1983.

[4] C. A. Gross, *Power System Analysis*, John Wiley & Sons, Hoboken, NJ, 1979.

[5] S. J. Chapman, *Electric Machinery Fundamentals*, 5th ed., McGraw-Hill, New York, 2012.

[6] T. Gönen, *Electrical Machines with MATLAB*, 2nd ed., Taylor & Francis Group, Boca Raton, FL, 2011.

[7] W. L. Matsch and J. D. Morgan, *Electromagnetics and Electromechanical Machines*, 3rd ed., John Wiley & Sons, Hoboken, NJ, 1986.

[8] M. S. Sarma, *Electric Machines: Steady-State Theory and Dynamic Performance*, WMC Brown Publishers, Dubuque, IA, 1985.

[9] R. Ramshaw and R. G. van Heeswijk, *Energy Conversion: Electric Motors and Generators*, Saunders College Publishing, Philadelphia, 1990.

[10] G. R. Slemon, *Electric Machines and Drives*, Addison-Wesley Publishing, New York, 1992.

[11] S. Falcones, X. Mao, and R. Ayyanar, "Topology Comparison for Solid State Transformer Implementation," IEEE PES General Meeting, July 2010.

[12] B. L. Theraja and A. K. Theraja, *A Textbook of Electrical Technology*, 1st multicolour ed., S. Chand, New Delhi, 2005.

[13] S. J. Chapman, *Electric Machinery Fundamentals*, 1st ed., McGraw-Hill, New York, 2001.

[14] D. K. Rathod, "Solid State Transformer (SST): Review of Recent Developments," *Advance in Electronic and Electric Engineering* **4**, 1 (2014).

[15] X. Zhang, P. Wei, W. Deng, et al., "Emerging Smart Grid Technology for Mitigating Global Warming," *Int. J. Energ. Res.* 39, 13 (2015).

[16] Allied Market Research, *World Solid State (Smart) Transformer Market: Opportunities and Forecasts, 2014–2020*, www.alliedmarketresearch.com/solid-state-transformer-market.

[17] FREEDM Architecture, "Solid State Transformer (SST)," nd, freedm.wordpress.ncsu.edu/sst.

[18] J. H. Harlow, *Electric Power Transformer Engineering*, 2nd ed., CRC Press, New York, 2006.

[19] Flex-Core 4970 Scioto Darby Rd. #D, Hilliard, OH 43026.

[20] ABB Ltd., Affolternstrasse 44, 8050 Zurich, Switzerland.

[21] Hangzhou Smart Technology Co. Ltd., Room 2303, Unit 1, Building 5, Fuleide Plaza, No. 99 Wenze Road, Jinjikaifa District, Hangzhou, Zhejiang, China (310000).

[22] EREA Energy Engineering BVBV Ruggeveldstraat 1, 2110 Wijnegem, Belgium.

CHAPTER 5

INDUCTION MACHINES

5.1 INTRODUCTION

Induction machines are the most common form of electromechanical energy-conversion devices for industrial, commercial, and domestic appliances that operate at constant speed. The need for energy processing and conversion from electrical to mechanical energy to do mechanical work abounds in every sector of human endeavor. About one-third of the world's electricity consumption is used for running induction motors that drive pumps, fans, compressors, elevators, and machinery of various types to meet industrial, commercial, and domestic needs.

The AC induction motor is a common form of asynchronous motor whose operation depends on three electromagnetic phenomena:

(a) Motor action. When an iron rod (or other magnetic material) is suspended in a magnetic field so that it is free to rotate, it will align itself with the field. If the magnetic field is moving or rotating, the iron rod will move with the moving field so as to maintain alignment.

(b) Rotating field. A rotating magnetic field can be created from fixed-stator poles by driving each pole pair from a different phase of the AC supply.

(c) Transformer action. The current in the rotor windings is induced from the current in the stator windings, avoiding the need for a direct connection from the power source to the rotating windings. Consequently, the induction motor can be considered an AC transformer with a rotating secondary winding.

5.2 CONSTRUCTION AND TYPES OF INDUCTION MOTORS

The induction motor stator resembles the stator of a revolving-field, three-phase, synchronous generator. The stator or the stationary part consists of three-phase winding held in place in the slots of a laminated steel core enclosed and supported by a cast-iron steel frame, as seen in Figure 5.1 [5].

The phase windings are placed 120° electrical apart and may be externally connected in Y or delta for which six leads are brought out to a terminal box mounted

Energy Processing and Smart Grid, First Edition. James A. Momoh

Figure 5.1 Induction motor stator.

on the frame of the motor. When the stator is energized from a three-phase voltage, it produces a rotating magnetic field in the stator core.

Three-phase induction motors are of two kinds:

- cage-rotor or squirrel-cage induction motors
- wound-rotor or slip-ring induction motors

Cage Rotors

Cage-rotor machines (also called squirrel-cage machines) are the most common type of induction motors. In a cage-rotor design, solid conductors are in slots on the rotor. The ends of the conductors are short circuited at each end of the rotor using an "end-ring." For small to medium-sized machines (up to a few hundred horsepower) the rotor conductors are cast using aluminum. This construction makes the rotor relatively cheap to produce. In larger machines, rotors are usually made by manually hammering solid copper bars into the rotor slots, then manually brazing an end-ring into place. Fabricated rotor cages are significantly more expensive that cast rotor cages. Figure 5.2 [5] shows a cage-rotor induction motor.

Wound Rotors

Wound-rotor induction machines have a three-phase winding, similar to the stator winding, on the rotor as seen in Figure 5.3 [5]. The rotor is usually Y connected with the terminals of the three rotor phases connected to slip rings. In normal operation, the windings at the slip rings are short circuited to allow currents to flow. An advantage of wound-rotor machines is that external circuits can be connected to the rotor, allowing external control of the machine. While all induction machines can be controlled to operate at different torques and speeds, wound-rotor control is particularly attractive in some applications. However, wound-rotor induction machines are usually significantly more expensive than cage rotor machines. Possible applications for wound-rotor machines include:

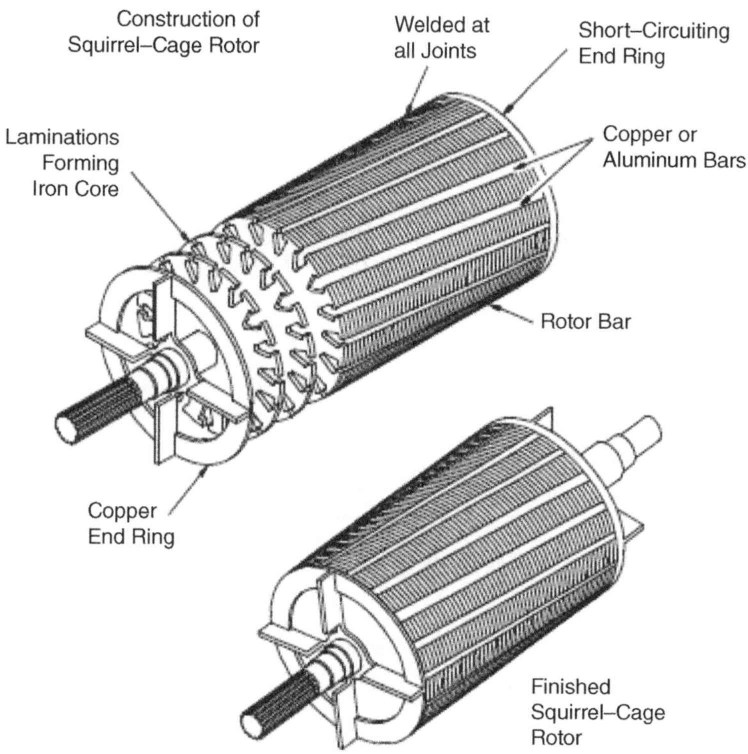

Figure 5.2 Cage-rotor induction motor.

- speed control of very large machines (multi-milliwatt)
- reduced cost control of large machines
- doubly fed induction generation (used in some wind turbines)

5.3 OPERATING PRINCIPLE

The basic idea behind the operation of an induction machine is quite simple. The three-phase stator winding is connected to a three-phase supply. Currents flow in the stator winding, producing an mmf and flux density that rotates at synchronous speed:

$$N_s = \frac{120f}{P} \tag{5.1}$$

The magnetic field passes conductors on the rotor and induces a voltage in those conductors. Since the conductors are short circuited, current flows in the rotor conductors, producing a second rotor magnetic field, which acts to oppose the stator

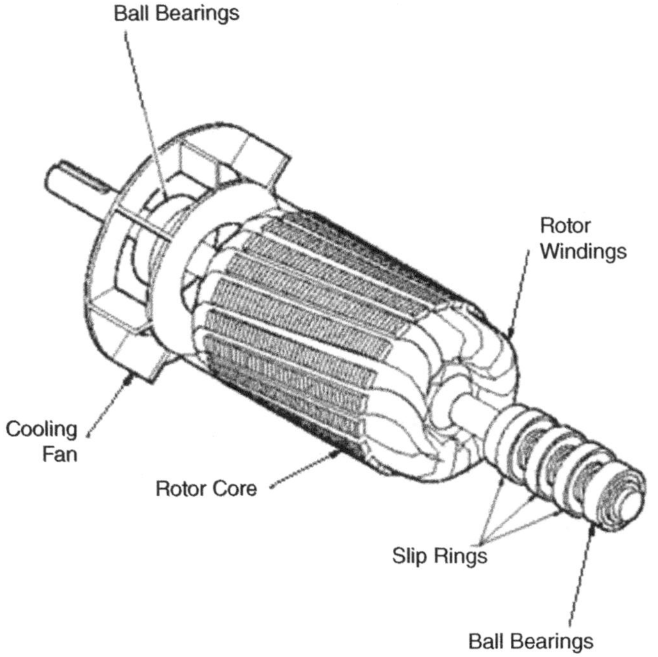

Figure 5.3 Wound-rotor induction motor.

magnetic field and rotates at synchronous speed. With two magnetic fields rotating at constant speed, a torque is produced.

5.4 BASIC INDUCTION-MOTOR CONCEPTS

The following induction-motor concepts are important to understanding the operation and application of these machines.

Leakage Reactance

The leakage reactance is for magnetizing the armature of the cylindrical-rotor machines. It is given by

$$X_{mu} = \frac{8f\mu_o mD_g L}{g_e}\left(\frac{K_{w1}N_{ph1}}{p_{a1}}\right)^2. \tag{5.2}$$

The mutual air-gap flux between the rotor and the stator is ϕ (weber, or Wb) per pole and rotates at an angular speed of ω radians per second. The flux linkage of N-turns of conductors will be at maximum when $\omega t = 0$ and also at $\omega t = 90°$. Assuming distributed winding, the flux linkage λ_a will vary as the cosine of the angle ωt. Thus, the flux linkage with coil a is

$$\lambda_a = N\phi \cos \omega t. \tag{5.3}$$

The induced voltage

$$e_a = -\frac{d\lambda}{dt} = \omega N\phi \sin \omega t \tag{5.4}$$

$$e_a = E_{max} \sin \omega t$$

$$e_a = E_{max} \cos\left(\omega t - \frac{\pi}{2}\right) \tag{5.5}$$

where

$$E_{max} = \omega N\phi = 2\pi f N\phi.$$

The rms value of the generated voltage is then

$$\frac{2\pi f N\phi}{\sqrt{2}} \tag{5.6}$$

$$E = 4.44 f N\phi. \tag{5.7}$$

In actual machines, the armature conductors are distributed in slots such that the phasor sum of the induced emf is less than the numerical sum. Hence, a factor to account for the difference is introduced, which we shall call the *winding factor* K_w (note that K_w varies between 0.85 and 0.95 for three-phase windings). Consequently, for distributed-phase windings, the rms value of the generated voltage is:

$$E = 4.44 K_w f N\phi \,(\mathrm{V}). \tag{5.8}$$

Let stator emf be E_1 and rotor emf E_r. Hence, we may write

$$E_1 = \frac{4.44 f K_{w1} N_{ph1} \phi_m}{a_1} \tag{5.9}$$

$$E_r = \frac{4.44 f K_{w2} N_{ph2} \phi_m}{a_2} \tag{5.10}$$

Since E_r was induced by transformer action, the induction motor behaves like a transformer with a rotating secondary winding. Hence, we may write an expression for the transformation ratio in terms of the stator and rotor parameters as:

$$a = \frac{E_r}{E_1} = \frac{K_{w2} N_{ph2} a_1}{K_{w1} N_{ph1} a_2} \tag{5.11}$$

where f is frequency in Hz, K_{w1} and K_{w2} are stator and rotor winding factors, N_{ph1} and N_{ph2} are the number of turns in the stator and rotor windings, and a_1 and a_2 are the corresponding number of current paths.

5.5 INDUCTION-MOTOR SLIP

Suppose that the rotor circuit is open and is rotated at a speed n rpm by some external means in the duration of the rotating flux ϕ_m. If n_{syn} is given in rpm, the rotational speed of ϕ_m is defined as the slip s

$$s = \frac{n_{syn} - n}{n_{syn}}. \tag{5.12}$$

When motor rotor is at slip s

$$sn_{syn} = n_{syn} - n \tag{5.13}$$

the rotor frequency is given by

$$f_r = sf. \tag{5.14}$$

This frequency also affects the resulting voltage E_s':

$$E_s' = sE_r. \tag{5.15}$$

5.6 ROTOR CURRENT AND LEAKAGE REACTANCE

If r_{22} is the resistance of the rotor winding in ohms/ϕ, with the corresponding leakage inductance L_{22}, the wound-rotor leakage inductance in henrys/ϕ leads to leakage reactance of the rotor, given by:

$$X_{22} = 2\pi f L_{22} \tag{5.16}$$

and with slip s accounted for,

$$sX_{22} = 2\pi sfL_{22}\Omega/\phi. \tag{5.17}$$

Therefore, rotor current

$$\begin{aligned} I_{22} &= \frac{sE_r}{\sqrt{r_{22}^2 + sX_{22}^2}} \\ &= \frac{E_r}{\sqrt{\left(\frac{r_{22}}{s}\right)^2 + X_{22}^2}}. \end{aligned} \tag{5.18}$$

This mimics the transformer-equivalent circuit from the secondary side comprising of $\frac{r_{22}}{s}$ in series with the reactance of the secondary side given as the equivalent circuit, as shown in Figure 5.4.

The primary side is equivalent to the stator-winding operating at stator frequency regardless of the slip, i.e.,

$$E_2 = aE_1. \tag{5.19}$$

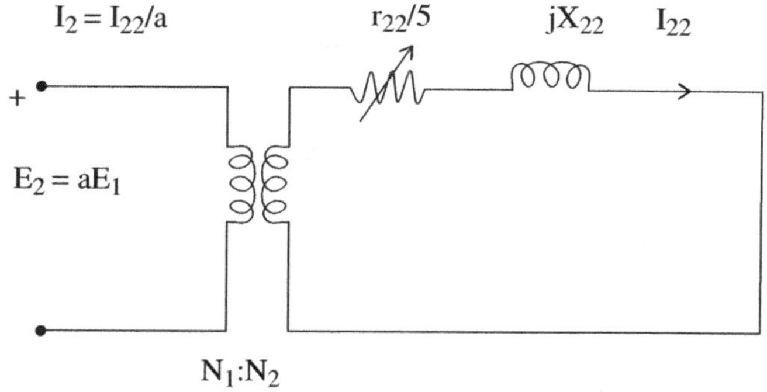

Figure 5.4 Induction-motor equivalent circuit.

This follows from the fact that the polyphase rotor currents at slip s have a synchronous frequency and therefore produce a rotor mmf that rotates at sn_{sync} rpm relative to the rotor in the same direction as the stator flux while the rotor is rotating at a speed of $(1-s)n_{syn}$ rpm in the same direction.

The resultant speed of the rotor mmf relative to the stator is the sum of those two speeds:

$$sn_{syn} + (1-s)\,n_{syn} = n_{syn}. \tag{5.20}$$

Synchronous speed $n_{syn} = sn$ of speed of stator at stator frequency f (Hz).

The ideal transformer may be used to model the equivalent circuit of an induction motor as shown in Figure 5.5.

Equivalent rotor resistance $\frac{r_2}{s}$ is shown in two parts, namely:

(a) Rotor resistance r_2 is dynamic resistance.

(b) Mechanical load on the machine is modeled as $\frac{(1-s)}{s}r_2$.

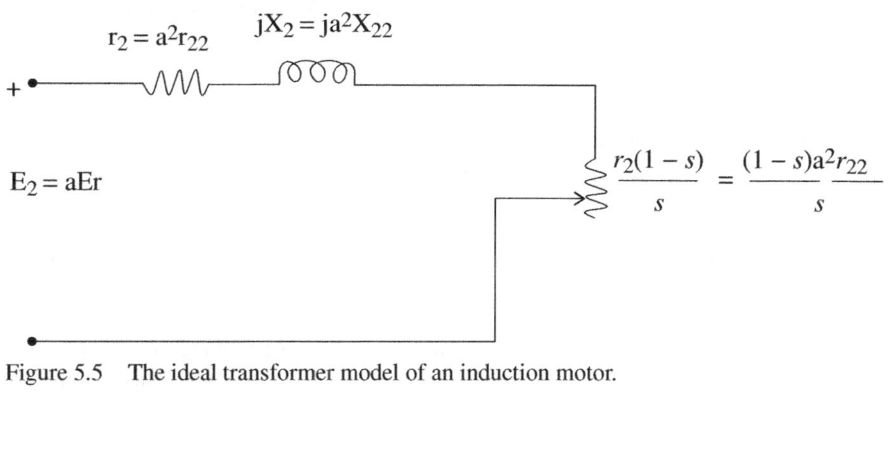

Figure 5.5 The ideal transformer model of an induction motor.

Example 1
A four-pole, three-phase induction motor operates from a supply whose frequency is 50 Hz. Calculate:

(a) speed at which the magnetic field of the stator is rotating
(b) speed of the rotor when the slip is 0.04
(c) frequency of the rotor current when the slip is 0.03
(d) frequency of the rotor currents at standstill

Solution:
Stator field revolves at synchronous speed, given by:

$$N_S = 120f/P = 120 \times 50/4 = 1,500\,\text{rpm}$$

rotor speed

$$N = N_S\,(1 - s) = 1,500\,(1 - 0.04) = 1,440\,\text{rpm}$$

frequency of rotor current

$$f' = sf = 0.03 \times 50 = 1.5\,\text{rps}$$

$$1.5\,\text{rps} \times 60\text{ seconds} = 90\,\text{rpm}$$

at standstill

$$s = 1$$

$$f' = sf = 1 \times f = 50\,\text{Hz}$$

Example 2
A 110 V, 50 Hz, delta-connected induction motor has a star-connected, slip-ring rotor with phase transformation ration of 3.8. The rotor resistance and standstill leakage reactance are 0.012Ω and 0.25Ω per phase, respectively. Neglecting stator impedance and magnetizing current, determine:

(a) rotor current at start with slip rings shorted
(b) rotor power factor at start with slip rings shorted
(c) rotor current at 4 percent slip with slip rings shorted
(d) rotor power factor at 4 percent slip with slip rings shorted
(e) external rotor resistance per phase required to obtain starting current of 100A in the stator supply lines

It should be noted that in a Δ/Y connection, primary-phase voltage is the same as line voltage. The rotor-phase voltage can be found by using the phase transformation ratio of 3.8, i.e., K = 1/3.8.

Solution:
Rotor-phase voltage at standstill = 1,100 × 1/3.8 = 289.5 V

(a) rotor impedance/phase = $\left(0.012^2 + 0.25^2\right)^{1/2} = 0.2503\Omega$

(b) rotor phase current at start = $289.5/0.2503 = 1,157\text{A}$

(c) power factor = $\text{R}_2/\text{Z}_2 = 0.012/0.2503 = 0.048$ lagging

(d) at 4 percent slip

$$\text{X}_r = s\text{X}_2 = 0.04 \times 0.25 = 0.01\Omega$$
$$\text{Z}_r = \left(0.012^2 + 0.01^2\right)^{1/2} = 0.0156\Omega$$
$$\text{E}_r = \text{sE}_2 = 0.04 \times 289.5 = 1158\text{V}$$
$$\text{I}_2 = \text{I}_1/\text{K} = 100 \times 3.8 = 380\text{A};$$
$$\text{E}_2 \text{ at standstill} = 289.5 \text{ V}$$
$$\begin{aligned}\text{Z}_2 = 289.5/380 = 0.7618\Omega,\ \text{R}_2 &= \left(\text{Z}_2^2 - \text{X}_2^2\right)^{1/2} \\ &= \left(0.7618^2 - 0.25^2\right)^{1/2} \\ &= 0.7176\Omega\end{aligned}$$

(e) external resistance required/phase = 0.7196 − 0.012 = 0.707Ω

5.7 ROTOR COPPER LOSS

As shown in Figure 5.4, for an induction-machine equivalent circuit, we can compute

$$(1-s)\,r_2 \tag{5.21}$$

$$P_{in} = \frac{I_2^2 r_2}{s}(\text{W/phase}) \tag{5.22}$$

$$P_{2(loss)} = I_2^2 r_2(\text{W/phase}) \tag{5.23}$$

from which we compute

$$s = \frac{\textit{Rotor Copper Loss}}{\textit{Rotor Real Power Input}}. \tag{5.24}$$

Given the value of P_{in} and P_{losses} for mechanical power

$$\begin{aligned}P_{mech} &= P_{2in} - P_{2loss} \\ &= \frac{I_2^2 r_2}{s} - I_2^2 r_2 \\ &= \frac{I_2^2(1-s)}{s} r_2.\end{aligned} \tag{5.25}$$

If we account for friction and stray losses ($P_{fw} - P_{stray}$), we have the net mechanical power as follows:

$$P_{mech} = P_{em} - \left(P_{fw} - P_{stray}\right) \tag{5.26}$$

where P_{fw} is power lost as friction and windage.

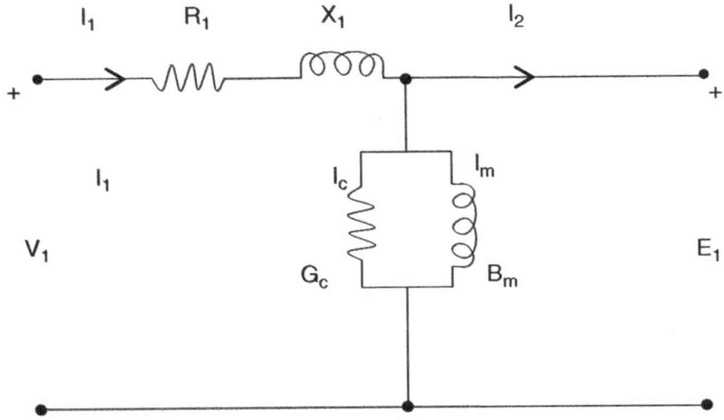

Figure 5.6 Induction-motor stator equivalent circuit-transformer analog model.

The induction motor can become a generator when I_{22} reverses its direction and

$$I_{22} = \frac{sE_r}{r_{22} - jsX_{22}}. \tag{5.27}$$

This is only possible when Q is supplied to the motor at the bus where it is connected.

5.8 DEVELOPING THE EQUIVALENT CIRCUIT OF POLYPHASE, WOUND-ROTOR INDUCTION MOTORS

Assuming the machine is three-phase connected and accounting for its losses in steady-state conditions, the currents and voltages align with neutral values at the stator side. The machine's rotating air-gap flux is defined by balanced, three-phase emf of the stator. This gives a terminal voltage, which differs from the stator emf by the voltage drop in the stator leakage impedance given as

$$V_1 = E_1 + I_1\left(R_1 + jX_1\right). \tag{5.28}$$

In terms of rotor emf

$$E_1 = E_2$$

where V_1 is stator terminal voltage, E_1 is counter emf generated by resultant field flux, I_1 is the stator current, R_1 is the stator effective resistance, X_1 is the stator leakage reactance.

As an analog to the transformer model, we have the stator equivalent circuit as shown in Figure 5.6.

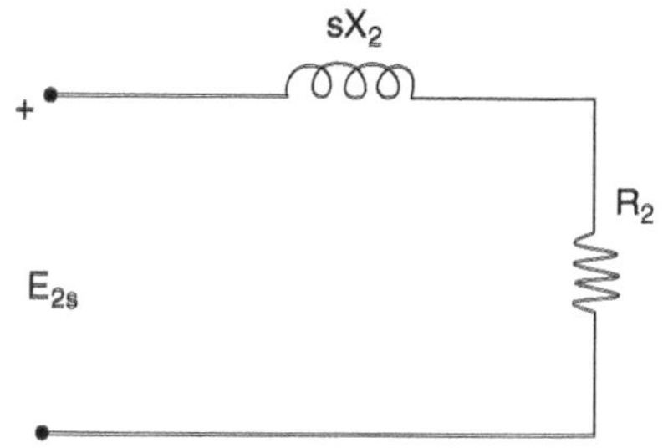

Figure 5.7 Rotor equivalent circuit of three-phase induction motor at slip frequency.

I_2 produces mmf linking mmf of the rotor current.

I_m is the stator current required to create the resultant air-gap flux as a function of E_1.

I_c is in phase with E_1 and magnetizing component I_{22} lagging E_1 by 90°.

These quantities are included in the circuit diagram in Figure 5.6.

As before, we can refer to the cage-wound rotor as a function of slip and then refer to the secondary side as seen in Figure 5.7.

$$E_{2s} = aE_{rotor} \tag{5.29}$$

$$sI_s = \frac{I_{rotor}}{a} \tag{5.30}$$

$$Z_{2s} = \frac{E_{2s}}{I_{rotor}}$$
$$= a^2 Z_{rotor}. \tag{5.31}$$

Thus, referring to the stator is essentially the same as referring the secondary quantities to the primary side of the static transformer.

Earlier, we defined as part of slip

$$\frac{E_{2s}}{I_{2s}} = R_2 + jsX_2 = Z_{2s} \tag{5.32}$$

where

Z_{2s} is the slip-frequency rotor-leakage impedance per phase referred to the stator, R_2 is the effective resistance, and sX_2 is the leakage reactance at slip frequency.

Thus

$$\frac{sE_1}{I_2} = \frac{E_{2s}}{I_{2s}} = R_2 + jsX_2 \tag{5.33}$$
$$= \frac{R_2}{s} + jX_2. \tag{5.34}$$

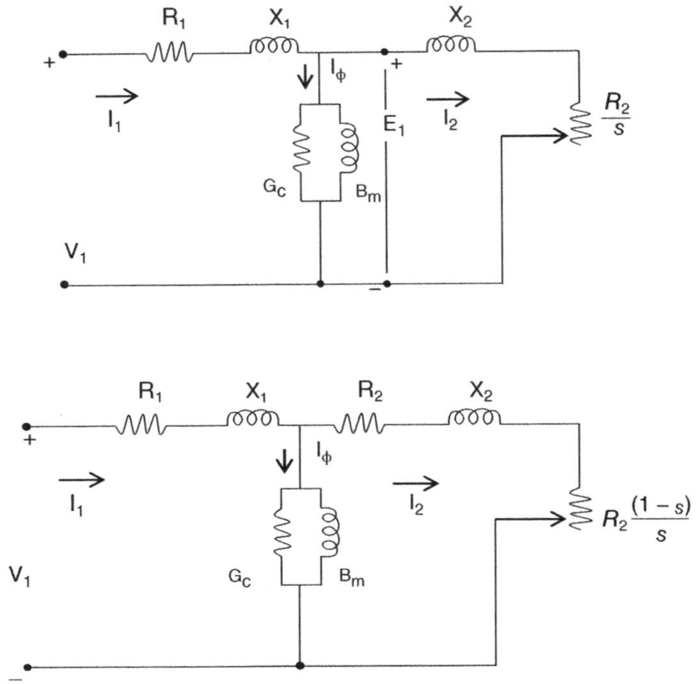

Figure 5.8 Equivalent circuits of three-phase polyphase induction motors.

The redrawn equivalent circuits for polyphase induction motors are shown in Figure 5.8.

The equivalent circuit is used to analyze the induction motor in general since V, I, losses, and torque are based on the model parameters.

We derive them simply as follows:

P_g transferred across the air gap from the stator is

$$P_{g1} = I_2^2 \frac{R_2}{s} a_1. \tag{5.35}$$

The total rotor loss I^2R can be written as $a_1 I_2^2 R_2$.

Net power is

$$P_{g1} - P_{loss} \tag{5.36}$$

$$= a_1 I_2^2 \frac{R_2}{s} - a_1 I_2^2 R_2$$

$$= a_1 I_1^2 R_2 \left(\frac{1}{s} - 1\right)$$

$$= a_1 I_1^2 R_L \left(\frac{1-s}{s}\right)$$

Motor mechanical power is

$$P = (1 - s) P_{g1}. \tag{5.37}$$

Total power delivered to the rotor is a fraction of $P_{g(mech)}$.

(1 – s) is converted into mechanical power because of P_{g1} (electrical power). The conversion (1 – s) mechanical is dissipated as rotor circuit I^2R loss.

The internal mechanical power per stator phase is equal to the power absorbed by the resistance

$$R_2\left(\frac{1-s}{s}\right). \tag{5.38}$$

Example

A 115 V, 60 Hz, three-phase, Y-connected, six-pole induction motor has an equivalent circuit consisting of stator impedance (0.07 + j0.3)Ω and an equivalent rotor impedance at standstill of (0.08 + j0.3)Ω. The magnetizing branch has G_c = 0.022 mho and B_m = 0.158 mho. Find:

(a) secondary current
(b) primary current
(c) primary power factor
(d) gross power output
(e) gross torque
(f) input power
(g) gross efficiency using equivalent circuit—assume a slip of 29 percent

Solution:

(a) Using the equivalent circuit shown in Figure 4.8,

$$\begin{aligned} I_2 &= \frac{V_1}{\left(R_1 + {}^{R_2}\!/_s\right) + j\left(X_1 + X_2\right)} \\ &= \frac{66.5\angle 0^\circ}{(0.07 + 0.08/0.02) + j(0.3 + 0.3)} \\ &= \frac{66.5\angle 0^\circ}{(4.07 + j0.6)} \\ &= 16.17\angle - 8.4^\circ. \end{aligned}$$

(b)

$$I_1 = I_\phi + I_2$$

$$\text{The exciting current } I_\phi = V\left(G_c + B_m\right)$$

$$= 66.5\,(0.022 - j0.158)$$

$$= (1.46 - j10.5)\text{ A}$$

$$\text{Therefore, } I_1 = I_\phi + I_2$$

$$= (1.46 - j10.5) + (16 - j2.36)$$

$$= 17.46 - j12.86 = 21.7\angle - 36.5^\circ.$$

(c) Primary power factor $= \cos 36.5^\circ = 0.804$.

(d) $P_g = 3{I_2}^2R_L = 3 \times 16.17^2 \times 3.92 = 3,075\text{W}$.

(e) Synchronous speed $N_s = 120 \times 60/6 = 1{,}200$ rpm.
Actual rotor speed $N = (1 - s)\,N_s = (1 - 0.02) \times 1,200 = 1,176$ rpm
Therefore, $T = 9.55\frac{P_m}{N} = 9.55 \times \frac{3{,}075}{1{,}176} = 24.97\text{Nm}$.

(f) Primary power input $= \sqrt{3}V_1 I_1 Cos\,\phi$
$\sqrt{3} \times 115 \times 21.7 \times 0.804 = 3,450\text{W}$.

(g) Gross sfficiency $= \dfrac{output\ power}{input\ power} \times 100\%$

$$= \frac{3,075}{3,450} \times 100 = 89.5\%.$$

5.9 COMPUTING CORRESPONDING TORQUE OF INDUCTION MOTORS

$$P = (1 - s)\,W_s T \tag{5.39}$$

$$P = \frac{1}{W_s} I_2^2 \frac{R_2}{s} \tag{5.40}$$

where

$$W_s = Synchronous\ velocity = \frac{4\pi f}{Poles}.$$

Note that torque and power do not include stray, frictional, windage, and load losses.

Example

A three-phase induction motor has a four-pole, star-connected stator winding and runs on a 220 V, 50 Hz supply. The rotor resistance per phase is 0.1Ω and reactance 0.9Ω. The ratio of stator to rotor turns is 1.75. The full load slip is 5 percent. Calculate for this load:

- **(a)** load torque in Nm
- **(b)** speed at maximum torque
- **(c)** rotor emf at maximum torque

Solution:

(a) K = rotor turns/stator turns = 1/1.75

Stator voltage/phase,

$$E_1 = 220/(3)^{1/2}V$$

Standstill rotor emf/phase,

$$E_2 = KE_1 = \frac{220}{\sqrt{3}} \times \frac{1}{1.75} = 72.6V$$

$$Z_r = \sqrt{R_2^2 + sX_2^2} = \sqrt{0.1^2 + (0.05 \times 0.9)^2} = 0.11\Omega$$

$$I_2 = sE_2/Z_r = 0.05 \times 72.6/0.11 = 33A$$

Rotor copper loss

$$P_{cu} = 3I_2^2R_2 = 3 \times 33^2 \times 0.1 = 327W$$

$$\frac{\textit{Rotor Cu Loss}}{\textit{Mech. Power Developed}} = \frac{s}{1-s}$$

$$\frac{327}{P_m} = \frac{0.05}{1-0.05}$$

$$P_m = 6,213W$$

$$T_g = 9.55 \times 6,213/1,425 = 41.6Nm$$

(b) For maximum torque,

$$s_m = \frac{R_2}{X_2} = \frac{0.1}{0.81} = \frac{1}{9}$$

$$N = N_S(1-s) = 1,500(1-1/9) = 1,333\,\text{rpm}$$

(c) Rotor emf/phase at maximum torque = 1/9 × 72.6

= 8.07 V

5.10 APPROXIMATION MODEL FOR INDUCTION MACHINES

To approximate the model further, the exciting branch is usually neglected to allow for 30 to 50 percent full-load current and because the leakage is also higher. This is drawn as either of the two diagrams in Figure 5.9.

Example

A three-phase, Y-connected, 220 V, $_{L\text{-}L}$, 10Hp, 60 Hz, six-pole, induction motor has the following constants in ohms:

$$R_1 = 0.294,\ R_2 = 0.144,\ X_1 = 0.503,\ X_2 = 0.209,\ X_\phi = 1.25.$$

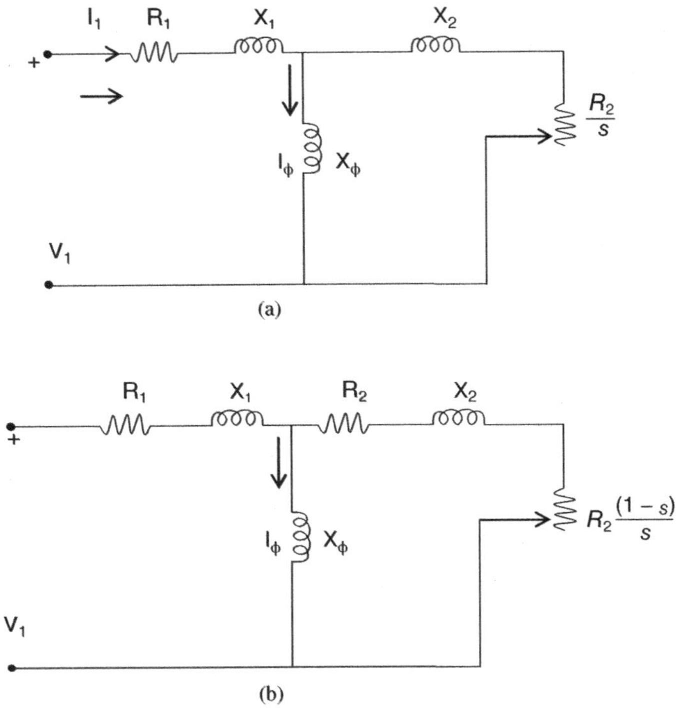

Figure 5.9 Approximate equivalent circuit of three-phase induction motor.

The total friction, windage, and core loss are assumed constant at 403 W, load slip of 2.0 percent.

Determine torque, power, I_{stator}, pf, and η of motor at rated frequency.

Solution:

The impedance Z_f is represented in Figure 5.10 as

$$Z_f = R_f + jX_f$$

$$= \frac{R_2}{s} + jX_2 \text{ in parallel with } X_\phi$$

$$Z_f = \frac{0.144}{0.02} + j0.209 \,||1.25$$

$$Z_f = \frac{0.144}{0.02} + j\left(\frac{0.209 \times 1.25}{0.209 + 1.25}\right) = 7.2 + j0.179$$

Sum of

$$R_f + jX_f + R_1 + jX_1 = 7.2 + j0.179 + 0.294 + 0.503$$

$$= 7.494 + j0.682$$

$$7.525\angle 5.2\Omega.$$

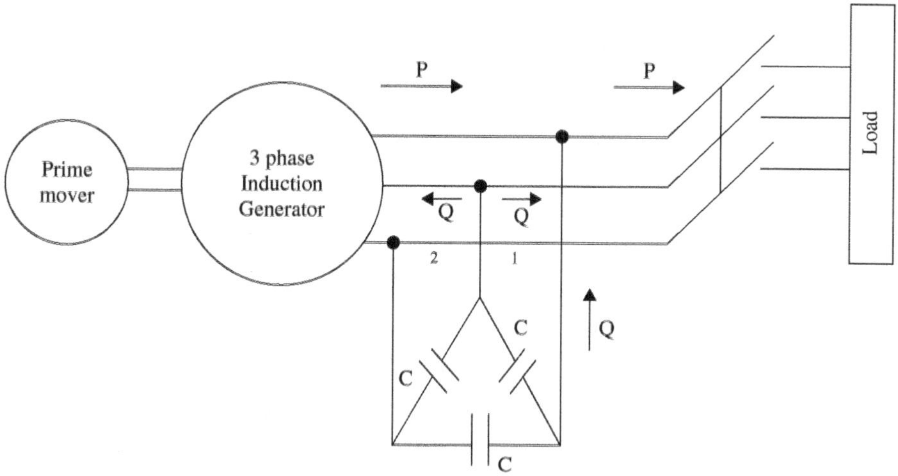

Figure 5.10 Induction generator in self-excited mode.

Applied voltage

$$V_{in} = \frac{220}{\sqrt{3}} = 127V$$

$$I_1 = \frac{220}{7.525\angle 5.2}$$

$$29.236\angle - 5.2$$

$$pf = Cos\,5.2$$

$$= 0.996$$

$$\text{Synchronous speed} = \frac{120f}{Poles}$$

$$= \frac{120 \times 60}{6}$$

$$1,200\,\text{rpm}$$

$$\omega_s = \frac{2\pi N_s}{60}$$

$$= \frac{2\pi \times 1,200}{60}$$

$$= 125.6\ rad/s$$

$$P_{g1} = I_2^2 \frac{R_2}{s}$$

$$3 \times 29.236^2 \times \frac{0.144}{0.02} = 18,462.5\text{W}$$

$$P = (1 - s)\ P_{g1}$$

$$0.98 \times 18,462.5 = 18,093\text{W}$$

Deducting losses of 403W

$$\text{Output power} = (18{,}093 - 403)\,\text{W}$$
$$= 17{,}690\text{W}$$
$$746\,\text{W} = 1\text{Hp}$$
$$\text{Output power} = 23.7\text{Hp}$$
$$Torque = \frac{P_{out}}{\omega_{rotor}}$$
$$= \frac{17{,}690}{0.98 \times 125.6}$$
$$\omega_{rotor} = (1 - s)\,\omega_s$$
$$= 143.72\text{Nm}$$
$$= 106.18\text{lb/ft}$$

Efficiency : Total stator I^2R loss $= 3(18.8)^2 \times 0.294 = 312$ W

$$\text{Rotor copper loss } I^2R = P_{g1} = \frac{I^2R_2}{s}$$
$$I^2R = sP_{g1} = 0.02\,(18{,}462.5)$$
$$= 369.25\text{W}$$

Friction + windage + core losses = 403W
Total = 403 + 369.25 + 312 = 1,084.25W
Output = 17,690

$$\therefore \quad Input = output + losses$$
$$= 18{,}774\text{W}$$
$$\eta = 1 - \frac{Losses}{Input\ Power}$$
$$1 - \frac{1{,}084.25}{18{,}774}$$
$$= 94.22\%$$

5.11 SPEED CONTROL OF INDUCTION MOTORS

To allow for multi-application of induction motors, engineers use many options to adjust the speed of induction motors, including:

(a) Pole changing motors. By simply changing coil connections, the number of poles can be changed in the ratio 2:1, typically for cage-rotor-induction motors.

(b) Line frequency control. Varying the line frequency can change the speed of the motor by using a wound-rotor induction motor as a frequency changer. The most effective and economic source of adjustable frequency must be determined for this method.

(c) Line-voltage control. This can be achieved by reduction of speed from n_1 to n_2 rpm by changing the control torque of the induction motor, which is proportional to the square of voltage at the primary side terminals.

(d) Rotor resistance control. This is done by adjusting the rotor-circuit resistance at the primary winding.

(e) Control of slip by auxiliary devices. By varying the slip s through changes in frequency, the speed of an induction motor can be controlled.

5.12 APPLICATION OF INDUCTION MOTORS

The following are some applications of induction motors.

(a) Squirrel-cage induction motors are more widely used in industrial applications than slip-ring induction motors since they are cheaper in cost, rugged in construction, and low maintenance. Squirrel-cage induction motors are suitable for applications where the drive requires constant speed, low starting torque, and no speed-control drives.

(b) Wound-rotor induction motors are suitable for loads requiring high starting torque and low starting current, and loads that have high inertia, which results in higher energy losses. They also are applicable for loads that require a gradual buildup of torque and speed control. Specific applications include in conveyors, cranes, pumps, elevators, and compressors.

5.13 INDUCTION-GENERATOR PRINCIPLES

The conditions under which polyphase induction machines behave as induction generators include the following.

(a) When the slip of a polyphase induction machine becomes negative, the rotor current and rotor emf attain negative values. The prime-mover torque becomes opposite to the electric torque. This condition can be achieved by coupling a prime mover whose speed can be controlled by the machine's rotor. If the speed of the prime mover is increased such that the slip becomes negative (i.e., the speed of the prime mover becomes greater than the synchronous speed) then the induction machine behaves like a generator.

(b) If the speed of the prime mover is further increased, such that it exceeds the negative maximum value of the torque produced, then the generating effect of the generator vanishes. Clearly the speed of the induction generator during the whole operation is not synchronous; therefore, the induction generation is also called *asynchronous generation*.

An induction generator is not a self-excited machine. In order to develop the rotating magnetic field it requires magnetizing current and reactive power. The induction generator obtains its magnetizing current and reactive power from the supply mains or from a synchronous generator connected in parallel with it. Thus, normally, the induction generator cannot work in isolation because it continuously requires reactive power from some external supply system.

The only way to make an induction generator self-excited in isolated mode is to use capacitor banks for reactive power supply instead of an AC-supply system.

5.13.1 Isolated Induction Generators

Isolated induction generators are also known as *self-excited generators*. A bank of capacitors is connected across the generator's stator terminals as shown in Figure 5.10.

The function of the capacitor bank is to provide the lagging reactive power to the induction generator as well as the load. So

$$Q = Q_1 + Q_2 \tag{5.41}$$

where Q is the total lagging reactive power supplied by the capacitor bank, Q_1 is the reactive power drawn by the induction generator, and Q_2 is the reactive power fed to the load.

5.13.2 Application of Induction Generators

The two main applications of induction generators are as follows:

- When externally excited, induction generators are widely used for regenerative breaking of hoists driven by three-phase induction motors.
- Self-excited generators help in conversion of unconventional energy sources into electrical energy and are used in wind turbines.

Illustrative Problems and Examples

1. A three-phase 60 Hz, induction motor runs at 715 rpm at no load and at 670 rpm at full load.
   - **(a)** How many poles does this motor have?
   - **(b)** What is the slip at rated load?
   - **(c)** What is the speed at one-quarter of the rated load?
   - **(d)** What is the rotor's electrical frequency at one-quarter of the rated load?

Solution:

(a) No-load speed $N_{NL} = 715\,\text{rpm}$
Full-load speed $N_{FL} = 670\,\text{rpm}$
$F = 60\text{Hz}$
$N_S = 120f/P$
$P = 120f/N_S$
$P = (120 \times 60)/715\,\text{rpm}$
Synchronous speed at $P = 10$ is $N_S = 120f/P$
$= 120 \times 60/10 = 720\,\text{rpm}$

(b) Slip at rated load

$$S = (N_S - N_L)/N_S$$
$$= (720\,\text{rpm} - 670\,\text{rpm})/720\,\text{rpm}$$
$$= 0.0694 \times 100\%$$
$$= 6.94\%$$

(c) $N_S = 720$ rpm

$$\text{Slip at } 1/4 \text{ load } = 0.25 \times 0.0644 = 0.01735$$
$$N = N_S - SN_S$$
$$= N_S(1 - N_S)$$
$$= 720(1 - 0.01735)$$
$$= 707.5\,\text{rpm}$$

(d) $\text{Frequency} = f \times \text{slip at } 1/4 \text{ load}$

$$= 0.01735 \times 60$$
$$= 1.041\,\text{Hz}$$

2. A 220 V, three-phase, six-pole, 50 Hz induction motor is running at a slip of 3.5 percent. Find:

(a) speed of the magnetic fields in revolutions per minute

(b) speed of the rotor in revolutions per minute

(c) slip speed of the rotor

(d) rotor frequency in hertz

Solution:

(a) 220V, six poles, 50Hz running at slip of 3.5 percent

$$N_S = 120f/P = (120 \times 50)/6$$
$$= 1,000\,\text{rpm}$$

(b) Speed of the rotor (S) = (NS – N)/NS

$$N = N_S(1 - S)$$
$$\text{where } S = 3.5\% = 0.035$$
$$N = 1,000(1 - 0.035)$$
$$= 965\,\text{rpm}$$

(c) Slip speed of the rotor $= N_S - N = 1,000 - 965\,\text{rpm} = 35\,\text{rpm}$

(d) Rotor frequency $= sf = 0.035 \times 50 = 1.75\text{Hz}$

5.14 CHAPTER SUMMARY

Principles of induction machines have been presented in this chapter. This class of electromechanical machines is important as it serves dual purposes when in an electric power system depending on its mode of excitation. We have found that, as an induction motor, it is a nonlinear load, which is of interest because of the power-quality problem of harmonic distortion it creates in an electric power system. As an induction generator, its application in wind energy systems is of importance. Consequently, irrespective of the configuration of an induction machine, its place as a load or an energy source in any power system is of utmost interest, especially in the design, operation, and management of a microgrid. Considerable resources have therefore been included in this chapter for the benefit of the student to facilitate overall knowledge of this machine in preparation for its applications in a microgrid system. Several exercises to illustrate computational modelling, performance, and speed regulation of induction machines have been presented in this chapter.

EXERCISES

1. A 50 kW, 440 V, 50 Hz, two-pole induction motor has a slip of 6 percent when operating at full-load conditions. At full-load conditions, the friction and windage losses are 520 W, and the core losses are 500 W. Find the following values for full-load conditions:

- **(a)** shaft speed in rpm,
- **(b)** output power in watts
- **(c)** load torque (load τ) in Newton meters
- **(d)** induced torque (ind τ) in Newton meters

2. A 480 V, 60 Hz, six-pole, three-phase, delta-connected induction motor has the following parameters: $R_1 = 0.461\Omega$, $R_2 = 0.258\Omega$, $X_1 = 0.507\Omega$, $X_2 = 0.309\Omega$, $X_m = 30.74\Omega$, and rotational losses 2,450 W. The motor drives a mechanical load at a speed of 1,170 rpm. Calculate:

- **(a)** synchronous speed in rpm
- **(b)** slip
- **(c)** line current
- **(d)** input power
- **(e)** air-gap power
- **(f)** torque developed
- **(g)** output power in horsepower
- **(h)** efficiency

3. A three-phase, six-pole, 10HP, 400 Hz induction motor has a slip of 3 percent at rated output power. Friction and windage losses are 300 W at rated speed. The rated condition total core losses are 350 W. $R_1 = R_2 = 0.05\Omega$ and $X_1 = X_2 = 0.05\Omega$. Using the approximate equivalent circuit for analysis, if the motor is operating at rated output power, speed, and frequency, find:

- **(a)** rotor speed
- **(b)** frequency of rotor currents
- **(c)** total power across the air gap
- **(d)** efficiency
- **(e)** applied-line voltage

4. A three-phase, 50HP, 460 V, 60 Hz, 865 rpm induction motor is operating at rated conditions and has $P_T = 43.75$ kW and $I_1 = 61$ KA. It is known that $R_1 = 0.15\Omega$ and rotational losses at rated speed are 1,050 W. Determine:

(a) full-load power factor
(b) full-load efficiency
(c) total rotor coil ohmic losses
(d) total core losses

5. A 208 V, four-pole, 60 Hz, Y-connected, wound-rotor induction motor is rated at 30Hp. Its equivalent circuit components are:

$R_1 = 0.100\Omega \quad R_2 = 0.070\Omega \quad X_m = 10.0\Omega$
$X_1 = 0.210\Omega \quad X_2 = 0.210\Omega$
$P_{mech} = 500\text{W} \quad P_{core} = 400\text{W}$

For a slip of 0.05, find:
(a) line current
(b) stator copper losses
(c) airgap power
(d) power converted from electrical to mechanical from induced torque
(e) power converted from electrical to mechanical from total load torque

6. The power crossing the air gap of a 60 Hz, four-pole, induction motor is 25 kW and the power converted from electrical to mechanical form in the motor is 23.2 kW.
(a) What is the slip of the motor at this time?
(b) What is the induced torque in this motor?
(c) If the mechanical losses are 300 W at this slip, what is the load torque of this motor?

7. An induction motor draws 50A from a 380 V, three-phase line at a lagging power factor of 0.9. If its rotor copper losses are 1,000 W and 500 W, respectively, its core losses are 650 W, the friction and windage losses are 200 W, and the stray losses are 250 W, determine:
(a) air-gap power
(b) mechanical power developed
(c) shaft output power in watts and horsepower
(d) efficiency of the motor

8. An induction motor draws 50A from a 480 V, three-phase line at a lagging power factor of 0.85. Its stator and rotor copper losses are 1,000 W and 500 W, respectively. Its core losses are 500 W, friction and windage losses are 250 W, and stray losses are 250 W. Determine:
(a) air-gap power
(b) mechanical power developed
(c) shaft output power in horsepower
(d) efficiency of the motor

9. The input to the rotor of a 208 V, three phase, 60 Hz, 24-pole induction motor is 20 kW. If the rotor current is 63.25A per phase and the rotor resistance is 0.05Ω per phase, and ignoring the rotational losses, determine:
(a) developed (electromagnetic) torque in Nm
(b) speed in rpm and rad/s
(c) horsepower output of the motor

BIBLIOGRAPHY

[1] Electrical4U, “Induction Motor,” www.electrical4u.com/electrical-engineering-articles/index.php?category=induction-motor.

[2] J. Casazza and F. Delea, *Understanding Electric Power Systems: An Overview of the Technology, the Marketplace, and Government Regulations*, John Wiley & Sons, Hoboken, NJ, 2003.

[3] H. Saadat, *Power System Analysis*, 3rd ed., PSA Publishing, 2010.

[4] A. E. Fitzgerald, C. Kingsley, and S. D. Umans, *Fitzgerald & Kingsley's Electric Machinery*, 4th ed., McGraw-Hill, New York, 1983.

[5] C. A. Gross, *Power System Analysis*, John Wiley & Sons, New York, 1979.

[6] S. J. Chapman, *Electric Machinery Fundamentals*, 5th ed., McGraw-Hill, New York, 2012.

[7] T. Gönen, *Electrical Machines with MATLAB*, 2nd ed., Taylor & Francis Group, Boca Raton, FL, 2011.

[8] W. L. Matsch and J. D. Morgan, *Electromagnetics and Electromechanical Machines*, 3rd ed., John Wiley & Sons, Hoboken, NJ, 1986.

[9] M. S. Sarma, *Electric Machines: Steady-State Theory and Dynamic Performance*, WMC Brown Publishers, Dubuque, IA, 1985.

[10] R. Ramshaw and R. G. van Heeswijk, *Energy Conversion: Electric Motors and Generators*, Saunders College Publishing, Philadelphia, 1990.

[11] G. R. Slemon, *Electric Machines and Drives*, Addison-Wesley Publishing, New York, 1992.

[12] B. L. Theraja and A. K. Theraja, *A Textbook of Electrical Technology*, 1st multicolour ed., S. Chand, New Delhi, 2005.

[13] Electrical Engineering (blog), “Squirrel Cage Motors,” http://electricalengineeringbasics.blogspot.com.

[14] Electropaedia, “Electric Drives: AC Motors (Description and Applications),” www.mpoweruk.com/motorsac.htm.

CHAPTER 6

SYNCHRONOUS MACHINES

6.1 INTRODUCTION

Synchronous generators are the main source of bulk power generation at large power stations. They may be driven by prime movers such as hydro turbines, wind turbines, gas turbines, or internal combustion engines. Today's most mature science of power generation rests on synchronous-generator technology. Generators are the largest single unit of electric machines in production, having power of some 1,000 MVA. In a large power system, more than a thousand machines may be interconnected through tie-lines supplying thousands of megawatts of power to consumers dispersed over a wide area. Each machine could be driven by prime movers.

Synchronous generators are scalable and may be built for smaller power-generation applications such as domestic, commercial, and small-scale industrial needs. These could have capacities ranging from 10 kW to 500 kW. Such machines are usually driven by diesel- or gasoline-powered internal combustion engines.

6.2 SYNCHRONOUS-GENERATOR CONSTRUCTION

Synchronous machines operate at 60 or 50 Hz for utility and for larger frequency in large aircraft. Almost all synchronous machines have stationary armatures commonly called the *stator* and rotating field structures called the *rotor*. The rotor may be constructed with salient or cylindrical (round with no polar projection) poles, as shown in Figure 6.1 [9]. The prime-mover speed influences the construction of the salient poles rotors. The armature carries three-phase winding in which AC emf is generated.

Generator stator windings are similar to that of polyphase induction machines. The stator core consists of windings (built in a segmented core) of high-quality laminates with slot-embedded, double-layer, top windings, as seen in Figure 6.2 [5].

The generator stators are connected in a three-phase Y model with 60° by two-power speed. The salient-pole synchronous machine is equipped with damper windings, which are usually a set of copper or brass conductors set in pole-face slots and connected for stability, with several other useful functions such as starting synchronous machines. Their role is damping rotor oscillation, as shown in Figure 6.3

Energy Processing and Smart Grid, First Edition. James A. Momoh

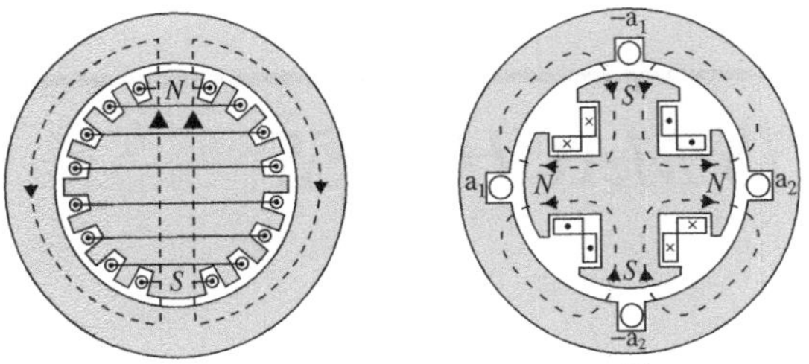

(a) (b)

Figure 6.1 Synchronous generator rotor types. (a) Cylindrical or round rotor (b) Salient-pole rotor.

[5], reducing overvoltage under short-circuit conditions, thus helping synchronize the machines.

6.3 EXCITERS

There are different types of excitation systems for synchronous machines. However, modern excitation systems use AC power sources through a silicon-controlled, solid-state rectifier. The DC winding on the rotating field is connected to the exciting system through slip rings and brushes as seen in Figure 6.4. A reasonable model of a modern exciter is linearized and only takes account of the major time constant of the exciter T_e, which is usually very small, and an exciter gain K_e, ignoring saturation and other nonlinearities. The model of such an exciter is given in the frequency domain as

$$\frac{V_f(s)}{V_r(s)} = \frac{K_e}{1 + T_e(s)} \tag{6.1}$$

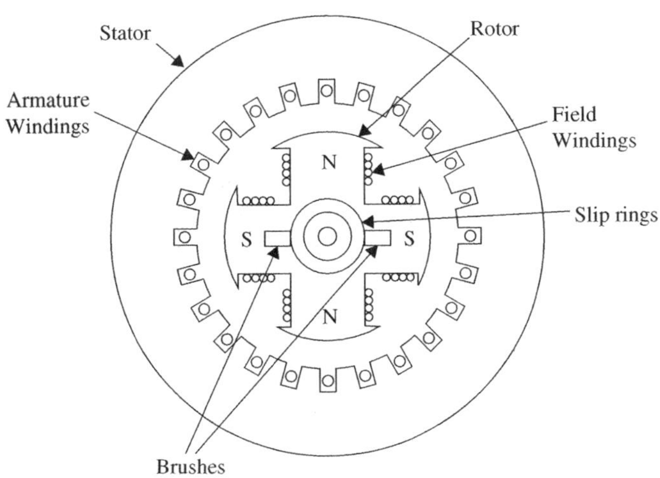

Figure 6.2 Cross-section of a synchronous machine.

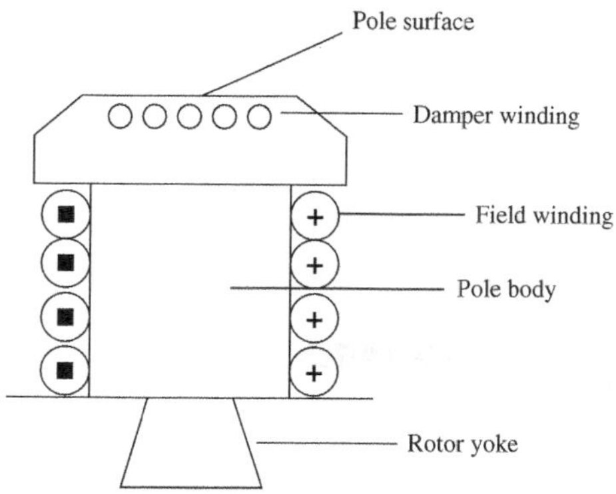

Figure 6.3 Rotor pole showing damper windings.

where V_f and V_r are the DC-field voltage and some reference voltages, respectively.

- For large hydro-generators (with low speed) there may not be self-excitation. But a self-starter or permanent-magnet-type exciter may be employed to activate the exciter.
- Silicon diodes and thyristor excitation are used to avoid wear and tear in classical excitation synchronous machines.
- Brushless systems have system-mounted rectifiers rotating with the rotor; the need for brushes is eliminated and no slip rings are needed.

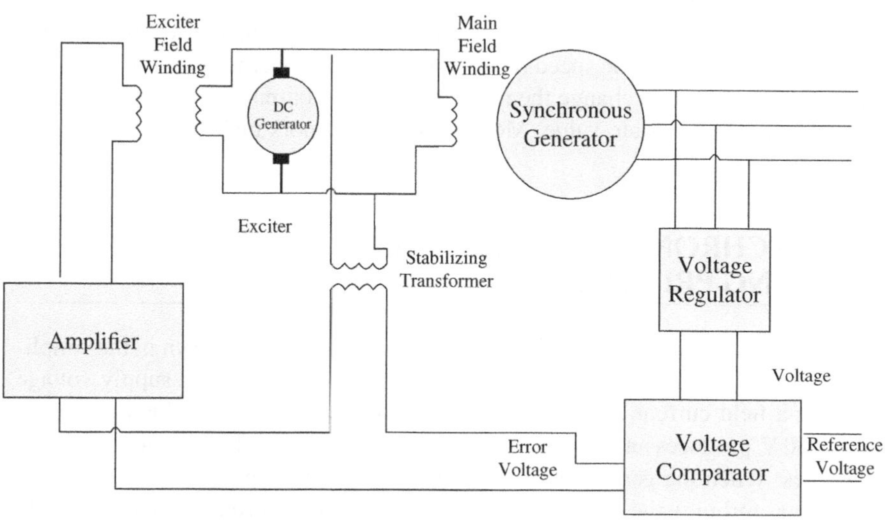

Figure 6.4 DC generator excitation system.

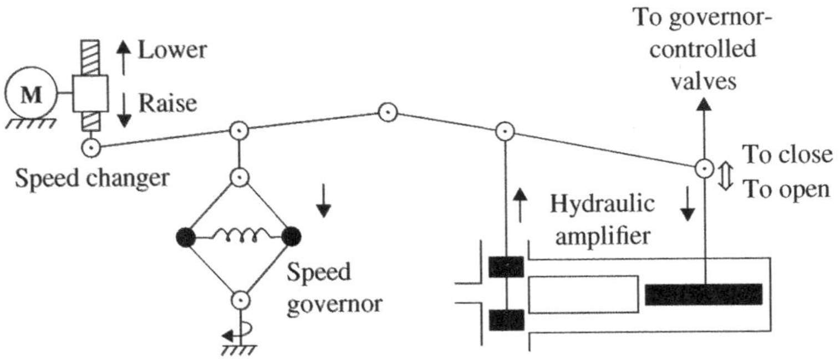

Figure 6.5 Mechanical governor system.

6.4 GOVERNORS

The mechanical system is an input to the electrical system and a shaft connects the two systems. When there is an increase in load, the generator applies a torque, which is in opposition to the torque from the mechanical system. This brings about a resultant reduction in the speed of the shaft of the system. This change in speed is an undesirable effect in the generator because it causes a variation in the frequency of the generated voltage. Thus, a governor is applied to the mechanical system (Figure 6.5) to bring about a predictable linear relationship between speed and power, so that reduction in speed can be controlled. Frequency is also controlled by the governor by increasing the no-load speed when there is a frequency reduction of the generated voltage.

Electrical power usually exceeds that of the input mechanical power when there is a sudden increase in electrical load, and the power difference is usually supplied by the kinetic energy stored in the rotating system. This results in the kinetic energy being reduced and thus makes the turbine speed decrease, which also results in the fall in frequency. The turbine speed reduction causes the governor to act by adjusting the turbine input valve to change the mechanical power output, which will bring the speed to a new steady-state value. Most modern governors make use of electronic speed sensors.

6.5 SYNCHRONOUS GENERATOR OPERATING PRINCIPLE

Consider a synchronous generator with wound salient rotors, as shown in the simplified diagram of Figure 6.6. The rotor winding is connected to a DC-supply voltage producing a field current, I_f. The external DC excitation voltage, which can be as high as 250 V, produces an electromagnetic field around the coil with static north and south poles. When the generator's rotor shaft is turned by the prime mover (hydro, steam, or gas turbine), the rotor poles also move, producing a rotating magnetic field as the north and south poles rotate at the same angular velocity as the turbine blades

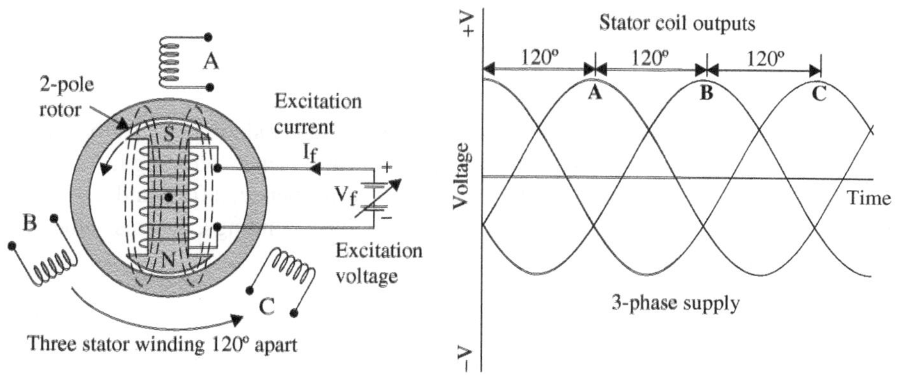

Figure 6.6 Synchronous generator operating principle. (a) Rotor-stator arrangement (b) 3-Phase AC induced voltage.

(assuming direct drive). As the rotor rotates, its magnetic flux cuts the individual stator coils one by one and, by Faraday's law, an emf and therefore a current is induced in each stator coil.

The magnitude of the voltage induced in the stator winding (Figure 6.6b) is a function of the magnetic field intensity, which is determined by the field current, the rotating speed of the rotor, and the number of turns in the stator winding. As the synchronous machine has three stator coils, a three-phase voltage supply corresponding to the windings, A, B, and C, which are electrically 120° apart, is generated in the stator windings.

The load is connected directly to the three-phase stator winding, and as these coils are stationary they do not need to go through large unreliable slip rings, commutators, or carbon brushes.

When the rotor is excited to produce an air-gap flux ϕ per pole and is rotating at an angular speed of ω radians per second, the flux linkage of N-turns of the conductor will be maximum when $\omega t = 0$ and also at $\omega t = 90°$. Assuming distributed winding, the flux linkage λ_a will vary as the cosine of the angle ωt. Thus, the flux linkage with coil a is

$$\lambda_a = N\phi \cos \omega t \tag{6.2}$$

and induced voltage

$$e_a = -\frac{d\lambda}{dt} = \omega N\phi \sin \omega t$$

$$e_a = E_{max} \sin \omega t$$

$$e_a = E_{max} \cos\left(\omega t - \frac{\pi}{2}\right) \tag{6.3}$$

where

$$E_{max} = \omega N\phi = 2\pi f N\phi.$$

The rms value of the generated voltage is then

$$E = 4.44 fN\phi. \tag{6.4}$$

In actual machines, the armature conductors are distributed in slots such that the phasor sum of the induced emf is less than the numerical sum. Hence, winding factor K_w is a factor used to account for the difference which varies between 0.85 and 0.95 for three phase windings. Consequently, for distributed-phase windings, the rms value of the generated voltage is

$$E = 4.44\, K_w fN\phi\,(\mathrm{V}). \tag{6.5}$$

6.6 EQUIVALENT CIRCUIT OF SYNCHRONOUS MACHINES

The fundamental component of field and armature fluxes $\emptyset_f$ and $\emptyset_a$, respectively, produces flux linkages with the armature, which induces corresponding components of armature voltage. The resultant $\emptyset_r$ of the field and the armature mmfs produce the net air-gap flux given by

$$\emptyset_r = \emptyset_f + \emptyset_{ar}. \tag{6.6}$$

This results in net armature flux linkages

$$\lambda_r = \lambda_f + \lambda_{ar}. \tag{6.7}$$

It should be noted that all the quantities in Equations 6.6 and 6.7 are phasors.

Correspondingly, each is responsible for a component of the armature voltage, represented as

$$E_r = E_f + E_{ar}. \tag{6.8}$$

The relationship between these quantities is shown in Figure 6.7.

The nature of leakage fluxes in a generator is similar to that of a transformer. The equivalent leakage flux ϕ_l and the flux ϕ_a of armature-magnetizing inductance are both in phase with the current in any phase of the armature winding. The magnetizing and leakage inductance L_m *and* L_l yield the synchronous inductance

$$L_s = L_m + L_l. \tag{6.9}$$

The corresponding synchronous reactance is given by

$$\omega L_s = \omega L_m + \omega L_l \tag{6.10}$$

$$X_s = X_m + X_l. \tag{6.11}$$

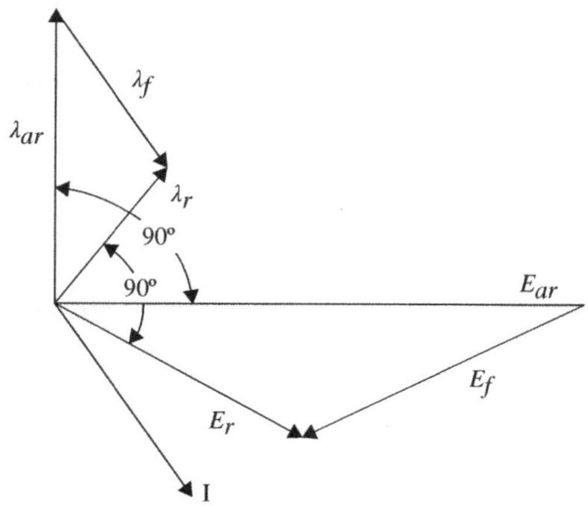

Figure 6.7 Phasor diagram relating flux linkage and voltage.

6.7 SYNCHRONOUS GENERATOR EQUIVALENT CIRCUITS

From the phasor diagram of Figure 6.8, if we assume E_r is the rotor voltage, $\vec{E}_f$ is the field voltage, and $\vec{E}_{ar}$ is the armature voltage, we can write the voltage in phasor form as:

$$E_f = E_{ar} + jI_a X_m. \tag{6.12}$$

Also,

$$E_{ar} - V = I_a R_a + jI_a X_l, \tag{6.13}$$

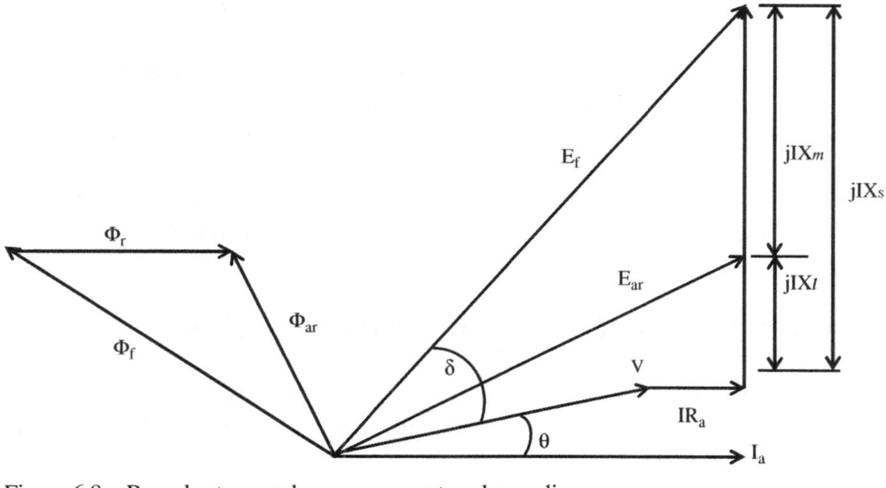

Figure 6.8 Round rotor synchronous generator phasor diagram.

so that we may write

$$E_{ar} = V + I_a R_a + jI_a X_l. \tag{6.14}$$

Substituting Equation 6.13 into 6.14 yields

$$E_f = V + I_a R_a + jI_a X_l + jI_a X_m.$$

But

$$X_s = X_l + X_m. \tag{6.15}$$

Hence,

$$E_f = V + I_a R_a + jI_a \left(X_l + X_m\right)$$

$$E_f = V + I_a R_a + jI_a X_s$$

$$E_f = V + I_a\left(R_a + jX_s\right). \tag{6.16}$$

The synchronous impedance of the machine may be represented by

$$Z_s = R_a + jX_s. \tag{6.17}$$

Hence

$$E_f = \mathrm{V} + \mathrm{j}I_a Z_s$$

$$\vec{E_f} = E_f \angle \delta.$$

Similarly,

$$\vec{V} = V \angle 0$$

$$\vec{I_a} = I_a \angle \theta.$$

where δ and θ are referred to as the *load* or *torque angle* and the *power-factor angle*, respectively.

The per-phase equivalent circuit of the synchronous generator may be derived from the phasor diagram relating E_f, terminal voltage V, and the drop across the synchronous impedance. This is given in Figure 6.9. If E_f is in series with synchronous impedance, I_a lags V_t by angle θ.

Further simplification of the phasor diagram of Figure 6.8 is presented in Figure 6.10 based on the model of the machine obtained in Equation 6.13.

6.8 OVER EXCITATION AND UNDER EXCITATION

With a magnitude of $E_f > V_t$, the machine is over excited. With a magnitude of $E_f < V_t$, the machine is under excited. The angle between excitation voltage E_f and V_t is known as the *torque angle* or *load angle*.

$$\textit{pu voltage regulation} = \frac{E_f - V_t}{V_t} \tag{6.18}$$

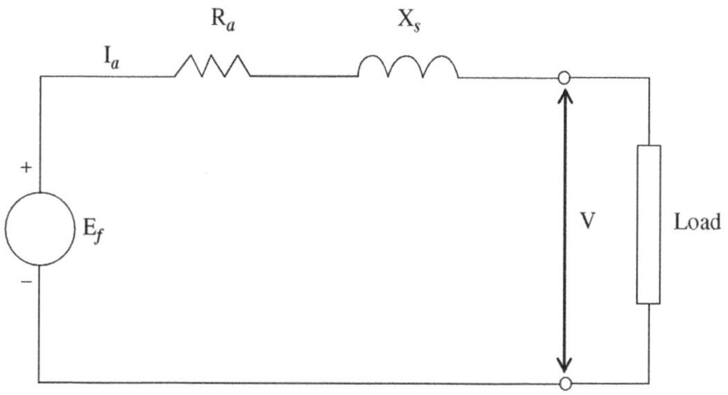

Figure 6.9 Synchronous generator equivalent circuit and phasor diagram.

6.9 OPEN-CIRCUIT AND SHORT-CIRCUIT CHARACTERISTICS

The process of determining the synchronous-machine parameters are done by carrying out an open-circuit test and a short-circuit test, as shown in Figures 6.11, Figure 6.12, and Figure 6.13 [9].

The open-circuit test is carried out with the terminals of the machine disconnected from any external circuit. The basic test procedure is:

(a) Open circuit the generator terminals.

(b) Drive the machine at synchronous speed using an external mechanical system.

(c) Slowly increase the field current and measure the open-circuit terminal voltage.

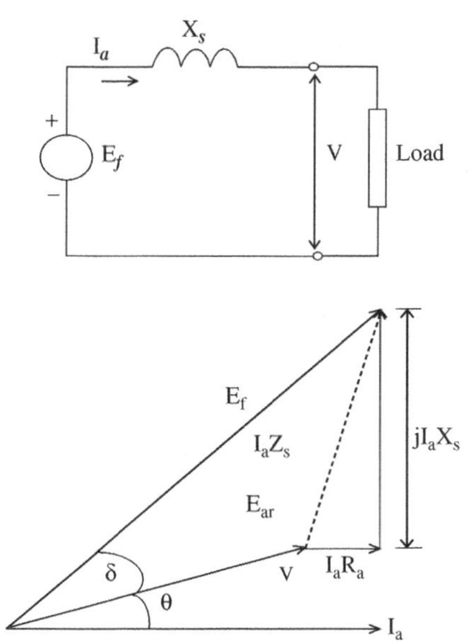

Figure 6.10 Per phase equivalent model and phasor diagram.

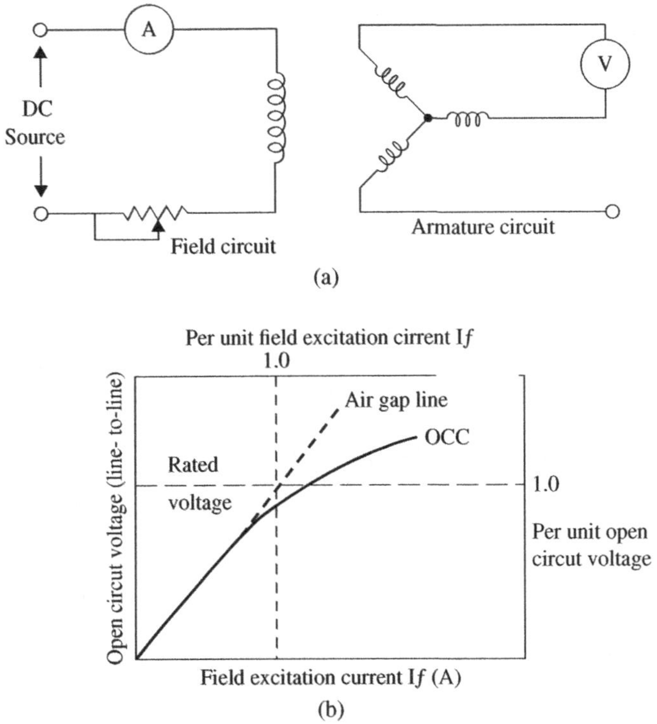

Figure 6.11 (a) Diagram of Open Circuit Test (b) Open-circuit test characteristics.

As there is no armature current, the measured terminal voltage is the induced voltage, from

$$E_f = V + I_a(R_a + jX_s)$$

$$E_f = V_{oc}.$$

The open-circuit characteristics or saturation curve can be obtained by plotting the armature terminal voltage (L–L) on an open circuit as a function of the field-excitation current when the machine is running at rated speed. The test produces a curve, as shown in Figure 6.11.

Saturation due to magnetic saliency during open-circuit test causes the open-circuit characteristics to change from a straight line to a set of curves.

From the curve, we can deduce Z_{oc} if $R \ll X$.

$$Z_{oc} = \frac{E}{\sqrt{3}I_{OC}} \tag{6.19}$$

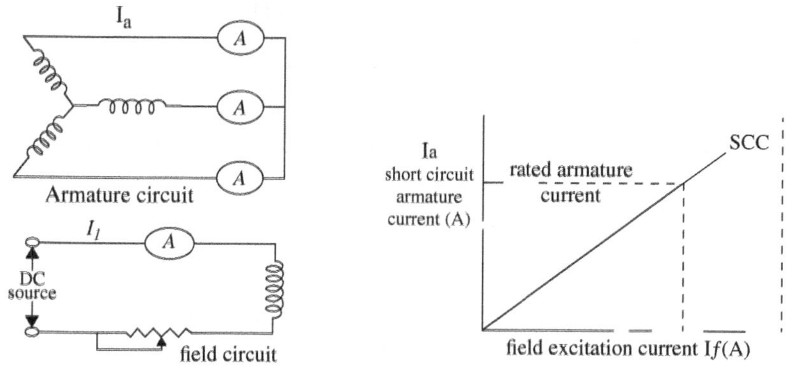

Figure 6.12 Short-circuit test: circuit connections for test and short-circuit characteristics.

Short-Circuit Test

As the name suggests, the short-circuit test is carried out with the terminals of the machine short circuited. The basic test procedure is as follows:

- **(a)** Set the field current to zero.
- **(b)** Short circuit the armature terminals.
- **(c)** Drive the generator at synchronous speed with the external mechanical system.
- **(d)** Slowly increase the field-winding current until the short-circuit armature current reaches the rated design value. Note that very low field-current levels are required to achieve rated short-circuit armature current.

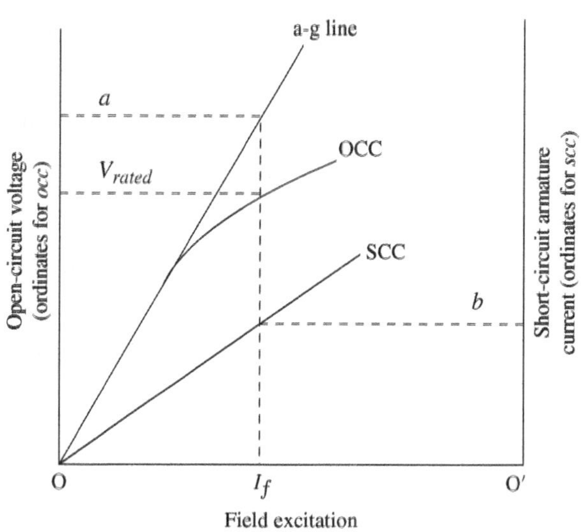

Figure 6.13 Open-circuit and short-circuit characteristics of a synchronous machine.

6.10 PERFORMANCE CHARACTERISTICS OF SYNCHRONOUS MACHINES

From the model of the synchronous generator E_f, V_t, and Z_s are determined. Analysis for describing the relationship between input power and output power is enabled by power angle or torque angle of rotor.

For a generator, complex power

$$S = P + jQ = V_t I_a^* \tag{6.20}$$

where V_t is terminal voltage of machine per phase, I_a is armature current for phase, and $I_a{}^*$ is its complex conjugate.

If R is neglected and $\vec{V}_t = V_t + j0$ and $\vec{E}_f = E_f\,(\cos\delta + j\sin\delta)$,

$$\text{armature current} = I_a = \frac{E_f - V_t}{jX_s} = \frac{E_f\,(\cos\theta + j\sin\theta) - V_t + j0}{jX_s} \tag{6.21}$$

$$= \frac{E_f\left(\cos\theta + V_t\right) + j\sin\theta}{jX_s} = \frac{E_f\left(\cos\theta + V_t\right) + jE_f\sin\theta}{jX_s} \tag{6.22}$$

but

$$I^* = \frac{\left(E_f\ \cos\delta - V_t\right) - j\left(E_f\sin\delta\right)}{-jX_s} = \frac{\left(E_f\sin\delta\right)}{X_s} + \frac{j\left(E_f\ \cos\delta - V_t\right)}{-X_s} \tag{6.23}$$

$$S = P + jQ. \tag{6.24}$$

Figure 6.14 shows the steady state power or torque angle characteristics of a cylindrical rotor machine.

$$P = \frac{V_t\ \left[E_f\sin\delta\right]}{X_s} = \frac{V_t\,E_f\sin\delta}{X_s} \tag{6.25}$$

$$Q = \frac{V_t E_f\ \cos\delta - V_t^2}{X_s} \tag{6.26}$$

Three-phase power is

$$3P_\phi = \frac{3V_t E_f\sin\delta}{X_s} \tag{6.27}$$

$$3Q_\phi = \frac{V_t E_f\cos\delta - V_t^2}{X_s} \tag{6.28}$$

$$\text{at }\sin\delta = 1,\quad \text{when }\delta = \frac{\pi}{2}$$

$$P_{max} = \frac{V_t E_f}{X_s}. \tag{6.29}$$

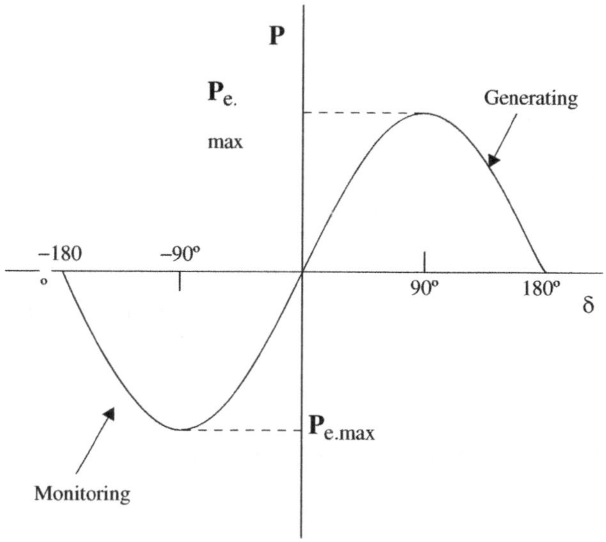

Figure 6.14 Steady-state power or torque-angle characteristics of a cylindrical rotor machine.

It is provided by maximum machine torque so

$$T_{max} = \frac{P_{max}}{w_m} = \frac{P_{max}}{2\pi\left(\frac{n_s}{60}\right)} \tag{6.30}$$

where n_s is the synchronous speed in rpm; thus we say at $\frac{\pi}{2} = \delta$, the machine reaches a steady-state stability limit.

Example 1

A 100 MVA, 11.5kV, 0.8 PF lagging, 50 Hz, two-pole, Y-connected, synchronous generator has a per-unit synchronous reactance of 0.8 and a per-unit armature resistance of 0.012.

(a) What are its synchronous reactance and armature resistance in ohms?

(b) What is the magnitude of the internal-generated voltage *EA* at the rated conditions? What is its torque angle at these conditions?

(c) Ignoring losses in this generator, what torque must be applied to its shaft by the prime mover at full load?

Solution:

The base phase voltage of this generator is $V_{\phi,\text{base}} = 11{,}500/\sqrt{3} = 6{,}640$ V. Therefore, the base impedance of the generator is

$$Z_{\text{base}} = \frac{3V_{\phi,\text{base}}{}^2}{S_{\text{base}}} = \frac{3\left(V_{\phi,\text{base}}\right)^2}{S_{\text{base}}} = 1.32\,\Omega.$$

(a) The generator impedances in ohm are:

$$R_A = (0.012)(1.32\,\Omega) = 0.0158\,\Omega$$
$$X_S = (0.8)(1.32\,\Omega) = 1.06\,\Omega.$$

(b) The rated armature current is

$$I_A = I_L = \frac{S}{\sqrt{3}V_r} = \frac{100\,\text{MVA}}{\sqrt{3}(11.5\,\text{kV})} = 5,020\,\text{A}.$$

The power factor is 0.8 lagging, so $I_A = 5,020\angle - 36.87°\,\text{A}$. Therefore, the internal-generated voltage is

$$E_A = V_\phi + R_A I_A + jX_S I_A$$
$$\begin{aligned} E_A &= 6,640\angle 0°\,\text{A} + (0.0158\,\Omega)\left(5,020\angle - 36.87°\,\text{A}\right) \\ &\quad + (1.06\,\Omega)\left(5,020\angle - 36.87°\,\text{A}\right) \\ &= 10,750\angle 23°\,\text{V}. \end{aligned}$$

Therefore, the magnitude of the internal-generated voltage $E_A = 10,750$ and $\delta = 23°\text{V}$.

(c) Ignoring the losses, the input power would equal the output power. Since

$$P_{out} = (0.80)(100\,\text{MVA}) = 80\,\text{MW}$$

and the applied torque would be

$$\tau_{app} = \tau_{ind} = \frac{80,\,\text{MW}}{\left(\frac{3,000\text{r}}{\text{min}}\right)\left(\frac{\frac{2\Pi\text{rad}}{\text{r}}}{\text{r}}\right)\left(\frac{1\text{min}}{60\text{s}}\right)} = 255,000\,\text{Nm}$$

Example 2

A three-phase, Y-connected, synchronous generator is rated 120 MVA, 13.8kV, 0.85 PF lagging and 60 Hz. Its synchronous reactance is 0.7 Ω and its resistance may be ignored.

(a) What is its voltage regulation?

(b) What would the voltage and apparent power rating of this generator be if it operated at 50 Hz with the same armature and field losses as it had at 60 Hz?

(c) What would the voltage regulation of the generator be at 50 Hz?

Solution:

(a) The rated armature current is

$$I_A = I_L = \frac{S}{\sqrt{3}V_r} = \frac{120\,\text{MVA}}{\sqrt{3}(13.8\text{kV})} = 5,020\,\text{A}.$$

The power factor is 0.85 lagging, so $I_A = 5,020\angle - 31.8°A$. Therefore, the internal generated voltage is

$$E_A = V_\phi + R_A I_A + jX_S I_A$$

$$E_A = 7,967\angle 0°A + (0.7\,\Omega)\left(5,020\angle - 31.8°A\right)$$

$$= 10,260\angle 16.9°V.$$

The resulting voltage regulation is

$$V_R = \frac{10,260 - 7,967}{7,967} \times 100\% = 28.8\%.$$

(b) If the generator is to be operated at 50 Hz with the same armature and field losses as at 60 Hz (so that the windings do not overheat), then its armature and field currents must not change. Since the voltage of the generator is directly proportional to the speed of the generator, the voltage rating (and hence the apparent power rating) of the generator will be reduced by a factor of 5/6.

$$V_{r,rated} = \frac{5}{6}(13.8kV) = 11.5kV$$

$$S_{rated} = \frac{5}{6}(120\,MVA) = 100\,MVA$$

Also, the synchronous reactance will be reduced by a factor of 5/6.

$$X_S = \frac{5}{6}(0.7\,\Omega) = 0.583\,\Omega.$$

(c) At 50 Hz-rated conditions, the armature current would be

$$I_A = I_L = \frac{S}{\sqrt{3}V_r} = \frac{100\,MVA}{\sqrt{3}(11.5kV)} = 5,020\,A.$$

The power factor is 0.85 lagging, so $I_A = 5,020\angle - 31.8°A$. Therefore, the internal generated voltage is

$$E_A = V_\phi + R_A I_A + jX_S I_A$$

$$E_A = 6,640\angle 0°A + (0.583\Omega)\left(5,020\angle - 31.8°A\right)$$

$$= 8,550\angle 16.9°V.$$

The resulting voltage regulation is

$$V_R = \frac{8,550 - 6,640}{6,640} \times 100\% = 28.8\%.$$

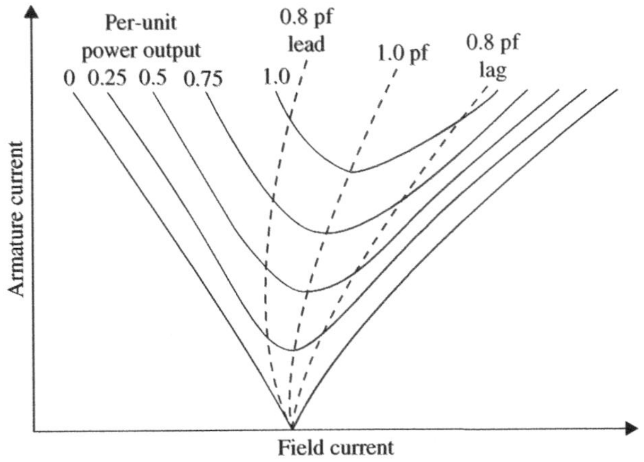

Figure 6.15 Synchronous machine V-curves.

6.11 GENERATOR COMPOUNDING CURVE

A synchronous machine has a power factor that is gradually controlled by the field current. In other words, changing its field excitation can control the power factor of the stator-line current.

The curve showing the relationship between the stator current and the field current at a constant terminal voltage with a constant real power is called a *synchronous machine V curve* because of the slope, as shown in Figure 6.15 [3]. The V curves can be developed for synchronous generators as well as for synchronous motors and will be almost identical.

6.12 SYNCHRONOUS GENERATOR OPERATING ALONE: CONCEPT OF INFINITE BUS

When a single synchronous machine operating in generating or motoring mode is connected to an infinite bus, the infinite-bus theory cannot be applied since there are no additional machines to compensate for field-excitation changes and prime-mover output to keep the terminal voltage and frequency constant. Since the terminal voltage changes drastically as the load changes, an automatic voltage regulator is required to control the field current in order to maintain a constant terminal voltage as the load varies.

Generally,

(a) addition of inductive loads causes terminal voltage to drop drastically

(b) addition of purely resistive loads has minimal effect on terminal voltage

(c) addition of capacitive loads causes the terminal voltage to rise drastically

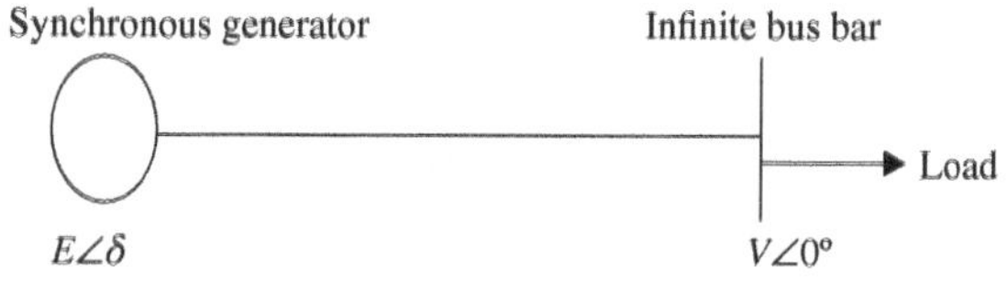

Figure 6.16 A single machine connected to an infinite bus bar.

Figure 6.16 shows the constant field current volt-ampere characteristics of a synchronous generator operating alone. The curves are for three different values of constant field currents and power factors.

Figure 6.17 shows a typical set of reactive power capability curves for a large, hydrogen-cooled turbine generator along with the effect of increased hydrogen pressure on allowable machine loadings. The rotor MVA of the synchronous machine is distorted by the stator current in terms of maximum allowable stator current. The upper and lower portions of the area inside the circle with a radius of maximum S represent the generator and motor. Figure 6.18 [3] shows the reliability power capability curves.

6.13 INITIAL ELEMENTARY FACTS ABOUT SYNCHRONOUS MACHINES

We must realize that:

$$\omega = 2\pi f(\text{electrical radians/sec}),\ \text{frequency } f = 60\,\text{Hz}$$

$$\omega_m = \frac{2}{P}\,(\text{radians})$$

or

$$n = \frac{120f}{P}\ \text{rpm.} \tag{6.31}$$

Thus, a polyphase winding excited by a balanced three phase produces an effect equivalent to that of a magnet or DC-rotating machine.

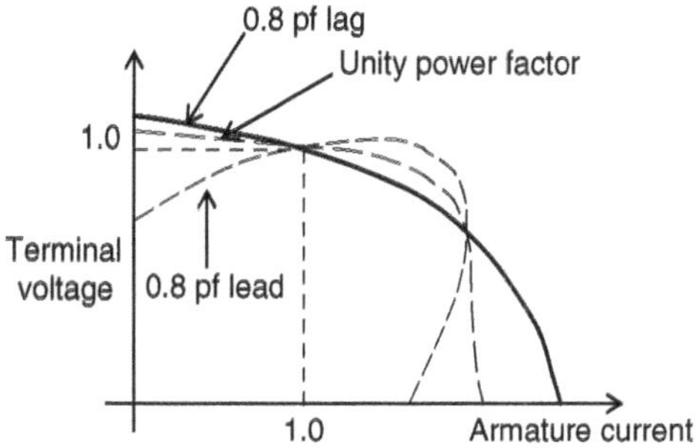

Figure 6.17 Synchronous generator armature field current curves.

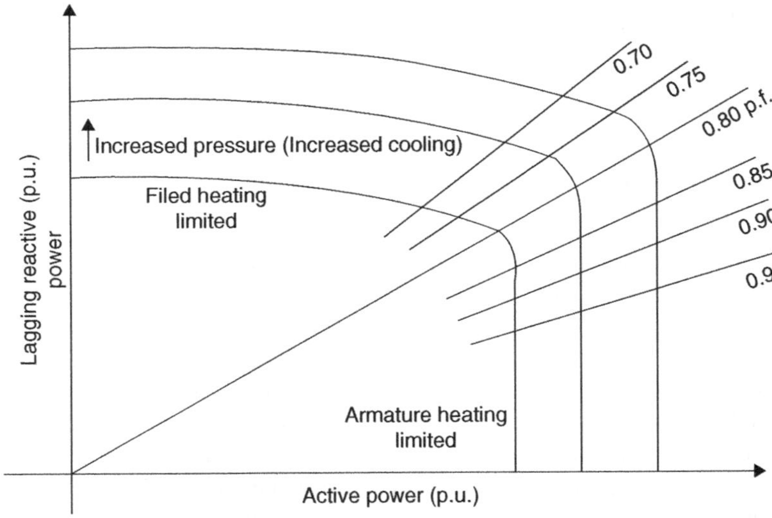

Figure 6.18 Reactive power capability curves.

For a synchronous machine, space distribution of radial gap flux density is given by

$$B = B_m \cos \beta \tag{6.32}$$

where

$$\begin{aligned} \phi &= \int_{-\frac{\pi}{2}}^{\frac{\pi}{2}} B_m lr \cos \beta d\beta \\ &= 2B_m lr. \end{aligned} \tag{6.33}$$

For a pole machine

$$\frac{2}{P} 2B_m lr$$

from which we can get the flux linkage λ as

$$\lambda = \frac{d\phi}{dt} = N\phi \cos \omega t \tag{6.34}$$

$$\begin{aligned} e &= -\frac{d\lambda}{dt} = \omega N\phi \sin \omega t = -\frac{Nd\phi}{dt} \cos \omega t \\ &= -N\frac{d\phi}{dt} \cos \omega t \end{aligned}$$

and

$$e = \omega N\phi \sin \omega t. \tag{6.35}$$

In terms of flux density

$$B = \phi A$$

$$e = B_{eff} l_{eff} r = (B_m \sin \omega t)\frac{2}{N}\left(r\omega_m\right)$$

$$e = \omega N \frac{2}{P} 2B_m lr \sin \omega t$$

$$e = \omega N \phi \sin \omega t \tag{6.36}$$

where e is a sinusoidal voltage with spatial distribution.

For a two-pole, synchronous machine under steady-state conditions, operation revolves at 60 rps or 3,600 rpm in order to produce voltage with frequency 60 Hz.

For a coil passing a complete cycle every two pairs of poles $\left(\frac{P}{2}\right)$ sweeping in each revolution,

$$f = \frac{P}{2}\frac{n}{60}\ \text{Hz}$$

$$n = \frac{120f}{P}\ \text{rpm}$$

$$\therefore \quad \omega = in\ radian = \frac{P}{2}\omega_m\ \text{(mechanical frequency)}.$$

6.14 CYLINDRICAL-ROTOR MACHINES FOR TURBO GENERATORS

The maximum value of induced voltage is

$$E_{max} = \omega N \phi = 2\pi f N \phi \tag{6.37}$$

$$E_{max} = \frac{2\pi}{\sqrt{2}} f N \phi = 4.44 f N \phi/\text{phase} \tag{6.38}$$

When connected with the equation of induced voltage in Equation 6.5, if we include the winding type in the salient-pole machine either in delta or star connection,

$$E_{rms} = 4.44\, k_w f N_{pu}\, \phi\, \text{V/phase} \tag{6.39}$$

where k_w is the winding factor.

Normally, $k_w < 1$ and is typically 0.85 – 0.95.

6.15 SYNCHRONOUS MACHINES WITH EFFECTS OF SALIENCY: TWO-REACTANCE THEORY

The saliency of a machine accounts for the reactance measurement at the terminal varying as a function of the rotor position. Armature current is resolved in two axes into the direct axis component I_d and the quadrature axis component I_q. I_d produces the armature reaction flux ϕ_{ar} along the axis of the field poles, while I_q produces an armature reaction flux ϕ_{aq}.

The sum of the two fluxes

$$\phi_{ad} + \phi_{aq} = \phi_{ar}. \tag{6.40}$$

E_f and I_a are out of phase by ϕ.

Similar to magnetizing reactance X_p of cylindrical-rotor theory, the inductive effects can be resolved into $X_{\phi d}$ and $X_{\phi q}$ given as

$$X_d = X_l + X_{\phi d} \tag{6.41}$$
$$X_q = X_l = X_{\phi q}. \tag{6.42}$$

X_l is same for direct and quadrature reactances. To account for X_d and X_q in X_s we have

$$E_f = V_t + I_a R_a + jI_d X_d + jI_q X_q. \tag{6.43}$$

Given the phase shift between I_d and I_q we can determine the torque angle from

$$Tan\ \delta = \frac{I_q X_q\, cos\, \phi}{V_t + I_a X_q\, sin\, \phi}. \tag{6.44}$$

Let us show how the saliency affects the power angle.

The figure and related equations are derived as follows

$$I_a X_a = V_t\ sin\, \delta \tag{6.45}$$
$$I_d X_d = E_f - V_t\ cos\, \delta \tag{6.46}$$
$$I_d = I_a \sin\ (\phi + \delta) \tag{6.47}$$
$$I_q = I_a \cos\ (\phi + \delta) \tag{6.48}$$

solving

$$I_a \cos\phi = \frac{E_f}{X_d}\sin\delta + \frac{V_t}{2X_q}\sin 2\delta - \frac{V_t}{2X_d}\sin 2\delta \tag{6.49}$$

$$P = V_t I_a \cos\phi = \frac{V_t E_f}{X_d}\sin\delta + \frac{1}{2}V_t^2\left(\frac{1}{X_q} - \frac{1}{X_d}\right)\sin 2\delta. \tag{6.50}$$

6.16 THE SALIENT-POLE MACHINE

The synchronizing power coefficient is

$$\frac{dP}{d\delta} = \frac{V_t E_f}{X_d}\cos\delta + V_t^2\frac{(X_d - X_q)}{X_d X_q}\cos 2\delta \tag{6.51}$$

$$\Delta P = \frac{\delta P}{\delta\delta}\Delta\delta. \tag{6.52}$$

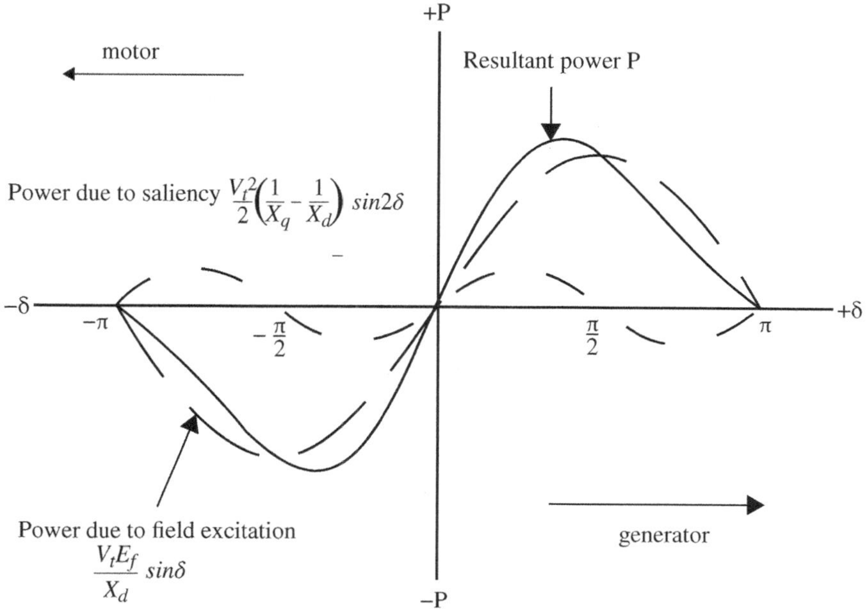

Figure 6.19 Power and load angle for saliency.

Figure 6.19 shows the power and load angle for saliency ΔP changes depend on load-angle changes $\Delta\delta$

$$\text{reactive power } Q = VI \sin\delta \tag{6.53}$$

if

$$I \sin\phi = I_d \cos\delta - I_q \sin\delta \tag{6.54}$$

$$V \cos\phi = E - I_d X_d \tag{6.55}$$

$$V \sin\phi = I_q X_q$$

$$Q = \frac{VE}{X_d}\cos\delta - \frac{V^2}{X_d}\cos 2\delta - \frac{V^2}{X_q}\sin 2\delta. \tag{6.56}$$

Efficiency Computation for Synchronous Generators

$$\eta_{pu} = \frac{Output = Power}{Input\ Power} = 1 - \frac{Losses}{Output + Losses} \tag{6.57}$$

For the case of a generator, power flow is shown in Figure 6.20.

Generator Power Computation

$$\text{pu } \eta = 1 - \frac{P_{cu} + P_f + P_s + P_{NL} + P_{wf}}{mP_o + P_{cu} + P_f + P_s + P_{NL} + P_{wf}} \tag{6.58}$$

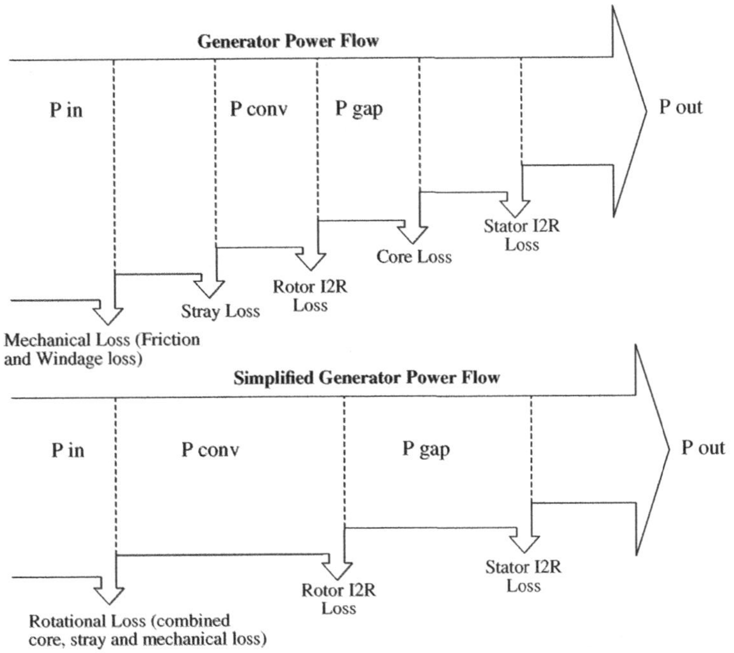

Figure 6.20 Synchronous generator power-flow diagram.

where m is the number of armature windings, P_o is the power output per phase (VI cos ϕ), P_{cu} is the armature-winding ohmic loss (mI^2R), P_f is the field-winding ohmic loss ($I_f^2R_f$), P_s is stray losses, and P_{wf} is the windage and frictional loss.

Note that windage and frictional losses are obtained from the open-circuit test. From the short-circuit test, we compute the windage, friction, and armature-winding ohmic loss to give the stray loss. Compute armature core losses + eddy current losses due to the armature-winding circuit.

6.17 SYNCHRONOUS MOTORS

Synchronous machines are reversible. If the rotor is driven by a prime mover, the machine is a generator, converting the mechanical energy input to electrical energy output. If the machine draws electrical energy from a supply, it will require a motor to drive a mechanical load connected to its shaft.

The motor is a constant-speed machine. It rotates at the synchronous speed n_s, given in terms of the frequency f and the number of pole pairs P. The generator theory is also applicable to synchronous motors.

As noted, the magnitude of the excitation of a generator connected to an infinite bus affects the power factor.

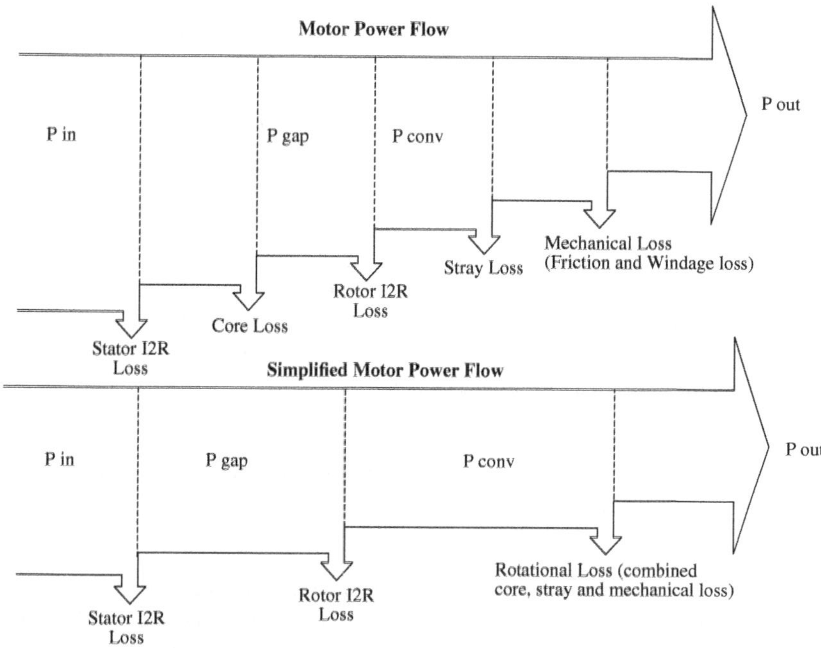

Figure 6.21 Synchronous motor power flow.

Applications

Synchronous motors are useful where constant speed is required for power-factor correction. However, synchronous motors are not self-starting; their applications are for low-speed drives.

The power-flow diagram for a synchronous motor is shown in Figure 6.21.

$$\eta = 1 - \frac{P_{cu} + P_f + P_s + P_{NL} + P_{wf}}{P_t} \tag{6.59}$$

Motor Analysis

For a round rotor motor

$$X_d = X_q = X_s \tag{6.60}$$

$$I = I_d + I_q. \tag{6.61}$$

The voltage equation for the round-rotor motor is given by

$$V = E + I(R_a + jX_s) \tag{6.62}$$

$$R_a \ll X_s$$

$$V = E + jX_s. \tag{6.63}$$

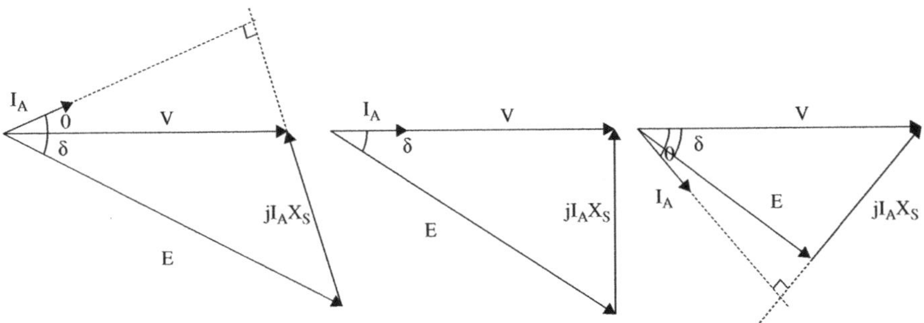

Simplified Leading Power Factor Phasor Diagram

Simplified Unity Power Factor Phasor Diagram

Simplified Lagging Power Factor Phasor Diagram

Figure 6.22 Phasor diagrams of a synchronous motor.

Figure 6.22 shows the phasor diagram of a synchronous motor for leading, unity and lagging power factor respectively.

Examples

(1) A hydraulic turbine running at 200 rpm is connected to a synchronous generator. If the induced voltage has a frequency of 60 Hz, determine the number of rotor poles.

Solution:

Number of poles

$$P = \frac{120f}{N}$$
$$= \frac{120 \times 60}{200}$$
$$= 36\,\text{poles}\,(18\text{ pole pairs})$$

(2) A 3,600 rpm, 60 Hz, 13.8kV, synchronous generator has a synchronous reactance of 20 Ω. The generator is operating at rated voltage and speed with the excitation voltage $E_f = 11.5\text{kV}$ and the torque angle $\delta = 15°$. Calculate:

(a) stator current

(b) power factor

(c) total output power

Solution:

(a) $$\bar{I}_a = \frac{\bar{E}_f - \bar{V}_{an}}{jX_s} = \frac{11{,}500\angle 15° - \dfrac{13{,}800}{\sqrt{3}}\angle 0°}{20\angle 90°} = 216.35\angle -46.54°\,\text{A}$$

$$I_a = 216.35\,\text{A}$$

(b) $PF = \cos\left(\angle\bar{V}_{an} - \angle\bar{I}_a\right) = \cos(46.54°) = 0.688$ lagging

(c) $P_T = \sqrt{3}V_L I_a PF = \sqrt{3} \times 13{,}800 \times 216.35 \times 0.688 = 3{,}557.8\,\text{kW}.$

(3) A 100 MVA, 13.8kV, 60 Hz, salient-pole, synchronous generator with eight poles has $X_d = 1.9\,\Omega$ and $X_q = 1.1\,\Omega$. The stator-coil resistance is negligible. The generator is supplying rated apparent power at a power factor of 0.866 lagging. Determine the value of excitation voltage E_f and torque angle δ.

Solution:

$$\bar{I}_a = \frac{S_T}{\sqrt{3}V_L}\angle - \cos^{-1}(\text{PF})$$
$$= \frac{100 \times 10^6}{\sqrt{3} \times 13,800}\angle\cos^{-1}(0.866) = 4,183.8\angle - 30°\,\text{A}$$
$$\overline{E'}_f = \bar{V}_{an} + jX_q\bar{I}_a = \frac{13,800}{\sqrt{3}}\angle 0° + \left(1.1\angle 90°\right)\left(4,183.8\angle - 30°\right)$$
$$\overline{E'}_f = 11,015\angle 21.21°\text{V}$$

Thus,

$$\delta = \angle\overline{E'}_f = 21.21°$$
$$\psi = (\theta + \delta) = (30 + 21.21) = 51.21°$$
$$I_d = I_a \sin\psi = 4,183.8 \sin(51.21°) = 3,261\text{A}$$
$$\bar{I}_d = I_d\angle(\delta - 90°) = 3,261\angle - 68.79°\text{A}$$
$$\bar{E}_f = \overline{E'}_f + \left(jX_d - X_q\right)\bar{I}_d$$
$$\bar{E}_f = 11,015\angle 21.21° + (0.8\angle 90°)(3,261\angle - 68.79°)$$
$$\bar{E}_f = 13,623.8\angle 21.21°\text{V}$$

$E_f = 13,623.8$V at a machine angle of 21.21°.

6.18 SYNCHRONOUS MACHINES AND SYSTEM STABILITY

Stability refers to the ability of the generators in a power system to operate in synchronism both under normal conditions and following disturbances. Three categories of instability are:

(a) steady-state instability

(b) transient instability

(c) dynamic instability

Steady-State Instability

Steady-state instabilty refers to the condition where the equilibrium of the generators connected to the power system cannot accommodate increases in power requirements that occur relatively slowly or when a transmission line is removed from service for maintenance.

Since the power flow from one point to another is proportional to the sine of the angular difference between the voltages at the two points and inversely proportional to the total impedance of the circuits connecting the two points, there is a maximum level of power flow, that is, the delivery level at which the angular difference is 90° and the sine is equal to one.

The fact that the power flow is dependent on the sine of the angular difference between the voltages has an important significance in that it defines the maximum amount of power that can be moved across the facilities connected by the impedance X12. If the power required by the customers at bus two is greater than the amount that can be delivered at a 90° separation in the voltages, the system is unworkable. When the customer load at bus two slowly increases and the angular spread responds until it reaches the 90° point and then goes beyond 90°, the system becomes unstable and will collapse.

If the net impedance is increased by removing a line, less power can be transmitted. If there are a number of lines connecting bus one with bus two, the loss or outage of any one of them will increase the impedance between the two buses and the system can again become unstable. Conversely, if the net impedance is reduced, more power can be transmitted. The value of the net impedance can be reduced by:

- building additional lines in parallel
- raising the design voltage of one or more of the existing lines
- decreasing the impedance of any of the existing lines by inserting a capacitor in series (remember XC cancels out XL)

Transient Instability

Transient instability refers to the condition where there is a disturbance in the system that causes a disruption in the synchronism or balance of the system. The disturbance can be a number of types of varying degrees of severity:

- opening a transmission line, increasing the XL of the system
- occurrence of a fault-decreasing voltage on the system (voltage at the fault goes to zero, decreasing all system voltages in the area)
- loss of a generator disturbing the energy balance and requiring an increase in the angular separation as other generators adjust to make up the lost energy
- loss of a large block of load in an exporting area

When there is a disturbance in the system, the energy balance of generators is disturbed. Under normal conditions, the mechanical energy input to the generator equals the net electrical energy output plus losses in the conversion process within the turbine generator and the power consumed in the power plant.

If a generator sees the electric demand at its terminal in excess of its mechanical energy input it will tend to slow down as rotational energy is removed from its rotor to supply the newly increased demand. If a generator sees an electric demand at its terminal less than its mechanical energy input, it will tend to speed up due to the sudden energy imbalance. This initial reaction is called the *inertial response*.

Disturbances may also change the voltage at the generator's terminals. In response, the generator's automatic voltage regulating system will sense the change and adjust the generator's field excitation, either up or down, to compensate.

Transient stability or instability considers that period immediately after a disturbance, usually before the generator's governor and other control systems have a chance to operate. In all cases, the disturbance causes the generator angles to change automatically as they adjust to find a new stable operating point with respect to one another. In an unstable case, the angular separation between one generator or group of generators and another group keeps increasing. This type of instability happens so quickly, in a few seconds, that operator corrective action is impossible.

If stable conditions exist, the generator's speed-governor system, sensing the beginning of a change in speed, will react to either admit more mechanical energy into the rotor to regain its speed or to reduce the energy input to reduce the speed. Directives may also be received by the generator from the company or area control center to adjust its scheduled output.

In addition to the measures noted above to improve steady-state stability, other design measures are available for selected disturbances to mitigate this type instability:

- improving the speed by which relays detect the fault and the speed by which circuit breakers operate to disconnect the faulted equipment sooner
- use of dynamic braking resistors, which, in the event of a fault, are automatically connected to the system near generators to reduce export from the generators
- installation of fast-valving systems on turbines, allowing rapid reduction in the mechanical-energy input to the turbine generator
- automatic generator tripping
- automatic load disconnection
- special transmission line-tripping schemes

Dynamic Instability

Dynamic instability refers to a condition where the control systems of generators interact in such a way as to produce oscillations between generators or groups of generators that increase in magnitude and result in instability, that is, there is insufficient damping of the oscillations. These conditions can occur either in normal operation or after a disturbance.

Results of Instability

In cases of instability, as the generator angles separate, the voltage and current angular relationships at points on the system change drastically. Some of the protective line relays will detect these changes and react as if they were due to fault conditions, causing the opening of many transmission lines.

The resulting transmission system is usually segmented into two or more electrically isolated islands, some of which will have excess generation and some that will be generation deficient. In excess-generation pockets, the frequency will rise. In generation-deficient pockets, the frequency will fall. If the frequency falls too far, generator auxiliary systems (motors, fans) will fail, causing generators to be automatically disconnected by their protective devices. Industry practice is to provide for situations where there is insufficient generation by installing under-frequency, load-shedding relays. These relays, keyed to various levels of low frequency, will actuate the disconnection of blocks of customer load in an effort to restore the load-generation balance.

In situations where the frequency rises because of excess generation, generators will be automatically removed from service by protective devices detecting an over-speed condition. If studies indicate potential excess-generation pockets, special selective-generation disconnection controls can be installed.

Illustrative Problems and Examples

1. At what speed must a six-pole, three-phase, synchronous generator run to generate 50 Hz voltage?
   Given f = 50 Hz, p = 6

$$N_s = 120 f/p$$
$$= 120 \times 50/6$$
$$= 1{,}000 \text{ rpm.}$$

2. An eight-pole, three-phase, synchronous generator is operated at 900 rpm. What is the frequency of the output voltage?
   Given $p = 8$, $N_S = 900$ rpm,

$$N_s = 120 \times f/p$$
$$f = pN_s/120 = (8 \times 900)/120 = 2 \times 30 = 60 \text{ Hz.}$$

3. A 60 kVA, three-phase, Y-connected, 440 V, 60 Hz, synchronous generator has a resistance of 0.15 Ω and a synchronous reactance of 3.5 Ω per phase. At rated load and unity power factor, find the internal excitation voltage magnitude and the power angle.
   Given $R_q = 0.15$ ohm, $X_S = 3.5$ ohm,
   since it is Y connected,

$$\frac{440 \times 10^3}{\sqrt{3}} = 254\angle 0°\text{V}$$
$$I = \frac{S}{\sqrt{3}V_L} = \frac{60 \times 10^3}{\sqrt{3} \times 440}$$
$$I = 78.72\text{A}$$

and since the power factor (pf) is unity,

$$pf = 1 = \cos\theta_L$$
$$\theta = \cos^{-1}1 = 0°$$
$$I = 78.72\angle 0°\,\text{A}$$

excitation voltage is

$$E_a = V + IZ$$

where

$$Z = 0.15 + j3.5$$
$$= 3.5\angle 87.55°\Omega.$$

Therefore,

$$\text{E} = 254\angle 0° + (78.72 < \angle 0°)(3.5\angle 87.55°)$$
$$= 254\angle 0° + 275.52\angle 87.55°$$
$$= 254 + j0 + 11.78 + j\,275.27$$
$$= 265.78 + j\,275.27$$
$$\text{E} = 382.64\angle 46°\text{V}$$

Hence, the excitation voltage is 382.64V and the power angle is 46°.

6.19 CHAPTER SUMMARY

The synchronous machine is at the heart of power generation in any conventional power system. Because of its scalability, it is also useful in small-scale power-generation applications. In this chapter, we presented the principles of this machine and its characteristics, which are essential for students to relate with such machines in a microgrid environment. The concepts include modeling, performance measures, and characteristics of the machines. Since the synchronous machine is a key energy source in any power system, the student also may want to explore its place in nontraditional power systems, especially in the presence of unconventional energy resources such as solar photovoltaics and wind energy. This chapter therefore prepares the student for future exercises in the application of this machine in an unconventional power network. Several working examples are given to illustrate the main features of synchronous machines under different conditions.

EXERCISES

1. A 1 MVA, 11kV, three-phase, Y-connected, synchronous generator supplies a three-phase load of 600 kW, 0.8 leading power factor. The synchronous reactance is 24 Ω per phase and the armature resistance is negligible. Find the power angle.

2. Calculate the excitation voltage for a three-phase, Y-connected, 2,500 kVA, 6.6kV, synchronous generator operating at full load and 0.9 power factor lagging. Per-phase synchronous reactance is 4 Ω and per-phase armature resistance is negligible. What will be the internal excitation voltage when the generator is operating at full load with 0.9 power factor leading? Explain whether the machine is over excited or under excited in each case.

3. A 75 kVA, 2.2kV, 60 Hz, three-phase, Y-connected, synchronous generator has a resistance of 0.2 Ω and a synchronous reactance of 6 Ω per phase, and is operated at full load. Draw a phasor diagram of the excitation voltage, load current, and terminal voltage when the load is:

   - 0.85 lagging power factor
   - unity power factor
   - 0.85 leading power factor

   Use the terminal voltage as the reference with the angle of zero degrees.

4. A 1 MVA, 11kV, three-phase, Y-connected, synchronous generator has a synchronous reactance of 5 Ω and a negligible armature resistance. At a certain field current the generator delivers a rated load at 0.9 lagging power factor at 11kV. For the same excitation, what is the armature current and power factor when the input torque is reduced such that the real power output is half of the previous case?

5. An 11kV, three-phase, Y-connected generator has a synchronous reactance of 6 Ω per phase and a negligible armature resistance. For a given field current, the open-circuit, line-to-line excitation voltage is 12kV. Calculate the maximum power developed by the generator. Determine the armature current and power factor for the maximum power condition.

6. Two three-phase generators (G1 and G2) supply a three-phase load through separate three-phase lines, as shown in Figure Q6. Each generator has a synchronous reactance of 3 Ω per phase and a negligible armature resistance. The three-phase, Y-connected load absorbs 30 kW at 0.8 power factor lagging. The line impedance is 1.4 + j1.6 Ω per phase between generator G1 and the load, and 0.8 + j1 Ω per phase between generator G2 and the load.

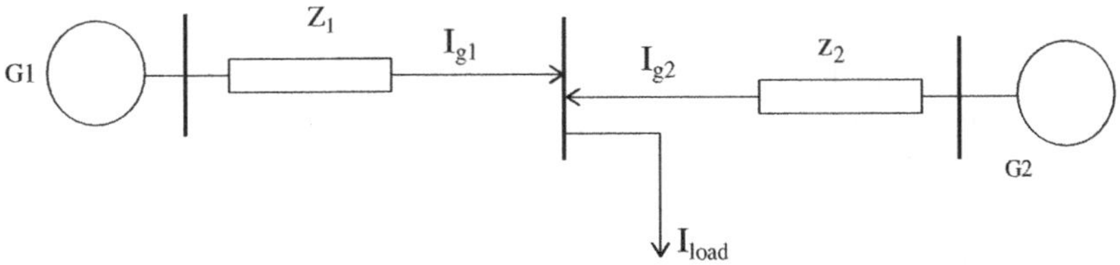

Figure Q6 Single line diagram of a Power System.

If generator G1 supplies 15 kW at 0.8 power factor lagging, with terminal voltage of 460 V line to line, assuming balanced operation, determine internal excitation voltage magnitude (per phase) and power angle of both generators. Use the terminal voltage of generator 1 as a reference angle.

7. An eight-pole alternator runs at 750 rpm and supplies power to a six-pole induction motor that has a full-load slip of 3 percent. Find the full-load speed of the induction motor and the frequency of its rotor emf.

8. An 11kV, three-phase, four-pole, 50 Hz, star-connected, synchronous generator has a synchronous impedance of 12ohms. Determine the emf per phase induced by the field current and the load angle when the machine supplies a load of 3 MW at 0.8 lagging to its bus bar.

BIBLIOGRAPHY

[1] J. Casazza and F. Delea, *Understanding Electric Power Systems: An Overview of the Technology, the Marketplace, and Government Regulations*, John Wiley & Sons, Hoboken, NJ, 2003.
[2] H. Saadat, *Power System Analysis*, 3rd ed., PSA Publishing, 2010.
[3] A. E. Fitzgerald, C. Kingsley, and S. D. Umans, *Fitzgerald & Kingsley's Electric Machinery*, 4th ed., McGraw-Hill, New York, 1983.
[4] C. A. Gross, *Power System Analysis*, John Wiley & Sons, Hoboken, NJ, 1979.
[5] S. J. Chapman, *Electric Machinery Fundamentals*, 5th ed., McGraw-Hill, New York, 2012.
[6] T. Gönen, *Electrical Machines with MATLAB*, 2nd ed., Taylor & Francis Group, Boca Raton, FL, 2011.
[7] W. L. Matsch and J. D. Morgan, *Electromagnetics and Electromechanical Machines*, 3rd ed., John Wiley & Sons, Hoboken, NJ, 1986.
[8] M. S. Sarma, *Electric Machines: Steady-State Theory and Dynamic Performance*, WMC Brown Publishers, Dubuque, IA, 1985.
[9] R. Ramshaw and R. G. van Heeswijk, *Energy Conversion: Electric Motors and Generators*, Saunders College Publishing, Philadelphia, 1990.
[10] G. R. Slemon, *Electric Machines and Drives*, Addison-Wesley Publishing, New York, 1992.
[11] B. L. Theraja and A. K. Theraja, *A Textbook of Electrical Technology*, 1st multicolour ed., S. Chand, New Delhi, 2005.
[12] D. P. Kothari and I. J. Nagrath, *Modern Power System Analysis*, 3rd ed., Tata McGraw-Hill, New Delhi, 2003.
[13] J. B. Gupta, *A Course in Power Systems*, 11th ed., S. K. Kataria & Sons, New Delhi, 2013.
[14] A. Knight, "Electric Machines," http://people.ucalgary.ca/~aknigh/electrical_machines/machines_main.html.
[15] "Synchronous Machine V-Curves," http://keysan.me.
[16] "Synchronous Generator Power Flow Diagram," http://people.ucalgary.com.

CHAPTER 7

DC MACHINES

7.1 INTRODUCTION

An electric machine is a link between electrical and mechanical systems. The conversion of electrical to mechanical systems is based on transferring voltage and current electrical quantities to mechanical quantities as torque and angular speed and vice versa. Generators and motors can be used interchangeably. Feeding a generator from the mains will convert it to a motor, while driving a motor with a prime mover will convert it to a generator. Generators or motors can be AC or DC machines.

DC and AC machines are subject to electromechanical energy conversion, which satisfies Lenz's, Faraday's or Lorentz's fundamental laws. We present herein a simple illustration of two principles on which such machines are developed:

- conductor moving in a uniform magnetic field
- current-carrying conductor in a uniform magnetic field

7.2 CONDUCTOR MOVING IN A UNIFORM MAGNETIC FIELD

The movement of a conductor in a magnetic field gives induced voltage in a conductor in the magnetic field, as shown in Figure 7.1 [1],

$$e = Blv \sin \theta \tag{7.1}$$

where B is the flux density in Tesla or Weber/m^2, l is the conductor length (m), and V is the velocity of its movement in the magnetic field (ms^{-1}).

7.3 CURRENT-CARRYING CONDUCTOR IN A UNIFORM MAGNETIC FIELD

When a current-carrying conductor is in a magnetic field of density B Tesla, as shown in Figure 7.2 [1], the conductor experiences a force F given by

$$F = BIl. \tag{7.2}$$

Energy Processing and Smart Grid, First Edition. James A. Momoh

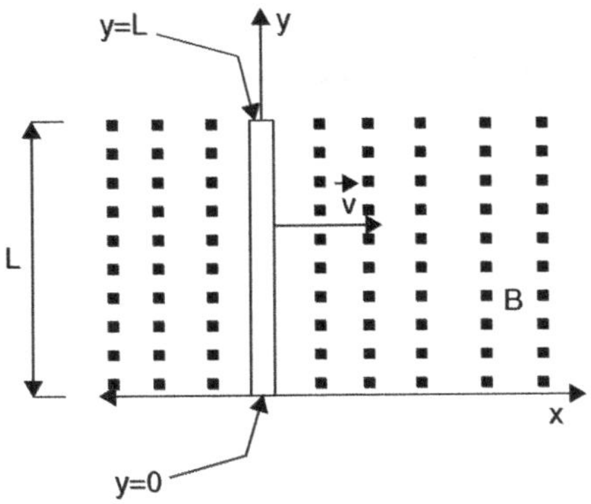

Figure 7.1 Conductor moving in a magnetic field.

A tuning force is impacted on the conductor, causing it to rotate in the magnetic field; this is also due to Lorentz's law.

Of all the rotating machines defined so far, synchronous, induction, and DC machines all use these fundamental principles to operate. To generalize these two principles, machines are designed based on two major components, namely a rotating (moving) part called the *rotor* and a stationary part called the *stator*.

The windings are made of copper, which allows current to flow through or allows for induced voltage based on the two principles. In general, the armature winding used to denote rotor and field winding is also used to denote stator winding for same class of machines.

Each of the machines by principle follows the concept—either electric-power conversion to mechanical torque, referred to as *motor action*, or mechanical-power conversion to electrical power, known as *generator action*.

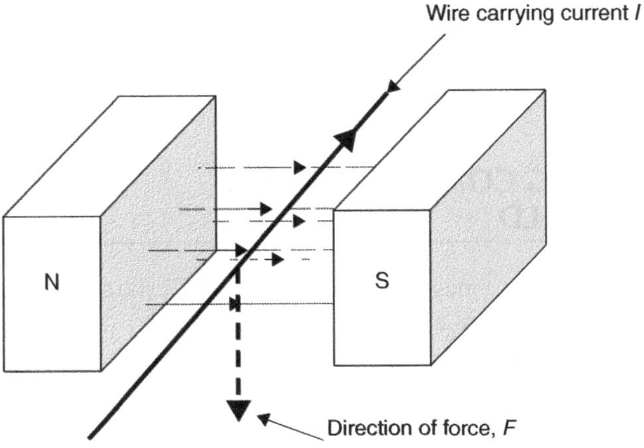

Figure 7.2 Current-carrying conductor in a magnetic field.

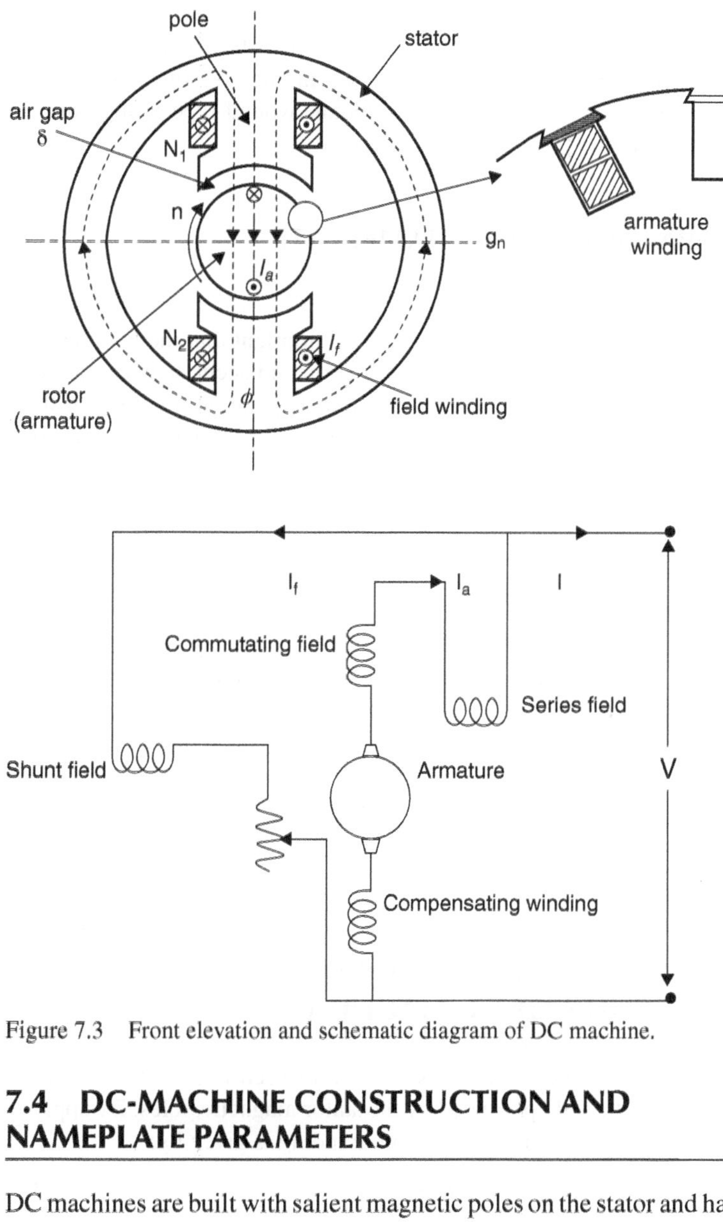

Figure 7.3 Front elevation and schematic diagram of DC machine.

7.4 DC-MACHINE CONSTRUCTION AND NAMEPLATE PARAMETERS

DC machines are built with salient magnetic poles on the stator and have commutating poles between the main poles. The coils are situated in the interpolar region with contact through carbon brushes. The windings are based on loop configuration and allow for pairs of poles in the wave winding. The brush holder holds the machines against the commutator surface by springs to maintain a constant pressure and smooth running speed.

A general schematic of a DC machine is shown in Figure 7.3. It is important to note that compensating winding is in the main poles, whereas the interpole is inserted in the commutating poles, which allows for reversal of current. So, we can claim here that a DC machine can be modeled in three configurations based on the location of the shunt or series winding of the interconnected windings of rotor and stator. Electric motors are described as shunt, series, or compound wound-type configuration depending on exciting current.

7.5 DC MACHINE PERTINENT NAMEPLATE PARAMETERS

In specifying machines for practical applications, it is important that the parameters are clearly spelled out to ensure easy equipment selection and accurate execution of tasks. Consequently, manufacturers usually include a nameplate on the body of the machine that spells out the machine's important parameters. Important parameters of DC machines that may be found on the nameplate include:

- armature and shunt field current (amps)
- armature resistance R_a (ohm)
- horsepower rating (HP)
- interpole winding and brushes resistance (ohm)
- machine type by winding (series, shunt, or compound wound)
- nominal power rating P (kW)
- number of poles P
- rotating losses
- speed N (rpm)
- terminal voltage V (volts)

7.6 DEVELOPMENT AND CONFIGURATION OF EQUIVALENT CIRCUITS OF DC MACHINES

From the principles of a rotating conductor, as detailed in Section 7.3, DC-machine armature rotates in a magnetic field produced by the stator magnetic poles. By applying Lenz's law, the voltage induced in the armature winding is given by:

$$e_{induced} = 2B(\theta)l\omega_m r \tag{7.3}$$

where l is the length of conductor in the slot of the armature, ϕ is the flux per pole ω_m is the mechanical speed, r is the distance of conductor from $= \frac{2\pi rl}{P}$, and A is the area per pole $= \frac{\phi P}{2\pi rl}$.

Consequently, we can recast Equation 7.3 as

$$e_{induced} = \frac{\phi P\omega_m}{\pi}. \tag{7.4}$$

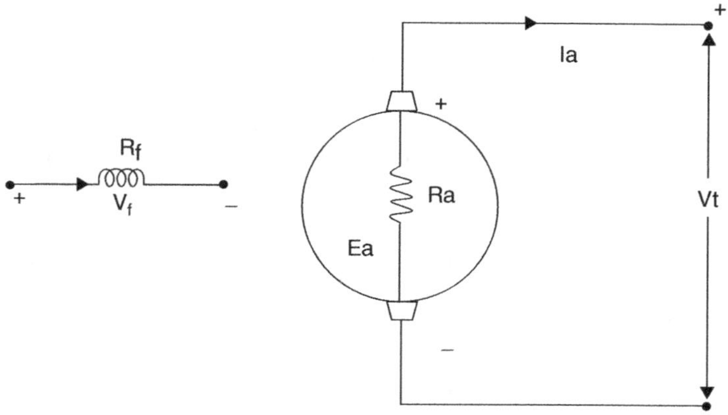

Figure 7.4 DC machine equivalent circuit.

This leads to computation of terminal voltage as a function of the number of turns connected in series with the armature, given as

$$E_a = \frac{Ne_i}{a} \tag{7.5}$$

where a is the number of parallel paths and N is the total number of turns in the winding.

$$E_a = \frac{N}{a}\left[\frac{\phi P \omega_m}{\pi}\right] \tag{7.6}$$

$$= \frac{NP}{\pi a}(\phi \omega_m) \tag{7.7}$$

Therefore, we observe that E_a is a function of ϕ and ω_m, given $\frac{NP}{\pi a}$ represented by a constant k for a specific machine.

So we may now write

$$E_a = k\phi\, \omega_m \tag{7.8}$$

where k varies with each machine configuration depending on N, P, π, and a.

Assume

$$NP = \frac{Z_p}{2} \text{ (number of pole pairs)}$$

then

$$k_a = \frac{Z_p}{2\pi a}. \tag{7.9}$$

E_a is now fully defined as voltage in generator operation and E_a is given as back emf in motor operation.

Typically, we develop the equivalent circuit (Figure 7.4) as follows under steady-state operations. As a generator (DC machine) we write

$$V_f = R_f\, i_f \tag{7.10}$$

and

$$V_t = E_a - I_a R_a \text{ (for DC generator)} \tag{7.11}$$

or simply

$$V_t = E_a + I_a R_a \text{ (for DC motor).} \tag{7.12}$$

Combining these two equations

$$E_a = k_a \, \phi \, \omega_m \tag{7.13}$$

where k_a depends on number of conductor and pole pairs, i.e.,

$$k_a = \frac{Z_p}{\pi P}. \tag{7.14}$$

Based on the second fundamental equation of machine operation (Equation 7.2), torque is developed based on resulting force on the conductor by Lorentz's equation. Torque is produced by armature winding with current in a magnetic-field winding of the stator.

The force armature conductor is calculated as

$$2B(\theta)\, l\omega_m r \tag{7.15}$$

$$F = \frac{B(\theta)\, lI_a}{a} \tag{7.16}$$

$$\text{torque} = \text{force} \times \text{r} \tag{7.17}$$

which gives the average torque

$$\text{T} = k_a \, \phi \, I_a \tag{7.18}$$

where k_a is a design constant that depends on $\frac{NP}{\pi a}$.

Therefore, in the case of motor operation:

- The terminal voltage generated is greater than emf and the electromagnetic torque produces rotation against a load.
- Alternatively, when the terminal voltage V_t is less than the generated voltage in the case of a generator, the electromagnetic torque opposes the torque applied to the shaft by the prime mover.

Simply, in the case of a motor

$$\text{electric power } P = E_a I_a = \text{mechanical power given as } P = T\omega_m$$

where $T\omega_m$ is the developed power within the mechanical system. The converse is also true for a motor or generator operation. So, we can write

$$\text{power } = T\omega_m = c\omega_m \phi I_a \tag{7.19}$$

where K_a, W_m, ϕ, and I_a are already defined.

Example 1

A shunt generator delivers 450A at 230 V and the resistance of the shunt field and armature are 50 ohm and 0.03 ohm, respectively. Calculate the generated emf.

Solution:

Current through shunt field winding

$$I_{sh} = 230/50 = 4.6A$$

Load current

$$I = 450A$$

Armature current

$$I_a = I + I_{sh}$$

Armature voltage drop

$$\begin{aligned} 450 + 4.6 &= 454.6A \\ &= I_a R_a = 545.6 \times 0.03 = 13.6V \\ &= E_g = V + I_a R_a \\ &= 230 + 13.6 = 243.6V \end{aligned}$$

Example 2

A four-pole DC machine has an armature of radius 12.5 cm, effective length of 25 cm. The poles cover 75 percent of the armature periphery. The armature windings consist of thirty-three coils each having seven turns. The coils are accommodated in thirty-three slots. The average flux density under each pole is 0.75T. If the armature is wound, determine:

- **(a)** armature constant k_a
- **(b)** induced-armature voltage when the armature rotates at 1,000 rpm
- **(c)** current and electromagnetic torque developed when the armature is 400A
- **(d)** power developed by the armature

If the armature is wave wound, repeat (a)–(d) above.

Solution:

$$\begin{aligned} k_a &= \frac{NP}{\pi a} = \frac{ZP}{2\pi a}, \\ Z &= 2 \times 33 \times 7 \\ &= 462 \\ a &= P = 4 \\ k_a &= \frac{462 \times 4}{2 \times 4 \times \pi} \\ &= 73.53 \end{aligned}$$

pole area

$$A_p = \frac{2\pi \times 0.125 \times 0.25 \times 0.75}{4}$$
$$= 36.8 \times 10^{-3} m^3$$
$$\phi = A_p B$$
$$\phi = 36.8 \times 10^{-3} \times 0.75$$
$$0.0276\text{Wb}$$
$$E_a = k_a \phi \omega_m$$
$$73.53 \times 0.0276 \times \frac{1,000}{60} \times 2\pi$$
$$= 212.5\text{V}$$
$$I_{coil} = \frac{I_a}{a} = \frac{400}{4}$$
$$= 100\text{A}$$
$$T = k_a \phi I_a$$
$$73.53 \times 0.0276 \times 400$$
$$= 811.8\text{Nm}$$
$$P_a = E_a I_a = 212.5 \times 400$$
$$85.0\text{kw}$$

or

$$T\omega_m = 811.8 \times \frac{1,000}{60} \times 2\pi$$
$$= 85.\text{kw}$$

wave wound

$$P = 4, \quad \therefore \text{poles of pairs} = 2$$
$$Z = 2 \times 33 \times 7 = 462$$
$$k_a = \frac{462 \times 4}{2 \times 2 \times \pi}$$
$$= 147.0$$
$$\omega_m = \frac{1,000}{60} \times 2\pi$$
$$= 104.67\text{rad/s}$$
$$E_a = 147.06 \times 0.0276 \times 104.67$$
$$= 425\text{V}$$
$$I_{coil} = 100\text{A}$$
$$I_a = 2 \times 100$$
$$= 200\text{A}$$
$$T = 147.06 \times 0.0276 \times 200$$
$$= 811.8\text{Nm}$$
$$P_a = 425 \times 200$$
$$85\text{kW}$$

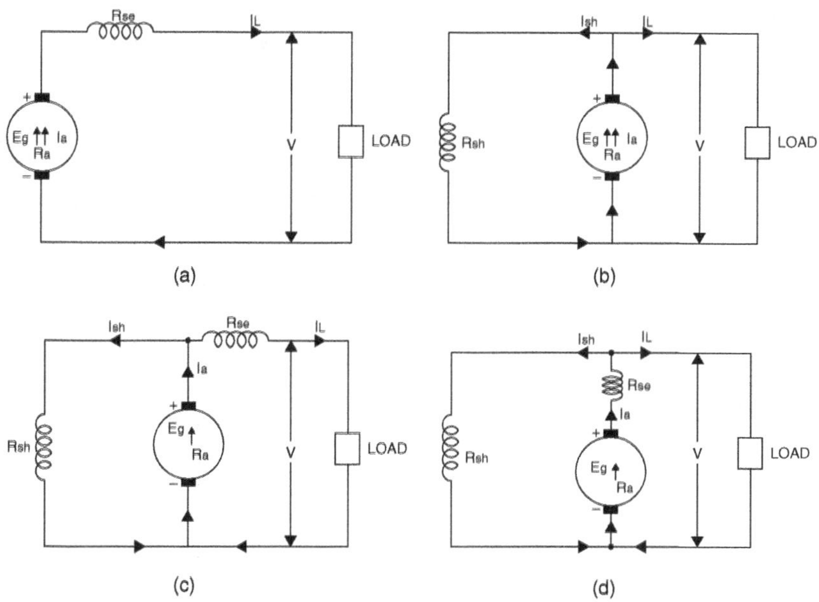

Figure 7.5 DC-machine configurations. (a) Series DC machine (b) Shunt DC machine (c) Compound (short shunt) (d) Compound (long shunt).

7.7 CLASSIFICATION OF DC MACHINES

The field current and armature current can be interconnected in many ways to provide various performance characteristics for DC-machine configuration.

1. Field poles in DC machines are configured by four methods (as shown in Figure 7.5):
   - **(a)** series-field winding (connected in series)
   - **(b)** shunt-field winding (connected in parallel)
   - **(c)** short-shunt compound winding
   - **(d)** long-shunt compound winding

The compound-machine configuration combines series and shunt winding and these two windings have different performance. The rheostat helps control the field current to achieve the desired performance.

Given these configurations, we now develop analysis for DC generators using the circuit equivalent.

Consider the machine to be separately excited (Figure 7.6), then the relevant equations are:

$$V_f = I_f R_f$$
$$E_a = V_t + I_a R_a$$
$$E_a = k_a \phi \omega_m$$
$$V_t = I_a R_L$$

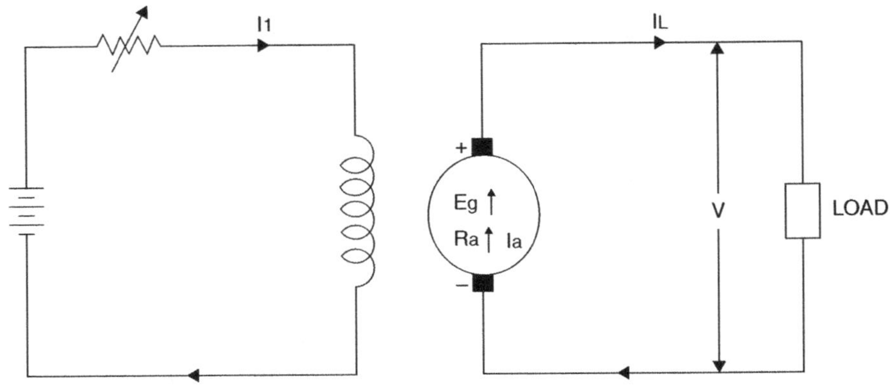

Figure 7.6 Separately excited DC machine.

and

$$I_a = I_t$$

where

$$V_t = E_a - I_a R_a$$

These equations define the termed or external V_t of the separately excited DC generator.

The relationship between V_t and I_t for different operating conditions is shown in Figure 7.7 [2].

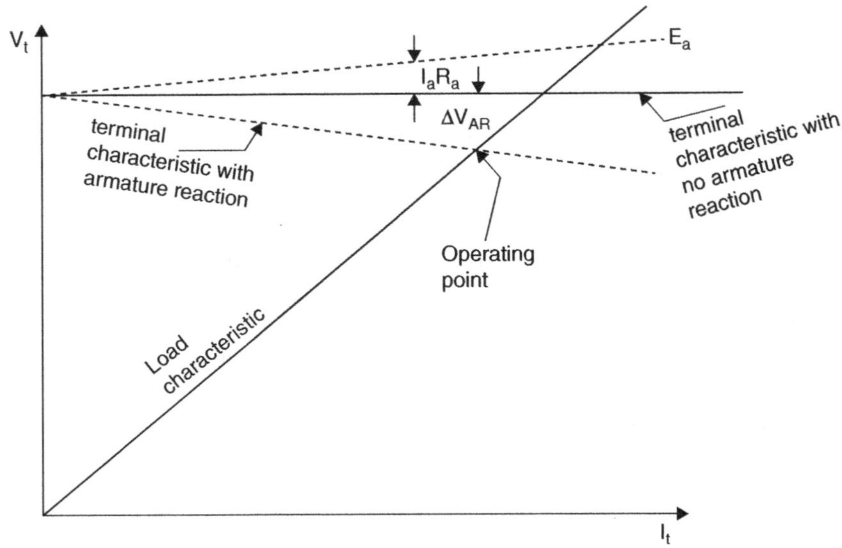

Figure 7.7 Terminal and load characteristics of separately excited generator.

From the terminal characteristics of a separately excited DC generator, there is a voltage drop due to armature reaction (ΔV_{AR}) from the machine internal characteristics to external characteristics. The operating point is the intersection between the generator's external characteristics and the load characteristics. The operating point gives the operating values of load current and terminal voltage.

Example 1

A shunt generator delivers 450A at 230 V and the resistance of the shunt field and armature are 50 ohm and 0.03 ohm, respectively. Calculate the generated emf.

Solution:

Current through shunt field winding = $I_{sh} = 230/50 = 4.6A$

Load current

$$I = 450A$$

Armature current

$$\begin{aligned} I_a &= I + I_{sh} \\ &= 450 + 4.6 = 454.6A \end{aligned}$$

Armature voltage drop

$$\begin{aligned} I_a R_a &= 545.6 \times 0.03 = 13.6V \\ E_g &= V + I_a R_a \\ &= 230 + 13.6 = 243.6V \end{aligned}$$

Example 2

A long-shunt compound generator delivers a load current of 50A at 500 V and has armature, series-field, and shunt-field resistances of 0.05 ohm, 0.03 ohm, and 250 ohm, respectively. Calculate the generated voltage and the armature current. Allow 1 V per brush for contact drop.

Solution:

$$I_{sh} = 500/250 = 2A$$

Current through armature and series winding = $50 + 2 = 52A$

Voltage drop on series field winding = $52 \times 0.03 = 1.56V$

Armature voltage drop = $I_a R_a = 52 \times 0.05 = 2.6V$

Drop at brushes = $2 \times 1 = 2V$

Now,
$$\begin{aligned} E_g &= V + I_a R_a + \text{series drop} + \text{brush drop} \\ &= 500 + 2.6 + 1.56 + 2 = 506.16V \end{aligned}$$

Example 3

A short-shunt compound generator delivers a load current of 30A at 220 V and has armature, series-field, and shunt-field resistance of 0.05 ohm, 0.3 ohm, and 200 ohm, respectively. Calculate the induced emf and armature current. Allow 1.0 V per brush for contact drop.

Solution:

$$\begin{aligned}
\text{Voltage drop in series winding} &= 30 \times 0.3 = 9\text{V} \\
\text{Voltage drop across shunt winding} &= 220 + 9 = 229\text{V} \\
\text{I}_{\text{sh}} &= 229/200 = 1.145\text{A} \\
\text{I}_{\text{a}} &= 30 + 1.145 = 31.145\text{A} \\
\text{I}_{\text{a}}\text{R}_{\text{a}} &= 31.145 \times 0.05 = 1.56\text{V} \\
\text{brush drop} &= 2 \times 1 = 2\text{V} \\
\text{E}_{\text{g}} &= \text{V} + \text{series drop} + \text{brush drop} + \text{I}_{\text{a}}\text{R}_{\text{a}} \\
&= 220 + 9 + 2 + 1.56 = 232.56\text{V}
\end{aligned}$$

Example 4

A 12 kW, 100 V, 1,000 rpm DC shunt generator has armature resistance $R_a = 0.1$ ohm, shunt-field winding resistance $R_f = 80$ ohm, and $N_f = 1,200$ turns/pole. Rated field current is 1A. Find V_t at full load if E_a is 98 V.

If armature resistance is at full load of 0.06A, determine full load at V_t.

Solution:

Given that

$$\begin{aligned}
V_t &= E_a - I_a R_a \\
&= 100 - (120 \times 0.1) \\
&= 88\text{V} \\
I_{ef} &= I_{rated} - I_f \\
V_t &= 98 - 120 \times 0.1 \\
&= 86\text{V}
\end{aligned}$$

Example 5

Determine the field current required to make the terminal voltage of $V_t = 100\text{V}$ at full load.

Solution:

$$\begin{aligned}
E_a &= V_t + I_a R_a \\
E_a &= 100 + 120 \times 0.1 \\
&= 112\text{V} \\
I_{eq} &= 1.4A
\end{aligned}$$

so

$$I_{f(actual)} = (1.4 + 0.6)A$$
$$= 1.46\text{A}.$$

7.8 VOLTAGE REGULATION

$$\%\text{ voltage regulation (VR)} = \frac{V_{t,NL} - V_{t,fl}}{Vt,fl} \times 100\% \tag{7.20}$$

$V_{t,NL}$ is the no-load terminal voltage, and $V_{t,fl}$ is the full-load terminal voltage.

Since E_a is equal to $V_{t,NL}$,

$$\%\text{ voltage regulation} \frac{E_a - V_{t,fl}}{Vt,fl} \times 100\%. \tag{7.21}$$

7.9 POWER COMPUTATION FOR DC MACHINES

For DC-machine operation, DC current must be provided. Use

$$P_{delivered} = E_a I_a \tag{7.22}$$

at armature reaction

$$P_{ta} = V_{ta} I_a \tag{7.23}$$

where

$$V_{ta} = E_a \pm I_a R_{total} \pm 2. \tag{7.24}$$

Finally, the total resistance

$$R = \sum R_a = R_a + R_{se} + R_{cp} + R_{cw} + R_{eq} + \cdots \tag{7.25}$$

This is used to determine the performance of the machine.

7.9.1 Loss Calculation

$$P_{cu} = I_a^2 \sum R_a \tag{7.26}$$

$$P_{brush} = I_a \tag{7.27}$$

$$P_{shaft} = P_D \pm P_{rotation} \tag{7.28}$$

where

$$P_{rotation} = P_{core} + P_{cu}.$$

7.9.2 Torque Computation

Assume a DC machine has V_a, I_a, E_a, K_{max}, then

$$P_D = E_a I_a.$$

This can be written as function of losses

$$P_D = P_o \pm P_{cu} \pm P_{brush} \tag{7.29}$$

and the corresponding torque

$$T_d = \frac{P_D}{\omega_n} = \frac{E_a I_a}{\omega_n} \tag{7.30}$$

$$T_d = \frac{(k_a \phi_d \omega_n) I_a}{\omega_n} \tag{7.31}$$

$$T_d = k_a \phi_d I_a \tag{7.32}$$

$$k_a = \frac{ZP}{2\pi a} \tag{7.33}$$

where

$$\omega_n = \frac{2\pi n_m}{60}$$

so that

$$T_d = \frac{E_a I_a}{\frac{2\pi n_m}{60}}. \tag{7.34}$$

n_m is in rad/s.

Using the loss formula, we can determine power flow and efficiency of the machine.

7.10 POWER FLOW AND EFFICIENCY

From the Figure 7.8 a and Figure 7.8b, we can compute efficiency

$$\eta = \frac{P_{out}}{P_{in}}$$

where

$$P_{out} = P_{in} - \sum P_{losses}$$

$$\eta = 1 - \frac{\sum P_{losses}}{P_{in}} \tag{7.35}$$

$$\eta = \left(1 - \frac{\sum P_{losses}}{P_{in}}\right) \times 100\%. \tag{7.36}$$

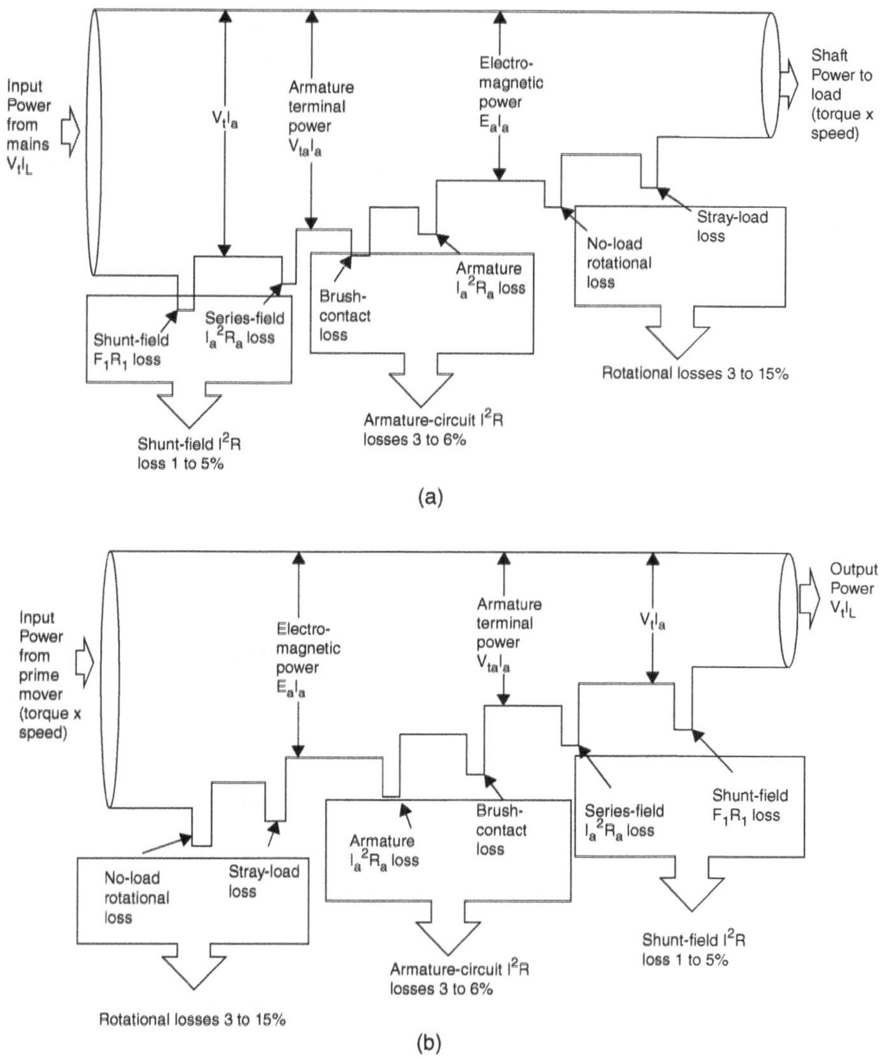

Figure 7.8 Losses and energy flow in DC generators and motors. (a) Energyflow in DC motor (b) Energy flow in DC generators.

Copper loss depends on armature-winding loss $(I_a^2R_a)$ or field-winding losses $\left(I_f^2R_f\right)$ when I_a and I_f are armature- and field-winding currents, respectively, R_a and R_f are armature and field resistances, respectively, brush loss $P_b = V_bI_a$, and V_b is the voltage drop across brushes, which is typically about 2 V.

7.10.1 Core Loss

The core loss is derived from the eddy-current and hysteresis losses, as discussed in Chapter 2. Core loss depends on the value of the magnetic flux density B and the speed of rotation of the machine. Mechanical losses can be given as a function of friction and windage. Stray losses account for approximately 1 percent of full-load loss. Power losses in DC machines are given or empirically computed.

Example 1

A shunt generator delivers 195A at terminal voltage 250 V. The armature resistance and shunt-field resistance are 0.02Ω and 50Ω, respectively. The iron and frictional losses equal 950 W. Find:

(a) emf generated

(b) copper loss

(c) output of the prime mover

(d) commercial, mechanical, and electrical efficiencies

Solution:

$$I_{sh} = \frac{250}{50} = 5A$$

$$I_a = 195 + 5 = 200A$$

$$\textit{armature voltage drop} = I_a R_a = 200 \times 0.02 = 4V$$

$$\text{Generated emf} = 250 + 4 = 254A$$

$$\text{Armature copper loss } I_a^2 R_a = 200^2 \times 0.02 = 800W$$

$$\text{Shunt copper loss } VI_{sh} = 250 \times 5 = 1,250W$$

$$\text{Total copper loss} = (1,250 + 800) = 2,050\text{W}$$

$$\text{Stray losses} = 950\text{W}$$

$$\text{Total losses} = 2,050 + 950 = 3,000\text{W}$$

$$\text{Output} = 250 \times 195 = 48,750\text{W}$$

$$\text{Input} = 48,750 + 3,000 = 51,750\text{W}$$

$$\text{Therefore, output of prime mover} = 51,750\text{W}$$

$$\text{Electrical power produced in armature} = 51,750 - 950 = 50,800\text{W}$$

$$\text{Mechanical efficiency } \eta_m = (50,800/51,750) \times 100 = 98.2\%$$

$$\text{Electrical or copper losses} = 2,050\text{W}$$

$$\eta_m = \frac{48,750}{48,750 + 2,050} \times 100 = 95.2\%$$

$$\text{and commercial efficiency } \eta_c = \left(\frac{48,750}{51,750}\right) \times 100 = 94.2\%.$$

7.11 DC MOTORS

Under a steady-state condition, the interaction between voltage and current for DC machines is given as

$$V_f = I_f R_f \tag{7.37}$$

$$V_t = E_a \pm I_a R_a \tag{7.38}$$

where we further suggest, as discussed in Section 7.12, that for a motor

$$E_a = k_a \phi \omega_m \tag{7.39}$$

from which it can be further shown that

$$T = k_a \phi I_a. \tag{7.40}$$

So, as usual

$$P_{em} = E_a I_a = T_e \omega_m. \tag{7.41}$$

Different configurations can be represented as follows:

(a) shunt motors, whose field current can be controlled by use of variable resistance in series with the field winding

(b) series motors, for which it is difficult to control the armature and field winding independently, so we treat flux as largely dependent on the armature current and field current

(c) compound motors, where the effect of armature current on the flux depends on different issues (short and long winding)

7.12 COMPUTATION OF SPEED OF DC MOTORS

From the fundamental expression

$$k_a \phi \omega_m = V_t - I_a R_a$$

$$\omega_m = \frac{V_t - I_a R_a}{k_a \phi} \tag{7.42}$$

which means that speed, ω_m, depends on the applied voltage V_t, armature current I_a, armature resistance R_a, and flux ϕ per pole.

The schematic diagram (as seen in Figure 7.9) of each DC-motor configuration with respect to control options is given as follows:

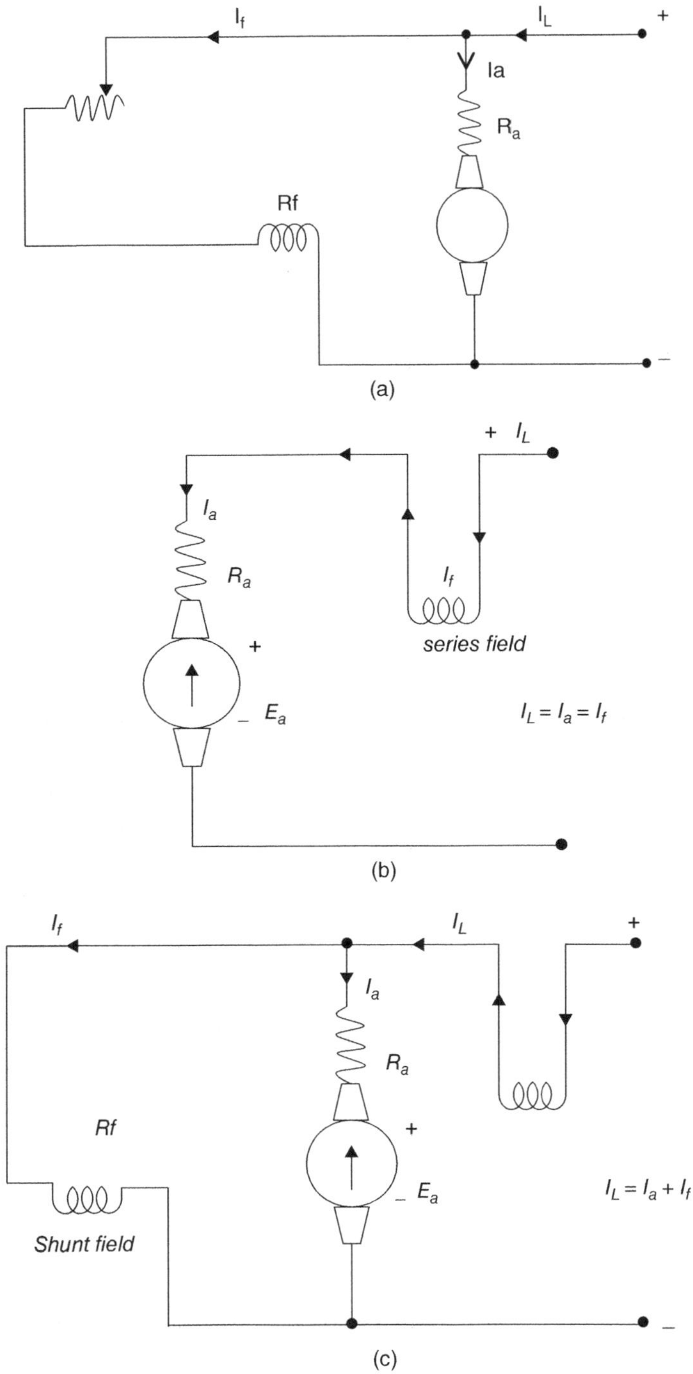

Figure 7.9 Configuration types of DC motors. Configuration (a) shunt DC motor (b) Series DC motor (c) Compound motor.

For configuration (a), shunt motor

$$T_e = k_a \phi I_a = k_m I_a \tag{7.43}$$

$$E_a = (k_a \phi)\omega_m = k_m \omega_m. \tag{7.44}$$

So, we can write

$$V_t = E_a + I_a R_a. \tag{7.45}$$

For configuration (b), series DC motor

$$T_e = k_a \phi I_a = k_m I_a \tag{7.46}$$

$$E_a = (k_a \phi)\omega_m = k_m \omega_m. \tag{7.47}$$

So, we can write

$$V_t = E_a + I_a R_a. \tag{7.48}$$

For configuration (c), compound motor

$$I_a = \frac{V_t - E_a}{R_a} \tag{7.49}$$

$$T_e = k_m I_a = k_m \left(\frac{V_t - E_a}{R_a} \right) \tag{7.50}$$

$$= \frac{k_m V_t - k_m E_a}{R_a}$$

But

$$k_m E_a = \omega_m k_m^2 \tag{7.51}$$

Then,

$$T_e = \frac{k_m V_t}{R_a} - \frac{k_m^2 \omega}{R_a} \tag{7.52}$$

$$P_{em} = T_e \omega_m = \frac{k_m V_t \omega_m}{R_a} - \frac{k_m^2 \omega_m^2}{R_a}. \tag{7.53}$$

The following figures namely Figures 7.10, 7.11, and 7.12 show the torque/armature current characteristics, torque/speed characteristics and speed/power characteristics respectively of the dc motor.

7.12.1 Connection Description

A shunt machine is a constant-speed machine with low-speed regulation. Speed can be controlled by field flux.

For a series motor, recall the equation for T_e given as

$$T_e = k_a \phi I_a = k_m I_a. \tag{7.54}$$

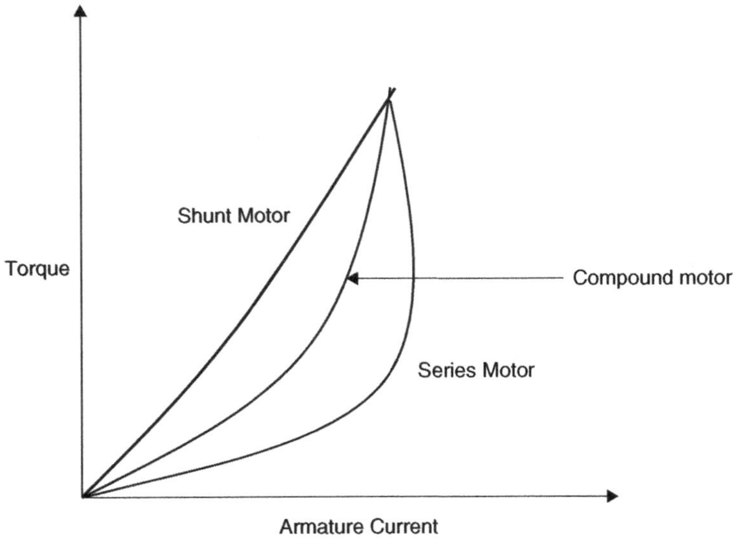

Figure 7.10 Torque/armature-current characteristics.

If $R_a \approx 0$, T_e is a function of flux ϕ, terminal voltage will be

$$V_t = E_a = k_m \omega_m = k I_a \omega_m \tag{7.55}$$

from which

$$T_e = k_a \frac{V_t}{k_a \omega_m^2}. \tag{7.56}$$

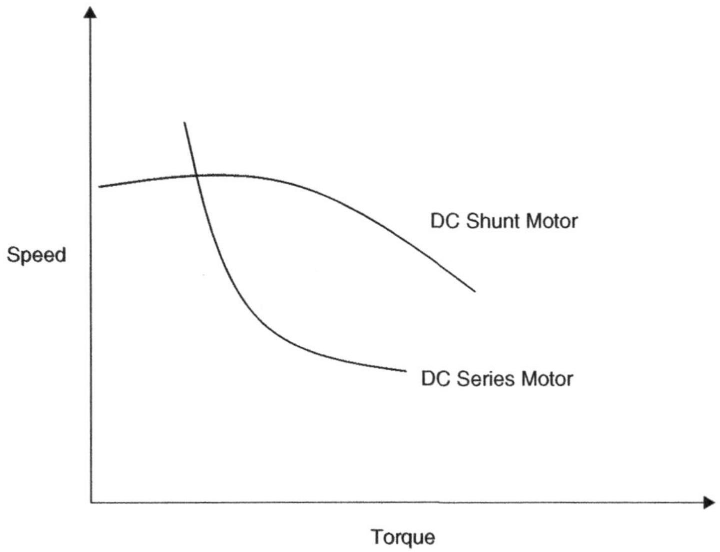

Figure 7.11 Torque/speed characteristics of a DC-motor configuration.

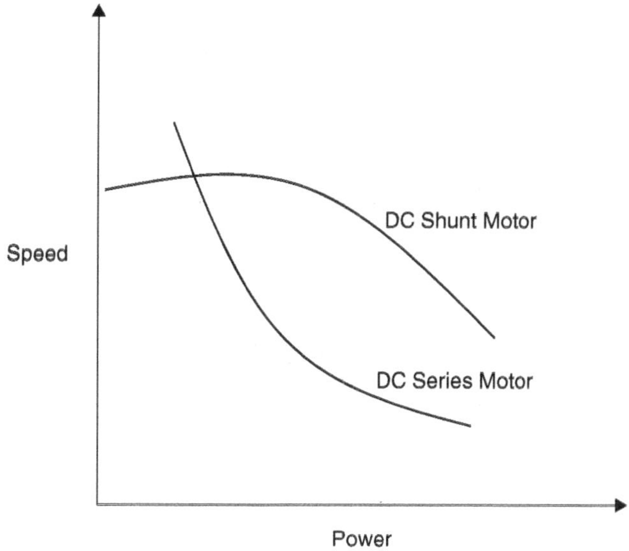

Figure 7.12 Speed/power characteristics of a DC-motor configuration.

Also

$$P_{em} = T_e \omega_m$$
$$= \frac{V_t^2}{k_a \omega_m} \quad (7.57)$$
$$P_{em} = \frac{V_t^2}{k_a \omega_m} = T_e \omega_m. \quad (7.58)$$

Thus, the speed/power relationship is as shown in Figure 7.13.

The relationship between speed and torque for each class of DC-machine configuration operating as a motor is as shown in Figure 7.14.

Example 1

A 440 V shunt motor has armature resistance of 0.8 ohm and field resistance of 200 ohm. Determine the back emf when giving an output of 7.46 kW at 85 percent efficiency.

Solution:

$$\text{Motor input power} = 7.46 \times \frac{10^3}{0.85}$$
$$\text{Motor input current} = 7,460/0.5 \times 440 = 19.95\text{A}$$
$$I_{sh} = 440/200 = 2.2\text{A}$$
$$I_a = 19.95 - 2.2 = 17.75\text{A}$$

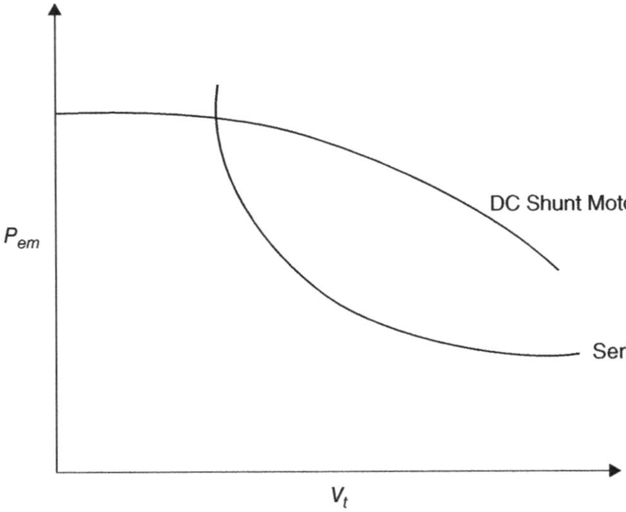

Figure 7.13 Speed–power relationship.

Now,

$$E_b = V - I_a R_a$$
$$E_b = 440 - (17.75 \times 0.8) = 425.8\text{V}$$

Example 2

The following data is for a DC shunt motor:

$$V = 220\,\text{V},\ \text{field resistance } 200\Omega,\ \text{armature resistance } 0.5\Omega.$$

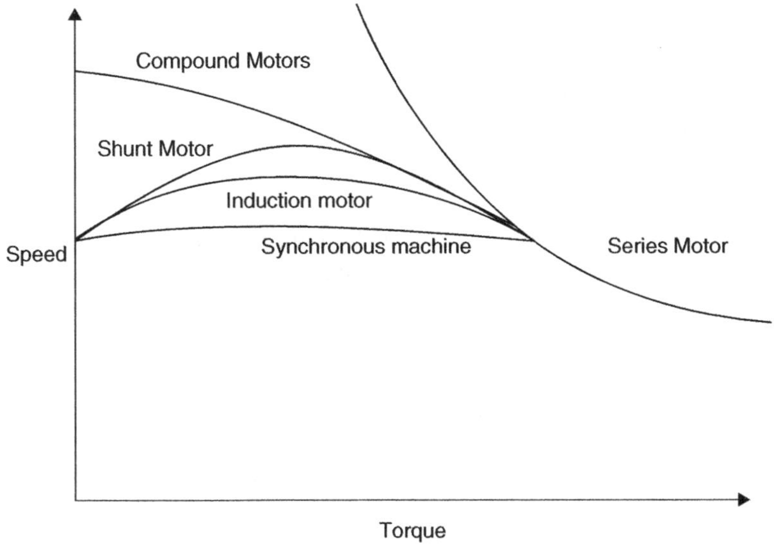

Figure 7.14 Speed-torque characteristics of electric machines.

At no load, armature resistance = 0.5Ω, speed = 1,000 rpm, armature current = 4A, torque = 10Nm.

(a) Find the armature current and the speed of the motor.

(b) If the motor is required to develop 10 hp at 1,200 rpm, compute the required value of the external series resistance in the field circuit.

Solution:
Assume field current

$$I_f = \frac{200}{200} = 1\text{A}$$

$$E_a = V_t - I_a R_a = 200 - (4 \times 0.5)$$

$$= 198\text{V}$$

$$E_a = k I_f \omega_m$$

$$k = \frac{E_a}{I_f \omega_m} = \frac{198}{\frac{1 \times 2\pi \times 1{,}000}{60}}$$

$$= 1.89$$

with machine on load

$$T_e = k_1 I_f I_a = 100$$

$$= 1.89 \times 1.0 \times I_a$$

$$I_a = \frac{100}{1.89 \times 1.0}$$

$$= 52.9.$$

We can now obtain

$$V_t = E_a - I_a R_a$$

$$V_t = 200 - 52.9 \times 0.5$$

$$= 173.55\text{V}.$$

Since

$$E_a = k_1 I_f \omega_m,$$

we can recompute

$$\omega_m = \frac{173.55}{1.89 \times 1.0}$$

$$= 918 \text{rad/s}$$

with corresponding speed

$$\frac{918 \times 60}{\pi}$$

$$= 876 \text{ rpm}.$$

For 10 Hp at 1,200 rpm,

$$T_e = \frac{10 \times 746}{\frac{2\pi}{60} \times 1,200}$$
$$= 59.34\text{Nm}.$$

So, given

$$59.24 = 1.89 I_f I_a$$
$$I_f I_a = 31.4$$
$$V_t = E_a + I_a R_a$$
$$= 1.89 \left(\frac{2\pi}{60} \times 1,200 \right) I_f + 0.5 I_a$$

or

$$200 = 237.6 I_f + 0.5 I_a$$
$$200 = 237.6 I_f + \frac{0.5 \times 31.4}{I_f}$$
$$I_f = 0.754A$$
$$I_a = \frac{31.4}{I_f}.$$

Example 3

The input to a 230 V DC shunt motor is 11 kW. Calculate:

(a) torque developed
(b) efficiency
(c) speed at this load

The particulars of the motor are: no-load current is 5A, no-load speed is 1,150 rpm, armature resistance is 0.5 ohm, and shunt-field resistance is 110 ohm.

Solution:
No-load input = 220 × 5 = 1,100 W

$$I_{sh} = \frac{220}{110} = 2A$$
$$\text{armature current at no-load} = 5 - 2 = 3\text{A}$$
$$\text{no-load armature copper loss} = 3^2 \times 0.5 = 4.5\text{W}$$
$$\text{constant losses} = 1,100 - 4.5 = 1,095.5\text{W}$$

with 11 kW input,

$$\text{input current} = 11{,}000/220 = 50\text{A}$$

$$\text{armature current} = 50 - 2 = 48\text{A}$$

$$\text{armature copper loss} = 48^2 \times 0.5 = 1{,}152\text{W}$$

$$\text{Total loss} = \text{armature copper loss} + \text{constant losses} = 1{,}152 + 1{,}095.5 = 2{,}248\text{W}$$

$$\text{output} = 11{,}000 - 2{,}248 = 8{,}752\text{W}$$

$$\text{efficiency} = 8{,}752 \times 100/11{,}000 = 79.6\%$$

$$\text{back emf at no load} = 220 - (3 \times 0.5) = 218.5\text{V}$$

$$\text{back emf at given load} = 220 - (48 \times 0.5) = 196\text{V}$$

$$\text{speed } N = 1{,}150 \times 196/218.5 = 1{,}031\ \text{rpm}$$

7.13 DC-MACHINE SPEED-CONTROL METHODS

Three methods are used mainly for DC machine control:

(a) field control

(b) armature-resistance control

(c) solid-state rectifier control

7.13.1 Field-Control Method

From

$$\omega_m = \frac{V_t - I_a R_a}{k_a \phi} \tag{7.59}$$

V_t is kept constant so ω_m depends on I_f only. ϕ decreases while speed increases and since I_f is directly proportional to ϕ, it follows that by varying I_f, the machine's speed can be varied. One way to do this is to insert a variable resistance in the field circuit to control I_f. This is one of the simplest but most efficient methods of DC-machine speed control.

7.13.2 Armature-Resistance Control Method

Terminal voltage and field current are maintained consistent at their rated values. Motor speed is easily controlled by varying armature resistance through an external rheostat. This control method is simple and inexpensive but has the disadvantage of low efficiency.

Again, ω_m changes depend on V_t. It is easy to compute voltage regulation or speed regulation using speed regulation

$$\frac{\omega_{NL} - \omega_{FL}}{\omega_{FL}} \times 100\% \tag{7.60}$$

The method of control here is applicable to shunt-, series-, and compound-motor configurations.

7.13.3 Solid-State Rectifier Control

Silicon-controlled rectifiers (SCRs) are one of the solid-state devices widely used in the control of DC motors. The SCR is used to rectify the voltage and control the armature current of the DC machine by adjusting the firing angle with the application of the appropriate trigger voltage to the gate of the SCR. The firing angle of the SCR determines the amount of DC current supplied to the motor's armature. Thus, by controlling the armature current, the machine is also controlled. This is consistent with the equation

$$V_t \cong E_a = k_a \phi \omega_m = kI_a \omega_m. \tag{7.61}$$

For a DC-series motor it can be shown that

$$V_t = E_a = kI_a \omega_m$$

from which we can see that

$$I_a \propto \frac{1}{\omega_m}.$$

Thus, increasing the armature current results in a decrease in motor speed, which can be achieved using the SCR.

7.14 WARD LEONARD SYSTEM

For a solid-state control rectifier using a DC machine, the speed-control system is called the Ward Leonard system. It is capable of controlling a DC motor from zero to maximum or maximum to zero.

The configuration of the Ward Leonard system is shown in Figure 7.15 [3]. AC source (induction or synchronous machine) serves as the prime mover for a

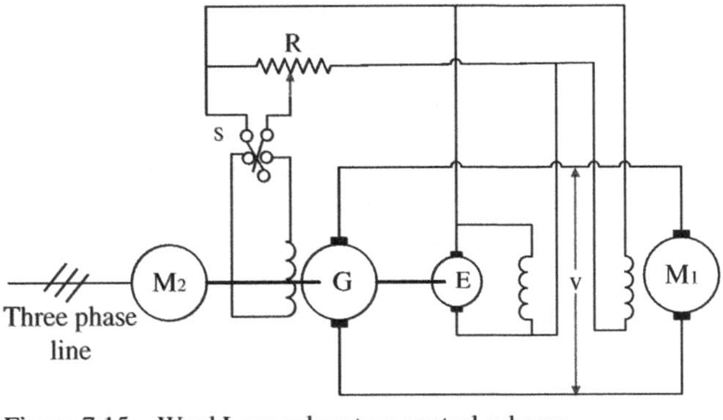

Figure 7.15 Ward Leonard system control scheme.

DC generator. The generator in turn supplies V_t, which supplies a DC motor for operation.

Varying the field current of the DC machine, which varies the speed of the motor, can consequently control the armature voltage in the motor. That is, the motor speed can be reduced or increased by the motor field current with changes in flux. Further, if the current of the generator is reversed, the polarity of the generator-armature voltage will reverse as well. These reversals also change or cause the motor direction of rotation to change. The speed variation occurs as the field current changes back and forth in the opposite direction for the Ward Leonard system.

Illustrative Problems and Examples

1. Calculate the flux in a four-pole dynamo with 722 armature conductors generating 500 V when running at 1,000 rpm, when the armature is (a) lap wound (b) wave wound.

Solution:
The generated emf is given by

$$E = \varphi NZP/A$$

(a) For lap wound C = 2P

$$E = \varphi NZP/60A$$

For lap wound,

$$P = A$$
$$500 = (\varphi \times 100 \times 722 \times 4)/60P$$

Now,

$$E = \varphi NZ$$
$$E = \varphi \times (1,000/60) \times 722$$
$$\Phi = (500 \times 60)/(722 \times 1,000)$$
$$\Phi = 41.6\,\text{MWb}$$

(b) For wave winding A = 2

$$500 = \varphi * 1,000 * 722 * 4)/(2 * 60)$$
$$\Phi = 20.8\,\text{MWb}$$

2. A DC generator generates an emf of 520 V. It has 2,000 armature conductors, flux per pole of 0.13 Wb, speed of 1,200 rpm, and the armature winding has four parallel paths. Find the number of poles.

Solution:

$$E = 520V$$
$$Z = 2,000$$
$$\Phi/\text{pole} = 0.13\text{Wb}$$
$$N = 1,200\,\text{rpm}$$

For four-parallel path

$$Z = 2,000/4 = 500$$
$$E = P\varphi N\,(Z/4)$$
$$520 = P \times 0.013 \times (1,200/60) \times 500$$
$$P = 4$$

3. When driven at 1,000 rpm with a flux per pole of 0.02 Wb, a DC generator has an emf of 200 V. If the speed is increased to 1,100 rpm and at the same time the flux per pole is reduced to 0.019 Wb per pole, what is the induced emf?

Solution:

$$E_1 = 2P\varphi NZ$$
$$200 = 2 \times (1,000/60) \times PZ \times 0.02$$
$$PZ = 6,300$$
$$E_2 = 2 \times 0.019 \times (1,100/60) \times 6,300$$
$$E_2 = 209V$$

4. A 230 V shunt motor has an armature resistance of 0.26Ω. Assuming a 2 V brush drop, calculate the back emf when the armature current is 36A.

Solution:

$$E_b = V_t - I_a R_a - \textit{brush drop}$$
$$= 230 - (36 \times 0.26) - 2$$
$$= 218.6V$$

7.15 CHAPTER SUMMARY

We have presented the concepts and fundamental principles of DC machines in this chapter. The internal structures were briefly discussed. Also, the various configurations and their important characteristics were discussed. We distinguished between DC generators and DC motors and discussed methods of speed control.

DC motors have found wide applications in control systems. As a rotating machine with an inertial mass, it is of interest to us as it applies to the microgrid. The behavior of DC machines as a generator or motor—for instance on a DC microgrid—should be of practical interest. The student is encouraged to learn these principles as an integral part of the knowledge relative to energy conversion in the smart grid and microgrid. Further, this chapter provided principles for studying the performance of DC-machine applications.

EXERCISES

1. A DC series motor operates at 750 rpm with a line current of 80A from the 230 V mains. Its armature circuit resistance is 0.14Ω and its field resistance is 0.11Ω. Assuming that the flux corresponding to a current of 20A is 40 percent of that corresponding to a current of 80A, find the motor speed at a line current of 20A at 230 V.

2. A DC shunt machine is connected across a 500 V supply. The total armature winding resistance is 0.05Ω and the field-circuit resistance is 100Ω. Calculate the ratio of speed as a generator to speed as a motor for the line current to be 200A.

3. A shunt motor draws 26.5A from a 120 V supply when it drives a load at 25 rps. The resistance of the field circuit is 120Ω and the armature-winding resistance is 0.2Ω. What is the electromagnetic torque developed?

4. When a motor is operating under load, the armature takes 8,600 W and its current is 38A. If the armature-circuit resistance, including brushes, is 0.4Ω, what horsepower is developed by the motor?

5. A 240 V, 20HP shunt motor has a resistance of 0.18Ω for the armature circuit, not including brushes (allow for a brush drop of 2 V), and 240Ω for the field circuit. The rated speed is 1,200 rpm and the full-load efficiency is 0.87. Calculate:
   - **(a)** torque
   - **(b)** stray losses
   - **(c)** rotational losses

6. The resistance of the armature circuit of a 240 V shunt motor is 0.1Ω, including brushes. When the field rheostat is adjusted so that the resistance of the shunt field circuit is 120Ω, the current taken by the motor is 82A and the speed is 900 rpm. Neglecting armature reaction, calculate:
   - **(a)** current
   - **(b)** speed if the load torque is increased by 50 percent

7. A 110 V shunt generator has full-load current of 100A, shunt-field resistance of 55Ω, and constant losses of 500 W.
   - **(a)** If full-load efficiency is 88 percent, find armature resistance.
   - **(b)** Assuming voltage to be constant at 110 V, calculate the efficiency at half full load and at 50 percent overload.
   - **(c)** Find the current.

8. A DC-series motor on full load takes 50A from 230 V DC mains. The total resistance of the motor is 0.22Ω. If the iron and friction losses together amount to 5 percent of the input, calculate the power delivered by the motor shaft. Total voltage drop due to the brush contact is 2 V.

9. A two-pole, DC shunt motor operating from a 200 V supply takes a full-load current of 35A, the no-load current being 2A. The field resistance is 500Ω and the armature has a

resistance of 0.6Ω. Calculate the efficiency of the motor on full load. Take the brush drop as being equal to 1.5 V per brush arm.

BIBLIOGRAPHY

[1] R. Ramshaw and R. G. van Heeswijk, *Energy Conversion: Electric Motors and Generators*, Saunders College Publishing, Philadelphia, 1990.

[2] E. H. Smith, *Mechanical Engineers Reference Book*, 12th ed., Butterworth Heinemann Ltd., Oxford, 1994.

[3] R. K. Rajput, *A Textbook of Electrical Machines*, Laxmi Publications Ltd., New Delhi, 2006.

[4] H. Saadat, *Power System Analysis*, 3rd ed., PSA Publishing, 2010.

[5] A. E. Fitzgerald, C. Kingsley, and S. D. Umans, *Fitzgerald & Kingsley's Electric Machinery*, 4th ed., McGraw-Hill, New York, 1983.

[6] C. A. Gross, *Power System Analysis*, John Wiley & Sons, New York, 1979.

[7] S. J. Chapman, *Electric Machinery Fundamentals*, 5th ed., McGraw Hill, New York, 2012.

[8] T. Gönen, *Electrical Machines with MATLAB*, 2nd ed., Taylor & Francis Group, Boca Raton, FL, 2011.

[9] W. L. Matsch and J. D. Morgan, *Electromagnetics and Electromechanical Machines*, 3rd ed., John Wiley & Sons, Hoboken, NJ, 1986.

[10] M. S. Sarma, *Electric Machines: Steady-State Theory and Dynamic Performance*, WMC Brown Publishers, Dubuque, IA, 1985.

[11] G. R. Slemon, *Electric Machines and Drives*, Addison-Wesley Publishing, New York, 1992.

[12] P. C. Sen, *Principles of Electric Machines and Power Electronics*, 2nd ed., John Wiley & Sons, Hoboken, NJ, 1997.

[13] Electronics Tutorials, "Magnetic Hysteresis," www.electronics-tutorials.ws/electromagnetism/magnetic-hysteresis.html.

CHAPTER 8

PERMANENT-MAGNET MOTORS

8.1 INTRODUCTION

A permanent-magnet motor is simply a machine whose field windings have been replaced by a permanent magnet. Permanent-magnet machines are important in that they have low eddy-current losses compared to conventional DC motors and synchronous motors. Although they have limitations, they are important and useful for a number of applications.

This chapter is divided into three major segments:

8.1. Permanent Magnet DC Motors

8.2. Permanent Magnet Synchronous Motors

8.3. Variants of Permanent-Magnet Synchronous Motors

For each of these subsections, the concepts and fundamentals, motivation and applications, design models, and materials are discussed. Solved problems are also included.

8.2 PERMANENT-MAGNET DC MOTORS

8.2.1 Permanent-Magnet DC Motor Fundamentals and Concepts

A permanent-magnet DC motor is like an ordinary shunt motor except that its field is provided by a permanent magnet instead of a salient-pole, wound-field structure. Although there are various types of DC motors, the motor shown in Figure 8.1 is suitable for understanding the basic principles. All kinds of DC motors work on the same principles.

For constructing a DC motor it is essential to establish a magnetic field. The magnetic field is established, obviously, by means of magnet. The magnet can by of many types including an electromagnet or permanent magnet. When a permanent magnet is used to create a magnetic field in a DC motor, the motor is referred to as a *permanent-magnet DC* or *PMDC motor*.

Energy Processing and Smart Grid, First Edition. James A. Momoh

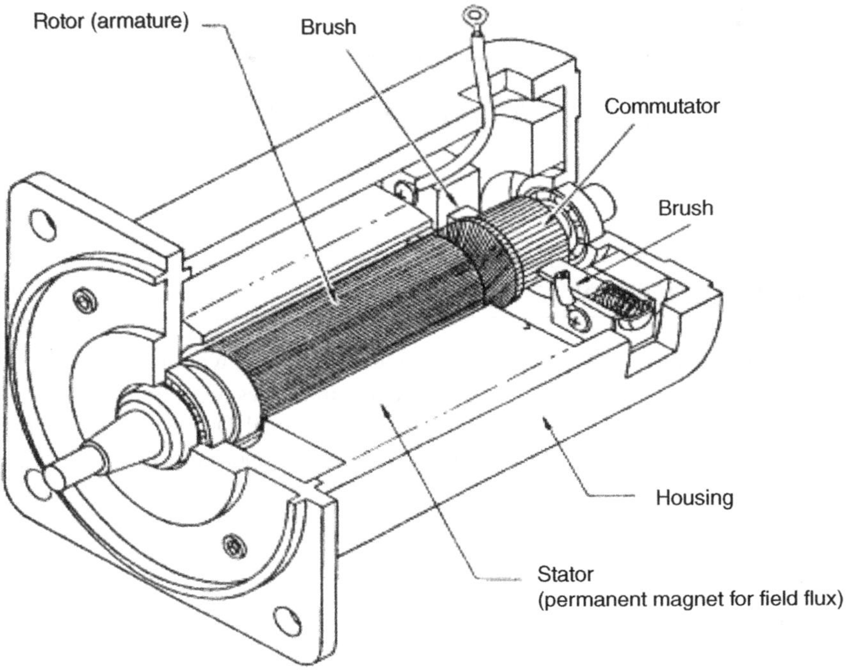

Figure 8.1 Permanent magnet DC motor.

As shown in Figure 8.1, the permanent magnets of a PMDC motor are supported by a cylindrical steel stator, which also serves as a return path for the magnetic field. Figure 8.2 shows the field system with a magnetic pole shoe.

Materials used for the permanent magnet are divided into diverse types. The magnet shown in Figure 8.1 has materials with high coercivity and residual flux density.

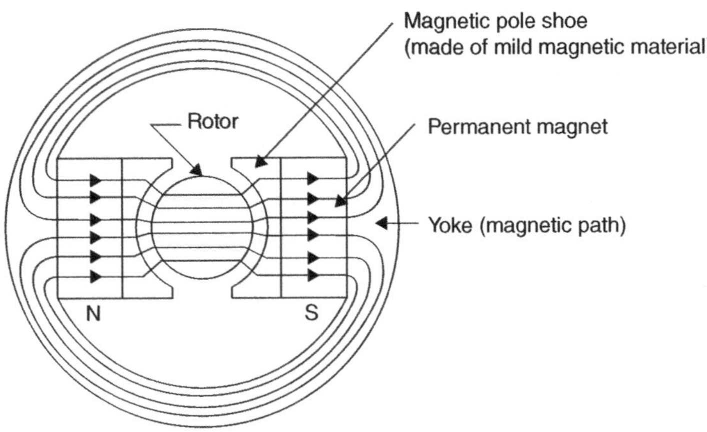

Figure 8.2 Movement of flux in PMDC.

TABLE 8.1 Motivations and Limitations of Permanent Magnet DC Motors

	Motivation	Limitations
Size	No field coil hence space for field coil is saved, which reduces the overall size of the motor	–
Performance	No need for a field-excitation arrangement; efficiency of these motors is generally higher than that of the wound-field motors	Weakened magnetic field strength due to the demagnetizing effect and non-compensation of its armature reaction Field in the air gap is fixed and limited and cannot be controlled externally; therefore, very efficient speed control of DC motor is difficult
Cost	Cheap and economical for fractional kW-rated applications	–
Environmental Factors	Low-voltage PMDC motors produce low air noise	Magnet sensitive to temperature
Losses	Eddy-current losses from a permanent magnet are lower than iron loss and rotor-cage loss Rotor is excited without current, which leads to no rotor-conductor loss or heat	–

8.2.2 PMDC Motivations and Applications

The motivations and limitations for using a PMDC motor are shown in Table 8.1.

PMDC motors are used extensively where small DC motors are required and very effective control is not required, such as in automobiles starters, toys, wipers, washers, hot blowers, air conditioners, and computer disc drives.

8.2.3 PMDC Design Models and Materials

Equivalent Circuit of a PMDC Motor

Since in a PMDC motor the field is produced by permanent magnet, there is no need to draw field coils in the equivalent circuit of a PMDC motor Figure 8.3. The supply voltage to the armature will have an drop in its armature resistance and the rest of the supply voltage will be countered by back emf of the motor. Hence the voltage equation of the motor is given as

$$V = IR \times E_b \tag{8.1}$$

where I is the armature current, R is the armature resistance of the motor, E_b is the back emf, and V is the supply voltage.

The armature is connected in series with the armature current and resistor R. This yields the back emf as

$$E_b = k_a \times \varphi_d \times \omega_m \tag{8.2}$$

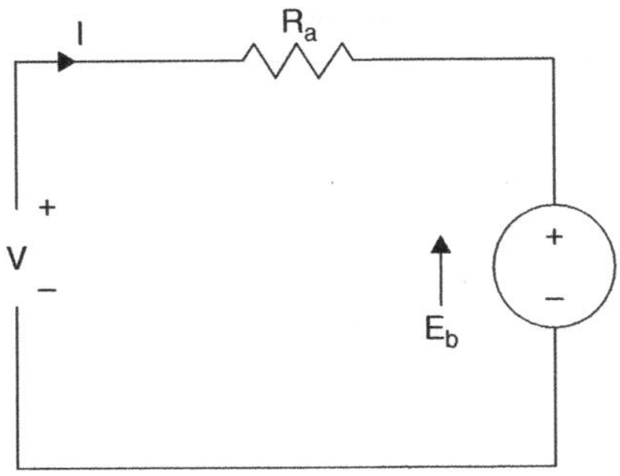

Figure 8.3 Equivalent circuit of a permanent magnet DC motor.

where k_a is the armature current, φ_d is the flux per pole, and ω_{m} is the flux speed in rad/sec.

Because of friction, the torque generated by the motor may be treated as being made up of frictional torque and usable torque:

$$\mathrm{T_d} = \mathrm{T}_l + \mathrm{T}_f \tag{8.3}$$

where $\mathrm{T_d}$ is the total torque of the permanent-magnet motor, T_l is the load torque or useful torque, and T_f is the frictional torque.

Mechanical power is defined as

$$P = T_d\,\omega \tag{8.4}$$

where ω is the rotational speed (rad/s).

The performance of a PMDC motor with respect to the current–torque relationship and flux-speed–current relationship is given in Figure 8.4.

In a PMDC motor, we have flux per pole defined as

$$\varphi_{\mathrm{d}} = \mathrm{k} \times \omega_{\mathrm{m}} \tag{8.5}$$

where $k = \mathrm{k_a} \times \varphi_{\mathrm{d}}$.

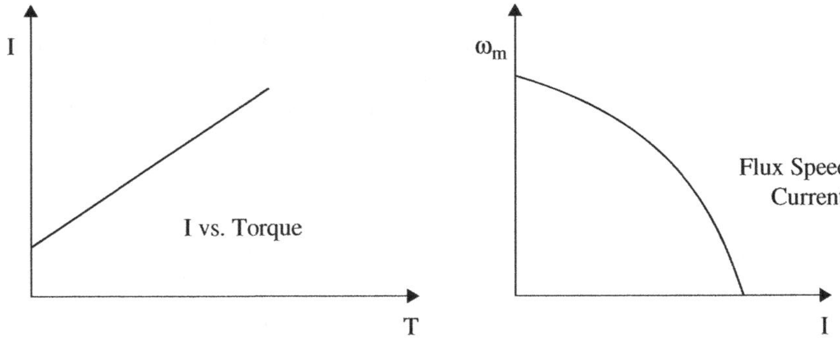

Figure 8.4 Graph of current (I) versus torque (T), and flux speed (ω_{m}) versus current (I).

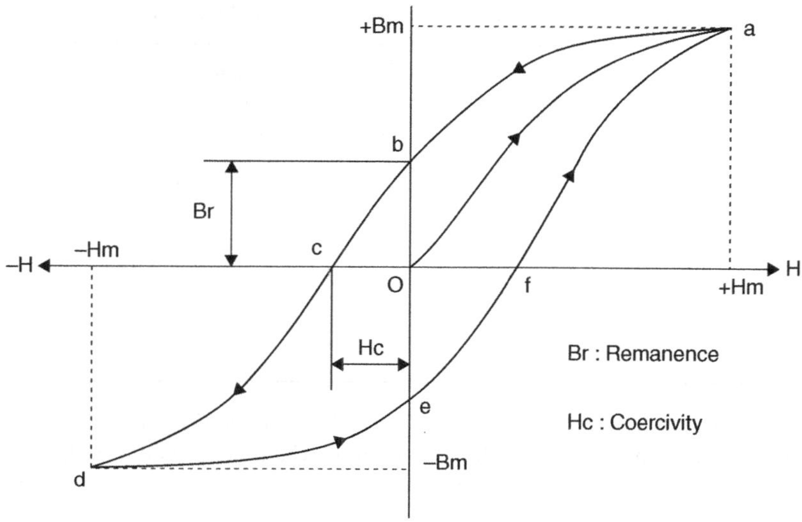

Figure 8.5 B–H curve.

So, back emf

$$E_b = k_a \times \varphi_d \times \omega_m \tag{8.6}$$

k_a = torque constant
T_d = total torque of permanent magnet motor and is

$$T_d = \frac{E_b \times I_a}{\omega_m} = kI_a \tag{8.7}$$

If V_t terminal voltage of the motor changes at no load, we can do performance analyses such as

$$Speed\ Regulator\ (SR) = \frac{\omega_{FL} - \omega_{NL}}{\omega_{NL}} \tag{8.8}$$

where ω_{FL} is the rotational speed at full load and ω_{NL} is the rotational speed at no load.

Materials for PMDC Motors

The advances in permanent magnet motors is due to special characteristics of new materials. The highest-performance magnets are brittle ceramics. Other desirable properties include chemical sensitivity, temperature sensitivity, and demagnetizing fields.

Materials used in permanent magnet motors all have B–H loops. Figure 8.5 [1] illustrates a typical B–H characteristic loop. It is divided into four quadrants, namely a magnetization quadrant, a demagnetization quadrant, a residual quadrant, and a coercive region.

TABLE 8.2 Summary of Properties of the Permanent Magnet Material Classes

Permanent Magnet	B_r (T)	H_c (kA/M)	Cost	Resistivity ($\mu\,\Omega - cm$)	Max. Working Temperature	Curie Temperature (°C)
Alnico	1.3	60	–	47	>500	–
Ceramic ferrite magnets	0.4	300	Low	>10,000	250	450
NdFeB (sintered)	1.1	850	Medium	150	80–200	310–350
Sm_2Co_7 (sintered)	1.0	750	Higher than NdFeB	86	250–350	700–800

The demagnetization quadrant is referred to as the *demagnetizing curve*.

The worst-case demagnetization shows at point D when flux density is reduced compared to the original line. It is recommended to keep the separating point away from the worst-case demagnetization condition.

The behavior or performance of a PMDC motor depends on the materials used in the permanent magnet. To understand the performance of a PMDC motor, it is important to know the properties of these materials:

1. **Alnico magnets** have good magnetization/demagnetization properties but very low coercive force and the B–H loop is too square shaped. The permanent demagnetization occurs too easily.
2. **Ceramic ferrite magnets** use barium and strontium. They are low cost and have a moderately high-service temperature at 400°C. This type of magnet is represented by a straight-line demagnetization curve B_r (residual). Flux density is low and hence the machine volume and size are typically large.
3. **Rare-earth magnets, using neodymium-iron-boron (NdFeB)**, have very good demagnetization properties except when at Curie temperature of 150°C. A range of B–H for each material and ordering measures for useful applications are determined. In general, rare-earth materials are expensive but cost-effective and useful in very small resistors.
4. **Samarium-cobalt (Sm_2Co_7) magnets** have very good demagnetization properties but are expensive because samarium is rare.

The combination of magnets with ferro-magnet materials is used to achieve high performance, high torque, and high efficiency at low voltage levels, and at a low cost.

A summary of the four magnet types is shown in Table 8.2.

Example 1

Consider a PMDC motor, where R_a is 1.2 Ω, V_c is 60 V, and ω is 1,950 rev/m. If $I_a = 1.5$ A at no load find:

(a) torque constant

(b) no-load rotational losses

(c) P_{out} in HP if speed changes to 1,500 rpm from a 50 V supply

Solution:

(a) Torque constant $k = k_a * \varphi_d$
if

$$E_a = k \times \omega_m$$

$$k = \frac{E_a}{\omega_m}$$

where

$$\begin{aligned} E_a &= V_{dc} - I_0 R_a \\ &= 60 - (1.5)\,1.2 \\ &= 58.2\,\text{V} \end{aligned}$$

thus

$$k = \frac{E_a}{\omega_m}$$

But

$$\begin{aligned} \omega_m &= 1,950\ \text{rpm} \times \frac{2\pi}{60} = 204.2\ \text{rad/sec} \\ k &= \frac{58.2}{204.2} \\ &= 0.285\ \text{V/rad} \end{aligned}$$

(b) At no load, rotational losses

$$\begin{aligned} P_{rotation\ loss} &= E_a I_a \\ &= 58.2 \times 1.5 \\ &= 87.3\,\text{W} \end{aligned}$$

if

$$\omega_m = 1,500\ \text{rpm},$$

then

$$\begin{aligned} \omega_m &= 1,500\ \text{rpm} \times \frac{2\pi}{60} = 157.08\ \text{rad/sec} \\ E_a &= k\,\omega_m \\ &= 0.285 \times 157.08 \\ &= 44.75\ \text{V} \\ E_a I_a &= (44.75)\,(4.38) \\ &= 196\,\text{W} \end{aligned}$$

(c)

$$P_{out} = P_{shaft} - P_{rotational\ loss}$$
$$= 196 - 87.3 = 108.7\ \text{W}$$

in horsepower [HP] (where 1 hp = 746 W)

$$P_{out}\ \text{in[HP]} = \frac{108.7}{746} = 0.1457$$

Problem 2

A motor with $k = 5.89\ \text{oz} \cdot \text{in/A}$ and a coil resistance of $1.76\ \Omega$ is driven with a supply voltage of 12 V. If the motor's friction torque is $1.2\ \text{oz} \cdot \text{in}$

Find:

(a) maximum torque available for driving a load

(b) how much current is flowing under these conditions

Solution:

(a) Using Equations 8.3 and 8.6, we can find the torque available at the output shaft

$$T_d = T_l + T_f = kI_a$$
$$T_l = kI_a - T_f$$

(b) Using Ohm's law to substitute for current

$$I = \frac{V}{R} = \frac{12V}{1.76\,\Omega} = 6.85\text{A}$$

$$T_l = \frac{kV}{R} - T_f = \frac{(5.89)(12)}{1.76\,\Omega} - 1.2$$

$$T_l = 39.0\ \text{oz} \cdot \text{in}$$

Problem 3

A permanent-magnet motor with a terminal resistance of $0.316\,\Omega$ and $KT = 30.2\ \text{Nm/A}$ is powered by a 12 V supply. Measurements show that the operating rotational speed is 3,616 rpm with a current flow of 1.79A. How much power does the motor generate under these conditions?

Solution:

Using Equations 8.3 and 8.4, full expression for motor-power output becomes

$$P_d = (T_l + T_f)\,\omega.$$

Here, T_f is assumed to be negligible, and simplifies the expression

$$P_d = (T_l)\,\omega = T_d\,\omega.$$

Rotational speed in radians/sec instead of rpm is obtained as:

$$3{,}616\,\text{rpm} \times 0.105\frac{\text{rad/s}}{\text{rpm}} = 380\,\text{rad/s}.$$

Substituting Equation 8.7 back to expression of power:

$$P_d = kI_a \omega$$

$$P_d = (0.0302\,\text{Nm/A}) \times (1.79\text{A}) \times (380\,\text{rad/s}) = 20.5\,\text{W}.$$

8.3 PERMANENT-MAGNET SYNCHRONOUS MOTORS

A permanent-magnet motor is a motor that has poles made up of permanent magnets. The magnets provide the flux in a synchronous machine. These magnets are generally placed on the rotor, eliminating the need for an external DC source. The rotor winding of synchronous machines is replaced by a permanent magnet. Its concepts, motivation and application, design, and materials are discussed in this section.

8.3.1 Permanent-Magnet Synchronous Motor Fundamentals and Concepts

A permanent-magnet synchronous motor (PMSM) or permanent-magnet motor (PMM) is a synchronous motor that uses permanent magnets rather than windings in the rotor.

The frequency of the excitation dictates rotor speed. The angular position of the PMM rotates at an angle ω_0, which is equal to angle ω_s of the synchronous machine and thus can produce AC currents. For a typical AC PMM, when energized, the machine produces torque with $P\omega_0 = T$.

First, we present the rotor construction of PMMs. In permanent magnet motors, there is no field-winding sector. The structure is smooth and cylindrical in shape. PMSMs can be built with different rotor configurations (as shown in Figure 8.6). Some PMSM types based on the arrangement of magnets on the rotor (shown in Figure 8.7 [2]) are as follows:

- surface-mounted, where magnets are mounted on the surface of the outer periphery of rotor lamination (Figure 8.7(a))
- surface-insert, where magnets are placed on the grooves of the outer periphery of the rotor lamination (Figure 8.7(b))
- interior transverse (Figure 8.7(c))
- salient poles with laminated-pole shoes and a cage-winding (Figure 8.6(a))
- interior-magnet rotors (Figure 8.6(b))
- surface-magnet rotors (Figure 8.6(c))
- inset-magnet rotors (Figure 8.6(d))
- rotor with buried magnets symmetrically distributed (Figure 8.6(e))
- rotor with buried magnets asymmetrically distributed (Figure 8.6(e))

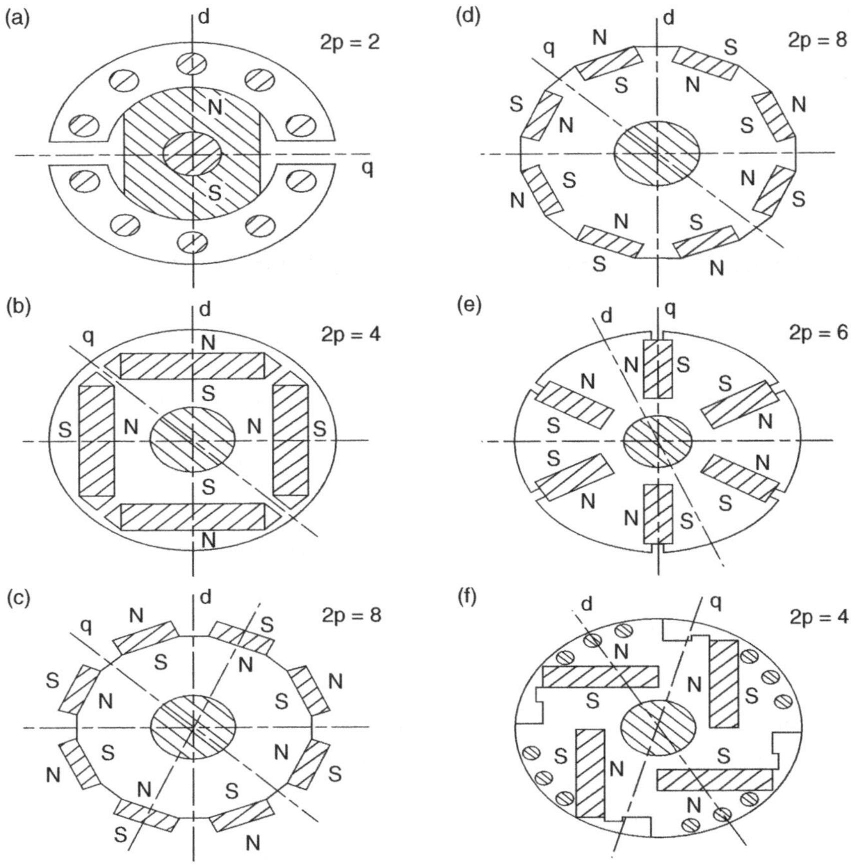

Figure 8.6 Rotor configurations for permanent-magnet synchronous motor.

In PMSMs, the stator contains 3Φ distribution windings, which is similar to the induction machine.

The arrangements for each magnet construction for permanent magnets in synchronous motors are shown in Figures 8.7 and 8.8 [3].

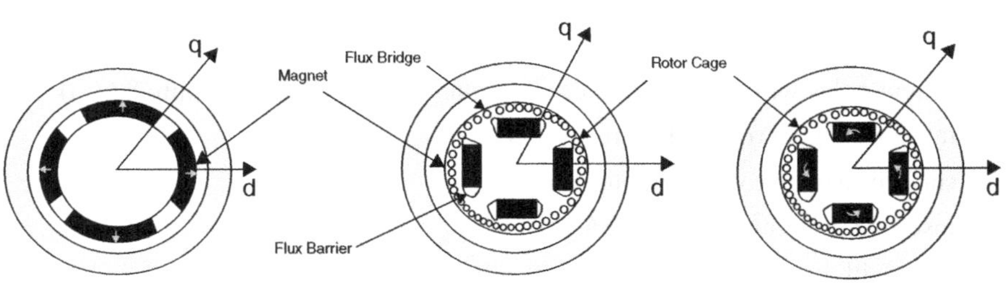

Figure 8.7 Arrangement of permanent magnets in synchronous motors.

TABLE 8.3 Motivations and Limitations of Permanent-Magnet Synchronous Motors

	Motivation	Limitations
Durability	–	Magnetic characteristics change with time.
Performance	• rotor inertia is low • large air gaps because of coercive-force densities • high efficiency and power density • simple variable-frequency control	• loss of flexibility of field-flux control • magnet sensitivity to demagnetization • loss of control leads to over speed, over voltage, and therefore fault-management issues
Cost	–	High cost of high-flux density permanent magnet
Environmental Factors	–	Loss of magnetization above core temperature
Losses	• copper losses eliminated	• always no-load spinning losses

8.3.2 PMSM Design Motivations and Applications

The motivations and limitations for using a PMSM are shown in Table 8.3.

PMSMs are used in air conditioners, heaters, blowers, windshield wipers, washers, power seats radios, etc. They are used at homes to operate electric stoves, electric toothbrushes, vacuum cleaners, electric knives, power tools, lawn mowers, etc. They are also used as starter motors for outdoor purposes, and in computers as tape drivers.

8.3.3 PMSM Design Models and Materials

Using the diagrams in Figures 8.7 and 8.8, we can derive the model equation for a permanent-magnet AC motor.

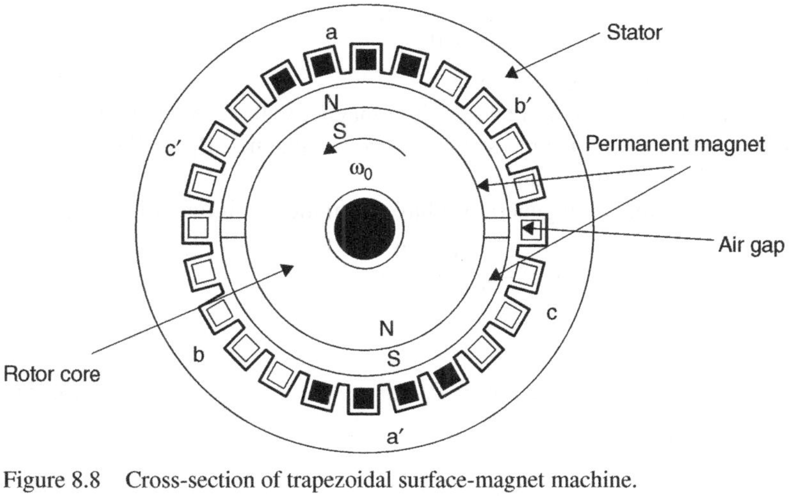

Figure 8.8 Cross-section of trapezoidal surface-magnet machine.

First let us discuss the permanent-magnet materials in the linear region of demagnetization. It has B_r (residual), with $B_m > B_r$ (critical view), so magnetic flux density

$$B_m = B_r + \mu_0 \, \mu_r \, H_m \tag{8.9}$$

where

$$B_m = \text{residual flux.}$$

We need to develop the PMM,

$$H \, l = I_{enclosed} \tag{8.10}$$

$$\frac{B}{\mu} l = I_{enclosed}. \tag{8.11}$$

This gives

$$l_m H_m + g_e \frac{B_g}{\mu_0} = 0 \tag{8.12}$$

using ampere circuital law, where, l_m is the toroidal radial length and g_e is the effective length of air gap.

If the flux is continuous, $B_m = B_g$, then combining Equations 8.7 and 8.10 gives

$$B_g = \frac{l_m}{l_m + \mu_r g_e} B_r. \tag{8.13}$$

Assume

$$B_{gap} \ll B_{residual}$$

to get,

$$B_g = \frac{l_m}{l_m + \mu_r g_e} B_r.$$

If μ_r = steel, by equation produce 0.9 Tesla.

For different materials, with a neodymium magnet with r = 5 mm and air gap 1 m, we get B = 0.9T. If, B_r is from a samarium–cobalt magnet, B_r = 0.857T. If the material is ferrite, the air-gap flux density ≤ 0.3T.

To build a circuit model in general we have from the recoil line, Figure 8.9

$$\text{slope(m)} = \frac{\Phi_r - 0}{F - (-F_0)} = P_{rc} \tag{8.14}$$

and range-slope equation

$$\therefore \quad \Phi_r = P_{rc} \Delta F. \tag{8.15}$$

If ΔF tends to zero (0),

$$\Phi_r = P_{rc} F_o \tag{8.16}$$

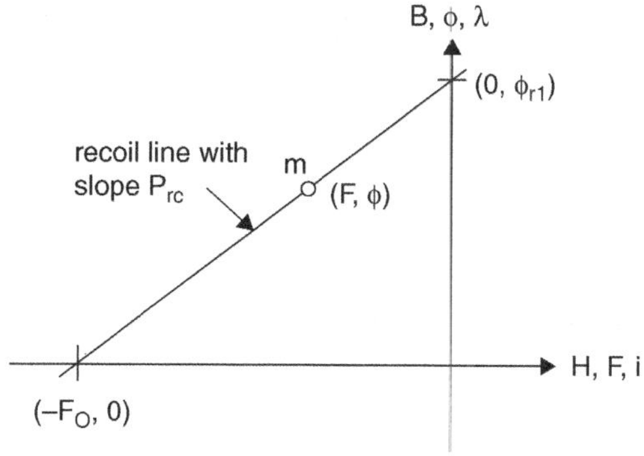

Figure 8.9 Recoil line graph.

so the equation of the circuit model is

$$F = F_O \frac{\Phi}{P_{rc}}. \tag{8.17}$$

Equation 8.17b suggests the equivalent circuit for a permanent magnet as shown in Figure 8.10.

It can easily be shown that source flux Φ is retarded or opposed by the reluctance in the magnetic circuit just as the electrical-equivalent circuits can be used to represent the magnetic circuit (mmf representing emf, flux representing current, and reluctance representing resistance), e.g., in source transformation shown in Figure 8.11.

The equivalent-circuit model of PMM is developed, and hence for each part of the machine it goes specifically as follows:

For the rotor with magnetic length l_m we derive,

$$\text{mmf} = \frac{B_r\, l_m}{\mu_r\, \mu_0}. \tag{8.18}$$

If the width of the magnet is 2γ of the magnet, then,

$$\text{mmf} = F\frac{4}{\pi}\frac{B_r\, l_m}{\mu_r\, \mu_0}\,\text{Sin}\,\gamma. \tag{8.19}$$

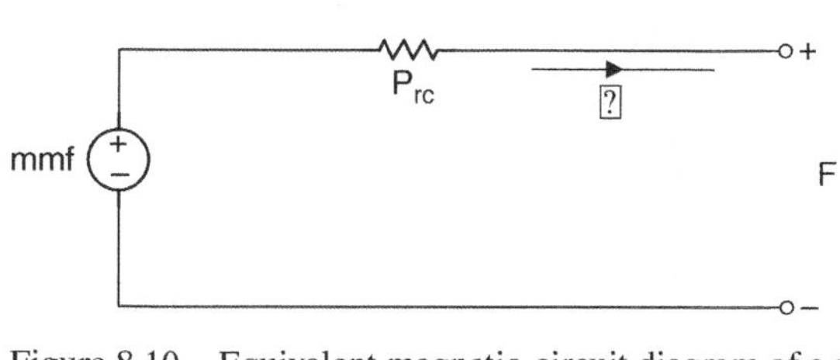

Figure 8.10 Equivalent magnetic-circuit diagram of a permanent-magnet motor.

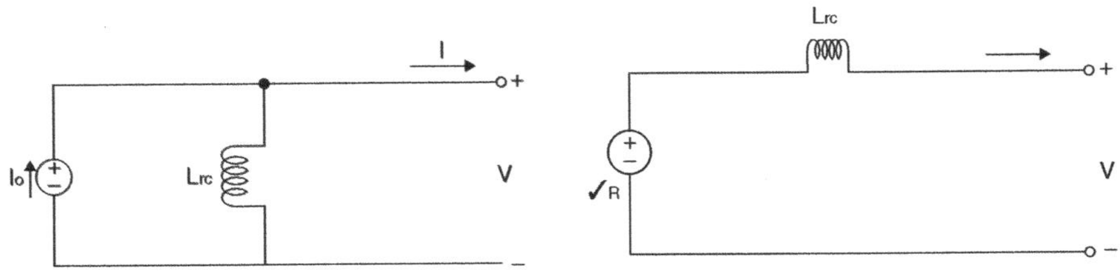

Figure 8.11 Norton's theorem current-to-voltage transformation.

Thus, the current value is determined from

$$\text{if} = \frac{4\text{f}}{3\text{N}_{se}} = \frac{16}{3\pi}\frac{\text{B}_r\, l_m}{\mu_r\, \mu_0}\frac{\text{Sin}\,\gamma}{\text{N}_{se}}\text{e}^{j\beta} \tag{8.20}$$

where

$$\beta = B_0 + wt.$$

For different materials, different values of β are determined, which give different values of L_m.

For the case of $\mu_r = 1$

$$\text{L}_m = \frac{3\pi}{8}\text{N}_{se}^2\frac{\mu_0\, l\, \text{r}}{\text{g}_e + l_m} \tag{8.21}$$

where N_{se} is the number of turns, l is the axial length, g_e is the gap length, l_m is the rotor-magnetic length, r is the radius of the outer coil, L_m is the magnetic inductance, and l is typically small, 0.25 to 0.4, based on the machine rating of two to five for induction machines.

From the B–H loop, the region of magnetic saturation is frequently of negligible effect. Thus, the mmf required for large machines is frequently of negligible effect.

Note that L_{ls} and L_m are constants. We may need to now provide the equivalent circuit of permanent magnet motor as shown in Figure 8.12.

$$\text{If } \bar{\lambda} = \text{L}\bar{\text{I}}, \text{ then L} = \frac{\bar{\lambda}}{\bar{\text{I}}} \tag{8.22}$$

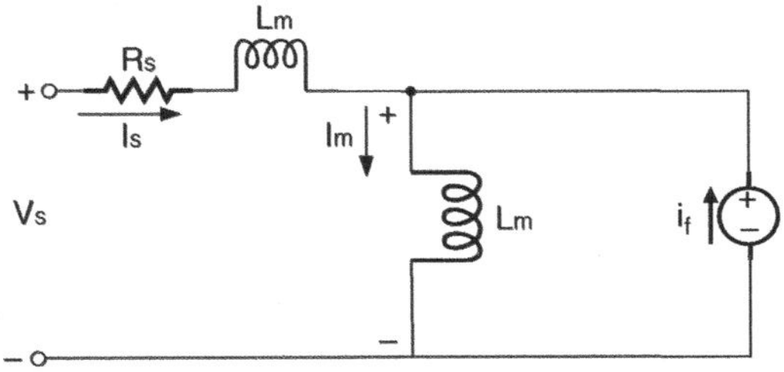

Figure 8.12 Equivalent circuit 1 of a permanent-magnet motor.

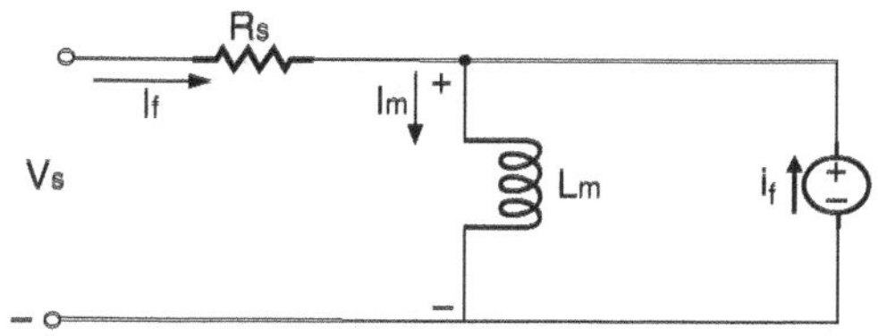

Figure 8.13 Equivalent circuit 2 of a permanent-magnet motor.

we can rewrite the equivalent diagram.

The inductive element function of current can be given as

$$L_m = L_{ls} + L_M \tag{8.23}$$

$$i_f' = \frac{L_m}{L_M}\vec{i_f} \tag{8.24}$$

which is the new system as shown in Figure 8.13

where i_f is the source current and i_f' is the output current.

From the equivalent circuit, as the PMM rotates we have $\omega_s = \omega_0$ at demagnetization region. The flux-density value B_d from the B–H loop = $\pm B_0$.

For different materials, e.g., ferrite, the value decreases with temperature. For samarium–cobalt (Sm-Co), B_d becomes negative:

$$B_{Samarium} = \frac{3N_{se}\mu_r\mu_0}{4(L_m + \mu_r g}\bar{i_s} \tag{8.25}$$

If $B_m < (B_g - \overline{B_s})$

$$\bar{i_s} = \frac{4}{3N_{se}\mu_r\mu_0}\left[L_m B_r - (L_m + \mu_r g_e\right) B_d. \tag{8.26}$$

You can convert to $j\omega$ domain and show i_s at different frequencies. Using $j\omega$ transformation

$$\bar{i_s} = \frac{-j\omega_s L_m}{R_S + j\omega_s\left(L_{ls} + L_m\right)} i_f\left(\varepsilon^{-j\omega t} - \varepsilon^{-t/\tau_s}\right) \tag{8.27}$$

Now $j\omega$ domain dominates.

We obtain

$$\bar{i_s} = \frac{L_m + L_{ls}}{R_S}. \tag{8.28}$$

The overall equivalent-circuit model of PMM is shown in Figure 8.14 [2]. The inductance L_{L_S} is lumped with the mutual inductance of the stator and damping windings. The combined mutual inductance is given by L_{md}. The magnetizing current $\overline{i_m}$ gives the equivalent magnetizing current for the PMM at the stator side.

The constant of the equivalent-circuit model for the PMM is given in Figure 8.14, defined into a direct quadrature axes reference frame.

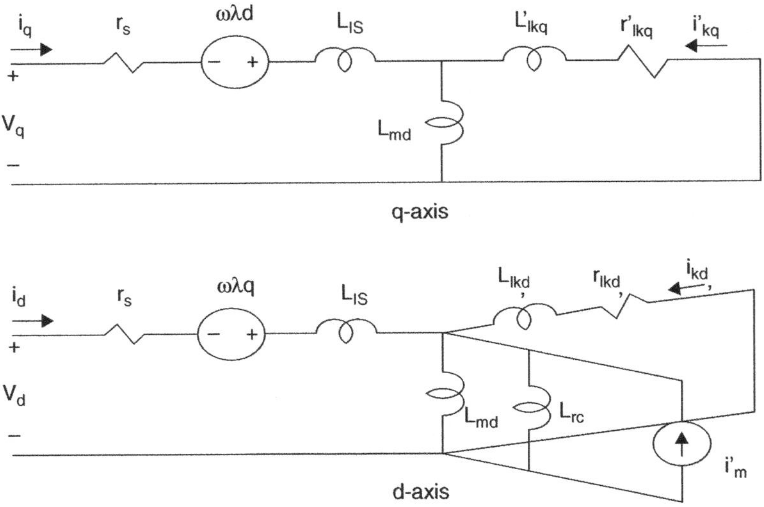

Figure 8.14 Equivalent-circuit model of permanent-magnet motor.

Figure 8.15 [3] shows a B–H curve with different operating points of a PMSM machine. At a no-load separating point B′ there is demagnetization. The shape is smaller with a larger air gap.

With current in the state, there is further demagnetization of the permanent magnet, which results in a different operation point at C′ at full load.

Refer to Section 8.1.3 for materials of a PMSM machine.

PMSMs provide a wide range of speed by using inverters on good magnetic types of materials such as neodymium–iron–boron or samarium–cobalt. Note that these materials with the same sizes may differ in efficiency under different

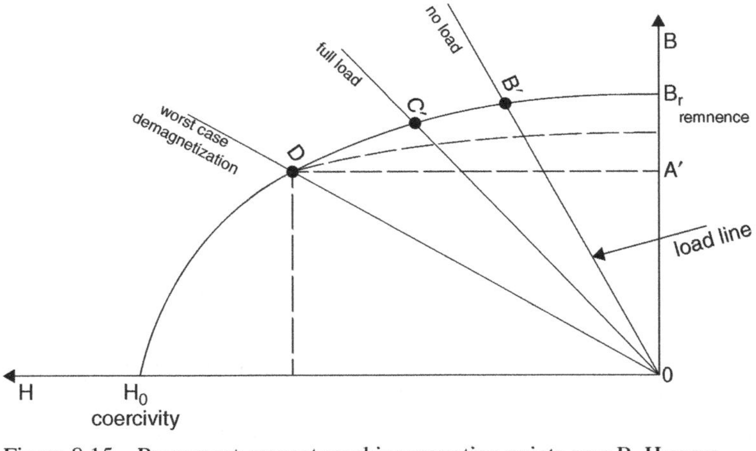

Figure 8.15 Permanent-magnet machine operating points on a B–H curve.

temperatures. The permanent-magnet materials used in this motor are same as that of a PMDC motor.

Example 1

The parameters of a star-connected, six-pole, 1.5 kW, 9.2 A, 1,500 rpm, 9.55 Nm, three-phase PMSM are:

$Rs = 0.513\,\Omega$, $Ld = 4.74$ mH, $Lq = 9.51$ mH.

Inverter input voltage = 285 V.

Find:

(a) normalized stator resistance

(b) normalized direct-axis inductance

(c) normalized quadrature-axis inductance

Solution:

Choose the base value as follows:

$$\mathrm{I_b} = 9.2\,\mathrm{A},\ \mathrm{P_b} = 1,500\,\mathrm{W},\ \mathrm{T_b} = 9.55\,\mathrm{Nm}$$

Base voltage,

$$\mathrm{V_b} = \mathrm{V}_b = \frac{\mathrm{P_b}}{3\,\mathrm{I_b}} = \frac{1,500}{3\times 9.2} = 54.35\,\mathrm{V}$$

Base speed,

$$\omega_\mathrm{b} = \frac{\mathrm{P_b}}{\mathrm{T_b}} = \frac{1,500}{9.55} = 157.068\ rad/s$$

Base impedance,

$$\mathrm{Z_b} = \frac{\mathrm{V_b}}{\mathrm{T_b}} = \frac{54.35}{9.2} = 5.91\,\Omega$$

Base impedance,

$$\mathrm{L_b} = \frac{\mathrm{Z_b}}{\omega_\mathrm{b}} = \frac{5.91}{157.07} = 37.6\,mH$$

Normalized stator resistance,

Ans. (a) $$\mathrm{R_{sn}} = \frac{\mathrm{R_s}}{\mathrm{Z_b}} = \frac{0.513}{5.91}\ 0.868\ \mathrm{pu}$$

Stator phasor voltage $\mathrm{V_s}$ is calculated as:

$$\mathrm{V_s} = 0.55\,\mathrm{V}_{dc} = 156.75$$

Normalized value,

$$\mathrm{V_{sn}} = \frac{\mathrm{V_s}}{\mathrm{V_b}} = \frac{156.75}{54.35} = 2.884\ \mathrm{pu}$$

Normalized direct axis inductance,

$$L_{dn} = \frac{L_d}{L_b} = \frac{4.74\ \text{mH}}{37.6} = 0.126\ \text{pu}$$

Normalized quadrature axis inductance,

$$L_{qn} = \frac{L_q}{L_b} = \frac{9.51\ \text{mH}}{3.76} = 0.2529\ \text{pu}$$

8.4 VARIANTS OF PERMANENT-MAGNET SYNCHRONOUS MOTORS

8.4.1 Trapezoidal Surface-Magnet Motors

A trapezoidal surface PMM is very similar to a sinusoidal PMM. The difference is that the trapezoidal surface-magnet motor has its 3Φ winding made of concentrated, full-pitch distribution instead of a sinusoidal distribution. Figure 8.16 represents the field winding and rotor.

The two-pole motor in Figure 8.16 [3] consists of a gap in the rotor magnets to reduce flux (magnetic losses), and has four slots per phase winding per pole. When the machine rotates, the flux linkage varies linearly at the demagnetizing region when it passes the magnetic gap. If the machine is driven by a prime mover, the stator three-phase voltages will have a trapezoidal wave shape as shown in Figure 8.17 [4].

Furthermore, an electronic inverter is required to establish a six-step current wave, which generates torque. A trapezoidal surface-magnet motor can serve as a brushless DC motor if it makes use of the inverter and an absolute-position sensor is mounted on its shaft. The sinusoidal-surface PMM can also serve as a brushless DC motor but the trapezoidal surface PMM gives closer DC-machine-like performance.

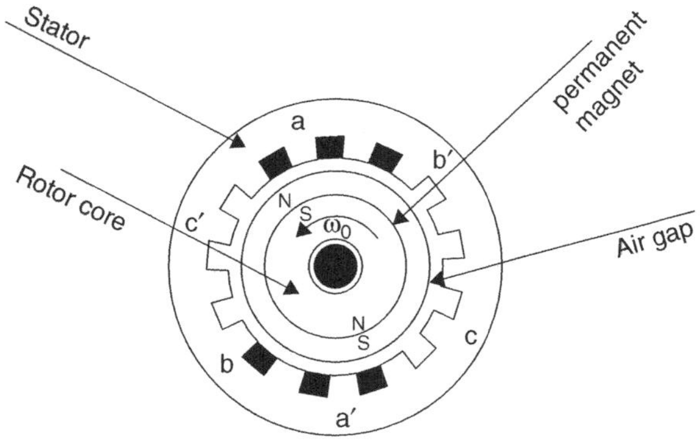

Figure 8.16 Cross-section of trapezoidal surface-magnet machine (two poles) with stator in its star form.

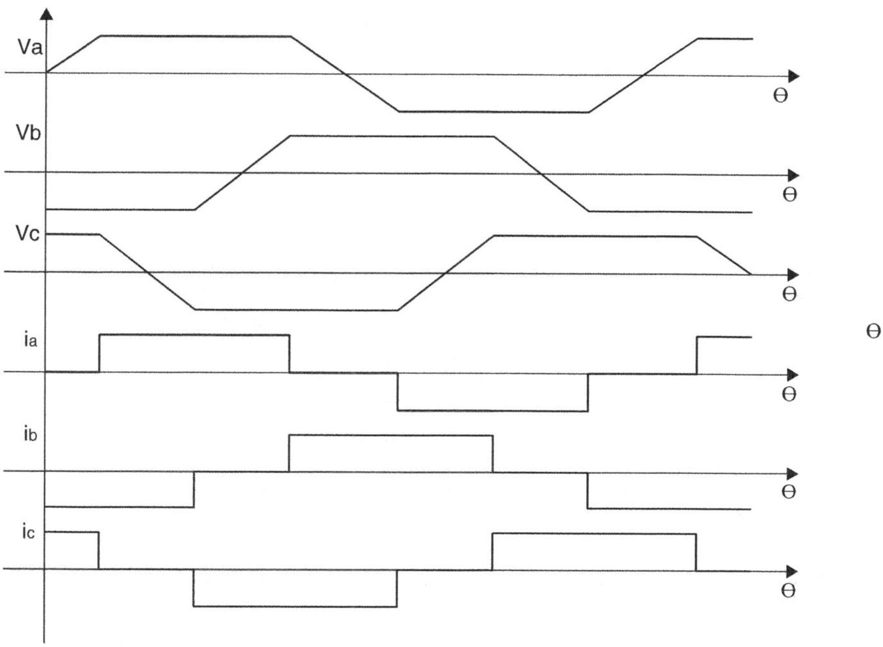

Figure 8.17 Stator-phase voltage and current waves in trapezoidal permanent-magnet machine.

8.4.2 Synchronous Reluctance Motors

A synchronous reluctance motor has the same structure as that of a salient-pole synchronous motor except that it does not have a field winding on the rotor. Figures 8.18 and 8.19 [5] show the cross-section and torque-angle curve, respectively, of a synchronous reluctance motor.

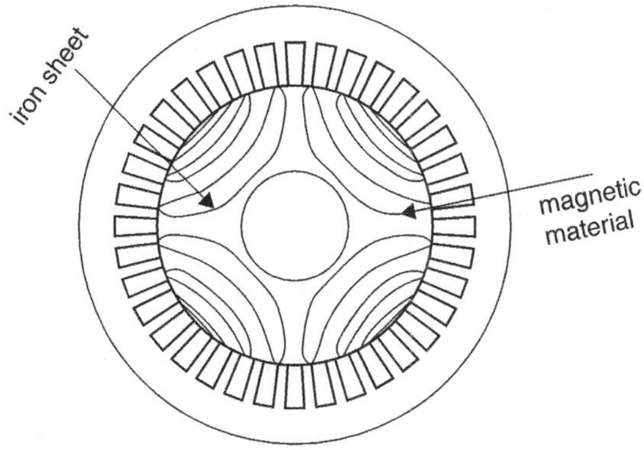

Figure 8.18 Cross-section of synchronous reluctance motor.

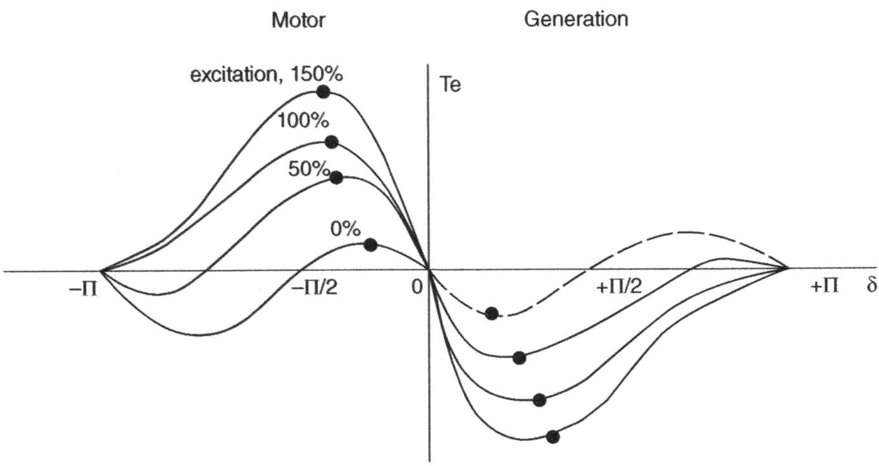

Figure 8.19 Torque-angle characteristics of a salient-pole machine.

The stator has a symmetrical winding, which creates a sinusoidal rotating field in the air gap. There is magnetic field induced in the rotor, causing it to align with the stator field in a minimum reluctance, which creates a reluctance torque on the rotor. The torque developed in this motor is given by

$$T_e = 3\left(\frac{P}{2}\right)\left(\Psi^2\frac{L_{ds} - L_{qs}}{2L_{ds}L_{qs}}\sin 2\delta\right) \tag{8.29}$$

where T_e is the torque developed in the motor, P is the pair of poles or pole pair, Ψ is the flux linkage, L_{ds} is the direct-axis inductance, L_{qs} is the quadrature-axis inductance, and δ is the rotor angle.

8.4.3 Variable Reluctance Motors

A variable reluctance motor has double saliency, i.e., both the rotor and stator have saliency. This motor type includes the stepper motor and the switched reluctance motor.

Stepper Motor

A stepper motor is a type of synchronous motor built to rotate a specific number of degrees in response to a digital input in the form of a pulse. This means that a digital signal is used to drive the motor, and every time it receives a digital pulse it rotates a specific number of degrees in rotation.

Stepper motors are excellent devices for accurate speed control or precise position control without any feedback. In such usage, the axis of the motor's magnetic field steps around the air gap at a speed that is based on the frequency of pulses. The rotor inclines to align itself with the axis of the magnetic field, which means that the

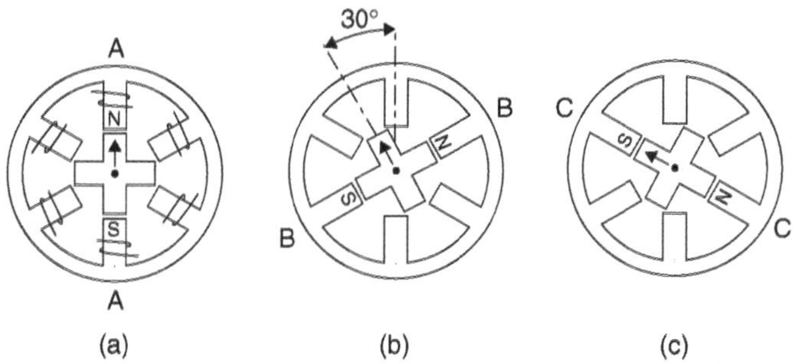

Figure 8.20 Principle of operation of 30° step variable-reluctance stepping motor.

rotor steps in sync with the motion of the magnetic field, thus it is called a stepper motor. Figure 8.20 shows operation of a 30° per step stepper motor.

Stepper motors are not suitable for variable-speed drives.

Switched Reluctance Motor

The structure of the switched reluctance motor, as shown in Figure 8.21 and Figure 8.22, has four stator-pole pairs and three rotor-pole pairs. The rotor has neither windings nor permanent magnets.

The stator poles have concentrated winding rather than sinusoidal winding. Each stator-pole pair is aligned and is then de-energized. The stator-pole pairs are sequentially excited using a rotor position encoder for timing.

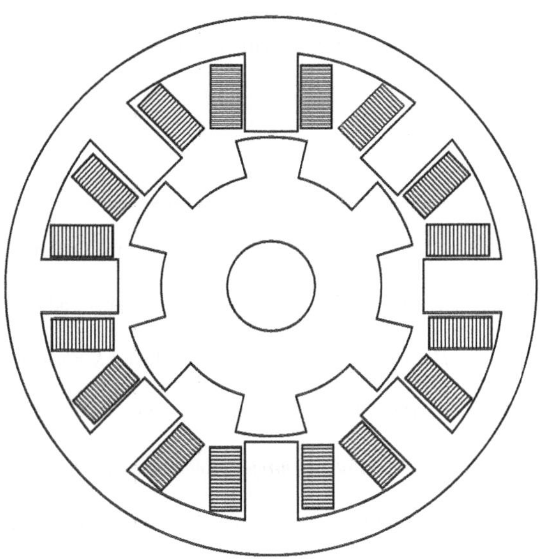

Figure 8.21 A typical switched reluctance motor.

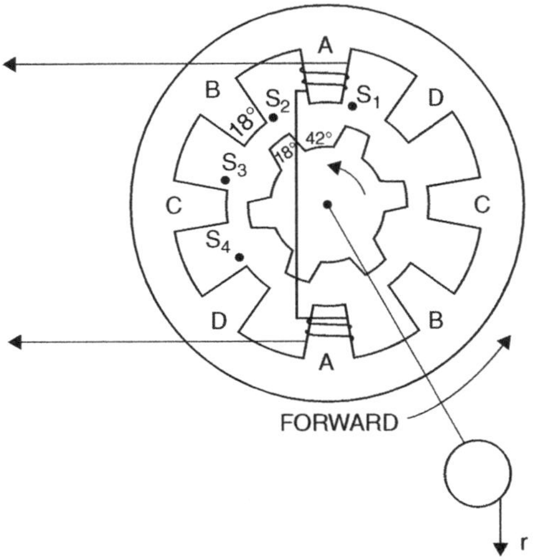

Figure 8.22 Structure of a switched reluctance motor.

8.5 CHAPTER SUMMARY

On this chapter permanent magnet operations & fundamentals are discussed. Applications of permanent magnet motor as DC motor and synchronous machine are discussed in detail. Worked examples and practices are provided on the concepts discussed.

Further, design, modeling and material usage are discussed. The chapter is concluded by discussing varients of permanent magnet synchronous motors.

BIBLIOGRAPHY

[1] E. H. Smith, *Mechanical Engineers Reference Book*, 12th ed., Butterworth Heinemann Ltd., Oxford, 1994.

[2] R. Krishnan, *Electric Motor Drives: Modeling, Analysis, and Control*, Prentice-Hall, Upper Saddle River, NJ, 2001.

[3] J. F. Gieras, *Permanent Magnet Motor Technology: Design and Applications*, 3rd ed., Taylor & Francis Group, Boca Raton, FL, 2009.

[4] R. Krishnan, *Permanent Magnet Synchronous and Brushless DC Motor*, CRC Press, Boca Raton, FL, 2010.

[5] G. Pellegrino, T. M. Jahns, N. Bianchi, et al., *The Rediscovery of Synchronous Reluctance and Ferrite Permanent Magnet Motors*, Springer, Berlin, 2016.

[6] R. Ramshaw and R. G. van Heeswijk, *Energy Conversion: Electric Motors and Generators*, Saunders College Publishing, Philadelphia, 1990.

[7] B. L. Theraja and A. K. Theraja, *A Textbook of Electrical Technology, Vol. 2: AC and DC Machines in S.I. Units*, S. Chand & Company, New Delhi, 2014.

[8] C. Ross, "Aspects of Permanent Magnet Machine Design," University of Illinois, https://pdfs.semanticscholar.org/d47d/031bc187875de4073d895590aa724892c8e4.pdf.

[9] PowerShow.com, Permanent Magnet Synchronous Motor, www.powershow.com/view/402a26-MTU0Z/Permanent_Magnet_Synchronous_Motors_powerpoint_ppt_presentation.

CHAPTER 9

RENEWABLE ENERGY RESOURCES

9.1 INTRODUCTION

An increasing demand for electric power in the twenty-first century and the need for more environmentally benign electric power systems are of critical concern to governments and stakeholders such as industries and end users. Electrical shortages, power quality, rotating outages, and increasing oil prices have motivated many utilities and consumers to look for alternative forms of highly reliable energy. Traditional utility ties have been deregulated, yielding room for new market structures and players. Regulatory commissions in different parts of the world have unbundled the vertical utility industry into separate business units that can be categorized into three broad categories: generation companies, which include utility and nonutility companies; transmission companies, which may be under state ownership; and distribution companies, which are privately owned business units.

9.2 DISTRIBUTED GENERATION CONCEPTS

Distributed generation (DG) is a relatively new approach to describe the new wave of generation at the customer side, which is less than that of the typical control power station in a competitive electricity market. DG has been given a variety of definitions relative to its rating, power delivery area, environmental impact, penetration level, and point of connection. Although these criteria are necessary, they are not sufficient. We provide here a practical definition for DG as "an electric power source connected directly to the power network, preferably at the customer side of the meter, sufficiently smaller than the controlling generating plant." This definition does not define the rating of the generation source, since the maximum rating depends on the voltage level of distribution on the sub-transmission network.

DG involves the use of technologies such as microturbines, stirling engines, fuel cells, and renewable energy such as photovoltaic, wind, and biomass systems. DGs can meet the needs of a wide range of users, from residential to commercial and industrial sectors.

Energy Processing and Smart Grid, First Edition. James A. Momoh

Note that the DG definition does not specify the technology options. Hence, DG can include any of the renewable sources; combined heat and power (CHP) applications are modular.

9.3 DG BENEFITS

DG can help reduce investments in transmission and distribution capacity. From a planning point of view, DG can be placed close to load centers, thus minimizing loss in operating networks and reducing costs for operations and controls. DG is especially favored to help reduce losses in distribution networks and serve as stand-alone or back-up generation. Other benefits include:

- **Grid support**: DG can contribute to the provision of ancillary services needed to maintain stable operation of the grid.
- **Environmental benefits**: DG can help improve or enforce environmental regulations.
- **DG combined generation and heat capacity**: DG has provided the so-called tri-generation to provide electricity, heat, and steam for different applications.
- **Fuel diversity**: DG uses different fuels at optimized prices, depending on the technology. Liberalization of electricity markets to an environment within a competitive market and due to various technologies makes it hard to generalize. Further, due to fuel diversity, any shortfall in fuel DG that is nonrenewable is considered risky and costly. As such, it may not help alleviate blackouts and invariably degrades the security of the power system, hence emphasizing the need for increased regulating (backup) power.

Policies are necessary to ensure that DG systems adhere to some quality of supply and to maintain system frequency. High voltage levels approved for DG connections relative to the utility company must also be known and properly controlled to achieve voltage security and respond to changing market conditions. This flexibility in construction of lines and centralized generation has made the DG market economically attractive.

Major policy issues surrounding DG include:

- high cost of implementing various DG technologies—a concern of policy makers and stakeholders
- less choice between costlier DG with expensive fuel supply—a fact that has discouraged the selection of DG compared with central generation
- policies to ensure economic efficiency of DG selection to liberalize the market

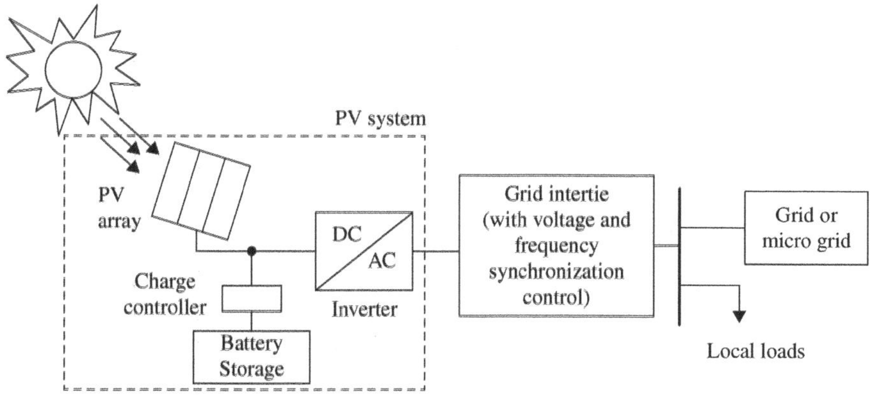

Figure 9.1 PV systems integration with microgrid.

9.4 WORKING DEFINITIONS AND CLASSIFICATIONS OF RENEWABLE ENERGY

Renewable energy is energy derived from natural sources that replenish themselves over a short period of time. These resources include sun, wind, hydropower, organic plant and waste material (biomass), and earth heat (geothermal). Whereas renewable resources can generate both electricity and heat, the term "green power" is used in a narrow sense to mean electricity products generated from renewable sources that are environmentally and socially acceptable.

9.4.1 Solar

Photovoltaic (PV) cells and modules are configurable from 1 to 5 MW. Figure 9.1 shows a typical modern PV system. In 1839, French physicist Edmond Becquerel was the first to discover that certain materials exposed to light produce current. Refinements at Bell Laboratory in 1954 led to the development of silicon-based PV cells, producing electricity conversion with more than 4 percent efficiency.

Following the energy crisis in recent years, the use of solar power has become more widespread. Commonly known as solar panels, PV modules are commercially available, provide no emissions, are an alternative to other energy sources, are reliable, and require minimum maintenance to operate. However, they are expensive compared with other renewable-energy options and up to four to six times larger than complicated alternative technologies. Today, advances in material science have led to the engineering and fabrication of solar panels with about 30 percent efficiency.

In the presence of sunlight, solar panels comprising discrete cells generate DC electricity, which after conversion to AC is connected to power the load or exported to the grid.

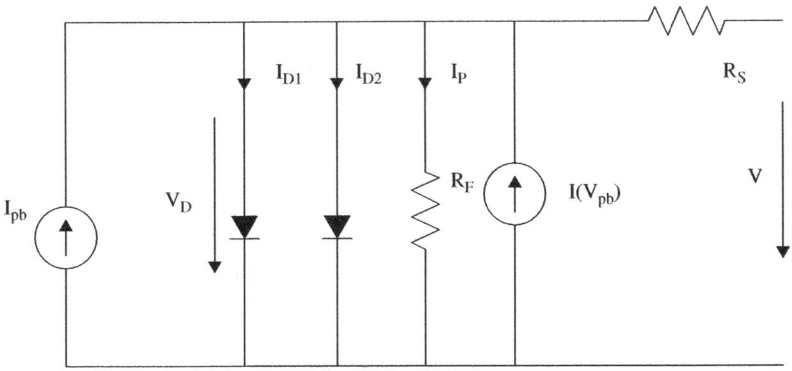

Figure 9.2 PV-equivalent circuit.

Application

PV cells can be part of a building, duplicating other building materials, and can have a wide range of application as DG ranging from residential and commercial users to remote power consumers (structures in school, homes, community facilities, and commercial buildings). The greatest potential for PV modules is as green power because they do not emit pollutants or CO_2. However, they are a poor fit as a peaking power application source, since the unit outputs are not easy to control. PV systems produce better power during daylight periods of maximum available solar radiation and require battery storage for backup. They are not a good fit as premium power due to the unpredictable nature of power from solar cells.

Cost Implications

Manufacturers continue to reduce the cost of installation while increasing efficiency as new technology is developed for manufacturing materials and other operational and management costs.

Modeling

PV systems need to be developed and verified to optimize the output of the system at the design stage for maximum energy production and peak-sharing applications. Three models of a PV-equivalent circuit are shown in Figure 9.2. The equivalent circuit consists of a diode and source, which are in parallel. A simplified model has the following *V-I* equations:

$$I = I_p - I_d = I_p - I_s \left(exp\frac{e(V + IR_s)}{mKT} - 1 \right) \tag{9.1}$$

$$I = I_{sc} - I_d = I_{sc} - I \left(e^{qV/KT} - 1 \right) \tag{9.2}$$

where I_p is the photon current, I_d is the diode current, I_s is the diode ideal factor, K is the Boltzmann constant, T is the absolute temperature, e is the charge of an electron, and I_{sc} is the solar insulation.

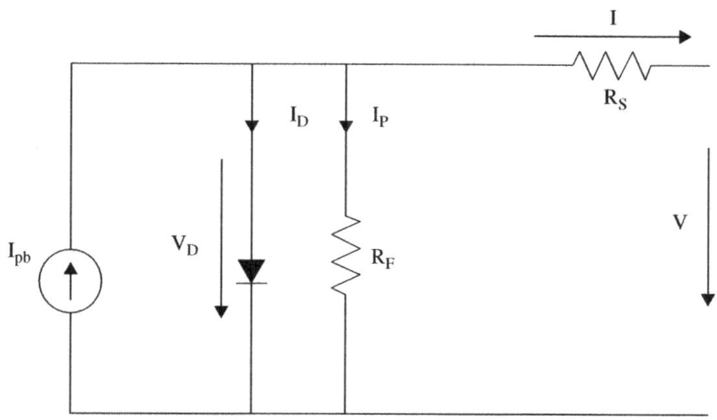

Figure 9.3 Modified equivalent circuit.

The open-circuit voltage, V_{oc}, of the PV is given as

$$V_{oc} = 0.0257 In\left(\frac{I_{sc}}{I_d} + 1\right). \tag{9.3}$$

At 25°C, Equation 9.2 becomes

$$I = I_{SC} - I\left(e^{38.9} - 1\right). \tag{9.4}$$

Modified Equivalent Circuit

The modified equivalent circuit (Figure 9.3) accounts for a real solar cell with external contacts, and voltage loss (drop) is accounted for through leakage currents and resistances. Using the V–I and Kirchoff's law

$$I_p = \frac{V_D}{R_P} = \frac{V + IR_s}{R_p} \tag{9.5}$$

$$I_{ph} - I_D - I_p - I = 0 \tag{9.6}$$

$$I_{ph} - I_s\left[exp\left(\frac{V + IR_s}{mV_+}\right) - 1\right] - \frac{V + IR_s}{R_p} - I = 0. \tag{9.7}$$

Also consider a simpler way (parallel connection):
for parallel resistance

$$I = I_{sc} - I_d - \frac{V}{R_p} \tag{9.8}$$

and for series connection

$$I = I_{sc} - I_D = I_{sc} - I_s\left(e^{\frac{qV_d}{KT}} - 1\right) \tag{9.9}$$

$$V_D = V + IR_s. \tag{9.10}$$

$$I = I_{sc} - I_d\left\{exp^{\left[\frac{q(V+IR_s)}{KT}\right]} - 1\right\} \tag{9.11}$$

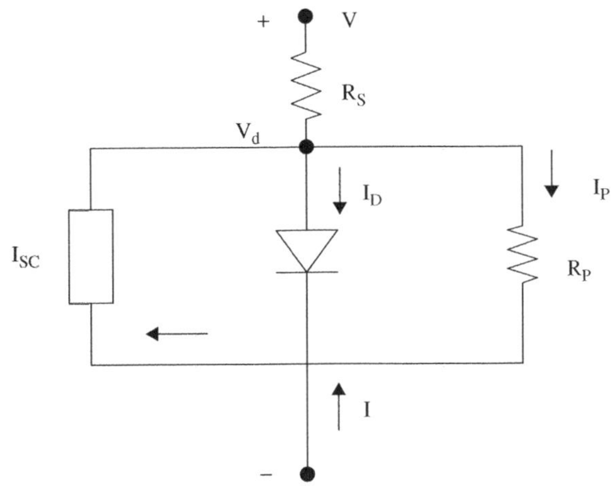

Figure 9.4 PV-equivalent circuit for both series and parallel resistances.

Example 1
Combine PV to series and parallel resistances R_s and R_p, respectively (Figure 9.4). The generalized PV equivalent circuit for both series and parallel resistances will be:

$$I = I_{SC} - I_d\left\{exp^{\left[\frac{q(V+IR_s)}{KT}\right]} - 1\right\} - \left(\frac{V+IR_s}{R_p}\right) \tag{9.12}$$

$$I = I_{SC} - I_d\left\{exp^{38.9(V+IR_s)} - 1\right\} - \frac{1}{R_p}\left(V+IR_s\right) \tag{9.13}$$

$$I_{SC} = I + I_d + I_p \tag{9.14}$$

$$I = I_{sc} - I_d\left(e^{38.9V_d} - 1\right) - \frac{V_d}{R_p} \tag{9.15}$$

Note that

$$V = V_d - IR_s \tag{9.16}$$

To obtain from cell to module, we use

$$V_{module} = n(V_d - IR_s) \tag{9.17}$$

where n is the number of cell modules.

PV Systems
A PV system can be connected to a grid (utility system), can stand alone, or can be integrated.

- A PV system connected to a grid sends DC power to a power condition unit converter, which converts DC to AC power.

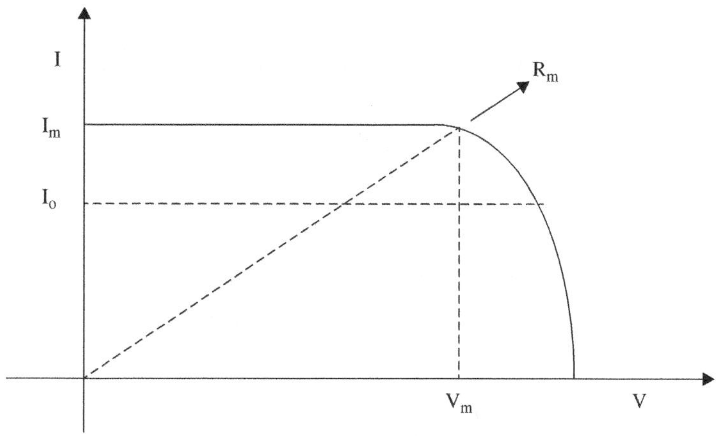

Figure 9.5 V-I characteristics: PV connected to resistive load.

- A stand-alone PV system off the grid is equipped with battery storage and a generator for backup power.
- An integrated PV system is directly coupled to its loads without battery or major power-conducting equipment.

Different types of loads affect the PV system, as shown from the performance characteristics of the V–I curves.

V–I Characteristics

PV systems connected to different loads exhibit different V–I characteristics, as seen in Figure 9.5. For a simple resistive load

$$V = IR \text{ or } I = \left(\frac{1}{R}\right) V \tag{9.18}$$

$$R_m = \frac{V_m}{I_m}. \tag{9.19}$$

To achieve R_{max}, we can use a maximum power tracker, which keeps a PV system operating at its highest efficiency point at all times.

Example 2

Consider a solar panel rated 20 W at 1.0 V. Determine:

(a) how many of these panels are needed to supply 1 A of current at 120 V

(b) how should they be connected

(c) value of the necessary load resistance

(d) total power that can be obtained from this array of solar cells

Solution:

(a) Since each panel produces 12 V, to get 120 V, the number of panels required in series is

$$
\begin{aligned}
&= \frac{Total\ voltage}{Volt/panel} \\
&= \frac{120}{10V/panel} \\
&= 12 \text{ panels}
\end{aligned}
$$

(b) The current rating of each panel can be found from the panel's power and voltage ratings, hence

$$
\begin{aligned}
I &= \frac{20W}{10V} \\
&= 20\text{A}.
\end{aligned}
$$

Thus, each path of panels connected in a series will provide 2 A. Therefore,

$$\text{no. of paths} = \frac{Total\ current}{Current/path} = \frac{10A}{2A/path} = 5$$

Hence, five parallel paths each having twelve panels in a series are required to form an array to meet the power requirement.

(c) The value of the required load resistance is found as:

$$R_L = \frac{120V}{10A} = 12\,\Omega.$$

The total power that can be obtained from the solar array is

$$
\begin{aligned}
\text{total power} &= (120)\,(10)\,\text{W} \\
&= 1.2\text{ kW}.
\end{aligned}
$$

Future Challenge of Solar Energy

The major challenge to solar energy is the cost of equipment and devices. The full cost of installation of a solar energy system is still three to six times as high as the installed cost of conventional nonrenewable generation. The cost of a PV cell module is expressed as the cost per peak watt output at standard insolation. Standard insolation (energy from sunlight on a flat surface perpendicular to the sun's rays) is defined as 1 kW per square meter. The key to developing solar-PV electricity generation is to reduce the cost of PV cells by increasing the production volume of cells so that the benefits of mass production can be realized.

9.4.2 Solar Concentrator Systems

Solar concentrator systems include power towers, linear, and dish/engine towers. These renewable-energy systems are based on energy conversion of solar to thermal energy to run conventional steam or gas turbines for power generation.

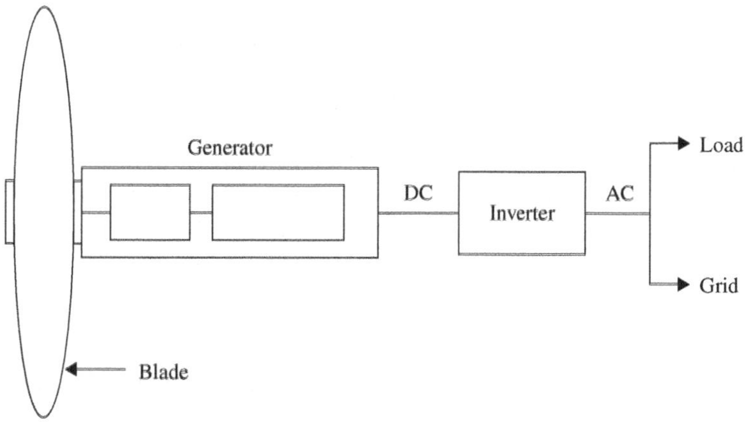

Figure 9.6 Basic components of a wind-turbine system.

9.4.3 Wind-Energy Systems

Windmills have been used for many years to harness wind energy for mechanical work such as pumping water for farms and ranches. Today, wind power is one of the fastest-growing sources of energy. After the energy crisis of the 1970s, wind energy was considered the most economically viable choice in the portfolio of available renewable-energy options. Wind turbines can produce electricity at an affordable cost without additional investments in infrastructure such as transmission lines. Wind turbines basically include the rotor, generator, turbine blades, and driver or coupling device as shown on Figure 9.6.

The operation is simple. Wind blows through the blades, with pressure exerted on the cross-sectional area of the blade. Aerodynamic force causes the blades to turn the rotor. The gearbox and the generator shown in Figure 8.8 are all in a single unit behind the machine blades. The output of the generator is passed through a unit for appropriate conversion from DC to AC. The windmill comes in different configurations, normally horizontal or vertical. The wind speed and the height of a pole-mounted windmill above the ground contribute to the power output of the wind-turbine system. The location of the system is equally important in sizing the output of the windmill. Wind turbines produce no emissions. They have a variety of sizes and applications and are classified into utility scale and individual scale. For large-scale utility projects, they can range from 1.5 to 5 MW loads. A small system can be as simple as a single pole and a blade.

Types of Wind Turbines

Wind turbines come in two basic types:

- horizontal-axis wind turbines (HAWTs), which have blades that rotate at about a horizontal axis parallel to the wind
- vertical-axis wind turbines (VAWTs), which have blades that rotates at about a vertical axis

Modeling of Wind Power

The mathematical model for wind power stems from aerodynamic power, given in

$$P = \frac{1}{2} p \pi R^2 v^3 C_p \tag{9.20}$$

where P is the air density, R is the radius of the circular area swept by turbine blade, in meters, v is the actual wind speed, and C_p is the turbine-power coefficient, which represents the power conversion efficiency of a wind turbine.

*T*ip speed ratio of the machine's blades to wind speed is

$$\lambda = \frac{\omega R}{v} \tag{9.21}$$

where $\omega =$ rotor rotational speed in radians/second is the rotor rotational speed in radians/second and C_p is maximum at $\lambda_{optimal}$.

The wind-turbine system uses induction generators that are independent of torque variation while speed varies between 1 to 2 percent.

Impact of Tower Height on Wind Power

A taller tower is expected to provide higher-speed winds to the turbine. Surface winds can also easily be affected by the irregularities or roughness of the earth's surface or forests/buildings. This is given as

$$Let \left(\frac{v}{v_0}\right) = \left(\frac{H}{H_0}\right)^{\alpha} \tag{9.22}$$

where v is the wind speed at height H, v_0 is the reference speed at reference height H_0, and α is the friction coefficient.

In the United States,

$$\alpha = \frac{1}{T} \tag{9.23}$$

whereas in Europe

$$\left(\frac{v}{v_0}\right) = \frac{ln\left(H/2\right)}{ln\left(H_0/2\right)}. \tag{9.24}$$

Example 3

If the average wind speed on an open plain is known to be 6 m/s at 10 m height, determine the wind speed at 50 m height if the roughness coefficient α is 0.15.

Solution:

We know that

$$\left(\frac{v}{v_0}\right) = \left(\frac{H}{H_0}\right)^{\alpha}$$

Thus

$$\left(\frac{v}{v_0}\right) = \left(\frac{50}{10}\right)^{0.15}$$

$$\left(\frac{v}{6}\right) = \left(\frac{50}{10}\right)^{0.15}$$

$$v = 6 \times \left(\frac{50}{10}\right)^{0.15}$$

$$= 7.6\,m/s.$$

Emission-Control Technologies for Wind Power

Wind turbines produce no emissions. As a green-power application, the efficiency of wind turbines is superb, since they do not emit CO_2 or pollutants. However, wind power has an obvious disadvantage. Because its output cannot be controlled, it is mostly suited for peaking applications, producing power only when there is sufficient wind. Wind power cannot serve as premium power because of its unpredictability. Moreover, it is unsuitable for CHP applications. Despite these drawbacks, windmills remain a subject of continuing research, with the principal focus on:

- lowering the minimum wind speed of operation
- developing voltage regulators to improve the turbine's ability to recharge its batteries while simultaneously producing electricity

Wind turbines are a relatively inexpensive way to produce electricity compared with PV, which is the most competitive green power to date.

Example 4

Determine the amount of power that is present in a 10 m/s wind striking a windmill whose blades have a radius of 5m.

Solution:

The area swept by the turbine blade is

$$A = \pi r^2$$

$$\pi(5)^2$$

$$= 78.54\ \text{m}^2.$$

Power present in wind is

$$P = \frac{1}{2}\rho A V^3$$

$$= \frac{1}{2}(1.24)(78.54)(10)^3$$

$$= 48,695\ \text{W}.$$

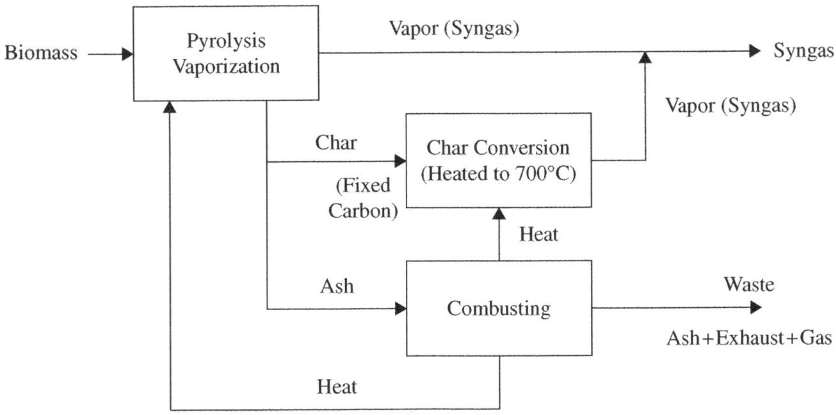

Figure 9.7 Biomass-gasification process.

If the turbine-power coefficient C_p is 0.2, then the amount that will be converted to usable electric power is

$$\begin{aligned} P &= 48,695C_p \\ &= 48,695\,(0.20) \\ &= 9,739\text{W}. \end{aligned}$$

9.4.4 Biomass Bioenergy

Bioenergy is the energy derived from biomass organic matter such as corn, wheat, soybeans, wood, and residues that can produce chemicals and materials that we normally get from petroleum. Bio-power is product of *biomass gasification*, which converts biomass to a gas that can be used to power a turbine and generate electricity. The biomass-gasification process is shown in Figure 9.7. The energy-conversion process for biomass also utilizes pyrolysis oil, produced when biomass is converted directly into fluid fuel. The most common fuels are ethanol alcohol or biodiesel derived from corn ethanol.

Biomass power plants are commercially available in the United States for up to 11 GW of installed capacity. Biomass power ranges from 0.5 to 3.0 GW using landfill gas and forest products, respectively. Biomass has traditionally been used for domestic cooking and heating, and such use is still widely practiced in developing countries.

Advantages and Disadvantages of Biomass Power

Biomass power is viable only when a sufficient quantity of bioproducts is available and a conversion process is done. Truly continuous applications are likely for biomass systems, and it appears to be a good fit for CHP applications. Since the output of these units cannot be controlled, they are not suitable for peaking applications. Biomass

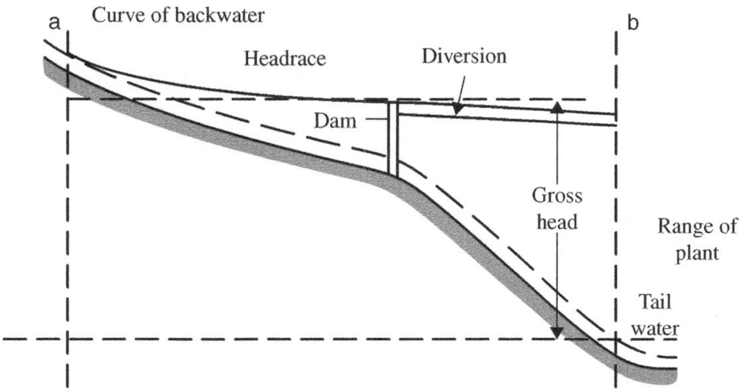

Figure 9.8 Diversion type hydropower plant.

power is not a premium power due to the limited availability of bioproducts. It is also not an ideal green power due to the emission of CO_2 and other pollutants. Several research efforts are under way to improve the quality of biomass power and reduce its environmental impacts.

The form of energy input is very inexpensive. However, the efficiency of biomass power is low (typically less than 20 percent), and is a relatively expensive way to produce electricity compared with PV. Several advances in technology are being used to reduce CO_2 emissions and improve the green-power nature of biomass fuels. The success of biomass energy depends on the continuity of fuel supplies.

9.4.5 Hydropower

Hydropower or waterpower is power derived from the energy of falling water or fast-running water, which may be harnessed for useful purposes. Since ancient times, hydropower from many kinds of water mills has been used as a renewable-energy source for irrigation and the operation of various mechanical devices, such as grist-mills, sawmills, textile mills, trip-hammers, dock cranes, domestic lifts, and ore mills. A trompe, which produces compressed air from falling water, is sometimes used to power other machinery at a distance.

There are three types of hydropower facilities.

- **Impoundment or dams**: Some hydropower plants use dams and some do not. Many dams were built for other purposes and hydropower was added later. In the United States, there are about 80,000 dams, of which only 2,400 produce power. The other dams are for recreation, stock/farm ponds, flood control, water supply, and irrigation.
- **Diversion**: A diversion, sometimes called a *run-of-river facility*, channels a portion of a river through a canal or penstock. It may not require the use of a dam. Diversion type plant is shown on Figure 9.8.

- **Pumped storage**: Pumped storage works like a battery, storing the electricity generated by hydropower and other sources (such as solar, wind, and nuclear) for later use to consumers during peak energy demand. Hydro-pump storage is explained further in Chapter 10.

Hydroelectric Plant Size

Hydropower plants range in size from small systems for a home or village to large projects producing electricity for utilities. The size of a hydroelectric power plant can also be categorized in terms of its power-generating capacity and the overall head of the system.

(a) Power-generating capacity

Facilities range in size from large power plants that supply many consumers with electricity to small and micro plants that individuals operate for their own energy needs or to sell power to utilities. Although definitions vary, the US Department of Energy has defined the various categories of hydropower plants as follows:

- Large hydropower facilities have a capacity of more than 30 MW.
- Small hydropower facilities have a capacity of 100 kW to 30 MW.
- Micro-hydropower plants have a capacity of up to 100 kW and may be designed for producing electricity for a home, farm, ranch, or village.

Once we have the flow rate and the head figures, we can roughly estimate the potential power available in kW with the following formula:

$$P_{avail} = \gamma \eta Q H \tag{9.25}$$

where Q is the discharge, m^3/sec, H is the net head, η is the system efficiency, and γ is the specific weight of water, kN/m^3.

(b) System overall head

Hydroelectric sites are broadly categorized as low- or high-head sites. Low head typically refers to a change in elevation of less than 10 feet (3 m). A vertical drop of less than 2 feet (0.61 m) will probably make a hydroelectric system unfeasible. A high flow rate can compensate for low head, but a larger and costlier turbine will be necessary.

Environmental and climatic factors as well as human activities in the watershed determine the amount and characteristics of stream flow on a day-to-day and seasonal basis. A storage reservoir can control flow but unless a dam already exists, building one can greatly increase cost and legal complications.

Problems Related to Hydropower

The following are some of the problems related to hydropower development:

- Unused rivers with good potentials are often far away from cities or in a low-population areas.

- Since water is highly essential to life, projects of this nature often generate disputes.
- The use of rivers or streams for hydropower often competes with recreational uses and environmental concerns.
- Displacement of wildlife is usually a fallout of hydropower projects.
- Flooding and submerged lands can become a problem if the project is not properly designed and managed.
- Siltation of the dam over time often limits the useful life of the dam reservoirs.

Example 5
During the design stage of a small hydropower station, the following data were collected for the site:

- average water flow rate = 23,000 tons/s
- net height of proposed impoundment = 75 m

If the combined efficiency of the water conveyor system from dam to turbine, turbine system, and generator is estimated at 0.77, estimate the capacity of the system. Let $g = 9.82\ m/s^2$.

Solution:

$$\begin{aligned}\text{Specific weight of water} = mg &= 1,000 \times 9.82 \\ &= 9,820\ \text{N/m}^3 \\ \text{Volumetric flow rate} &= \frac{23,000}{1,000} = 23\ \text{m}^3/\text{s} \\ P_{\text{avail}} &= \gamma\eta QH \\ P_{\text{avail}} &= 0.77 \times 9,820 \times 23 \times 75 \\ &= 13.04\ \text{MW}\end{aligned}$$

9.4.6 Fuel Cells

The fuel cell was first developed by Sir William Grove in 1839 and put to practical use in the 1960s by NASA to generate fuel for electricity needed by the Apollo and Gemini spacecraft. Fuel cells are quiet, clean, and highly efficient on-site generators of electricity that use the electrochemical process to convert fuel into electricity. In addition to generating electricity, fuel cells can also serve as a thermal-energy source for water and space heating or for cooling absorption.

Fuel cells can run using hydrogen, natural gas, methanol, or gasoline. The efficiency for conversion of fuel to electricity can be as high as 65 percent, as it does not depend on Carnot limits. This efficiency is what makes fuel cells environmentally friendly. Fuel cells come in a variety of different forms, all of which are under development. Examples include phosphoric acid fuel cells (PAFC), proton-exchange membrane (PEM) cells, solid-polymer molten carbonate fuel cells (MCFC), solid oxide fuel cells (SOFC), alkaline (a direct methanol) fuel cells, regenerative fuel cells,

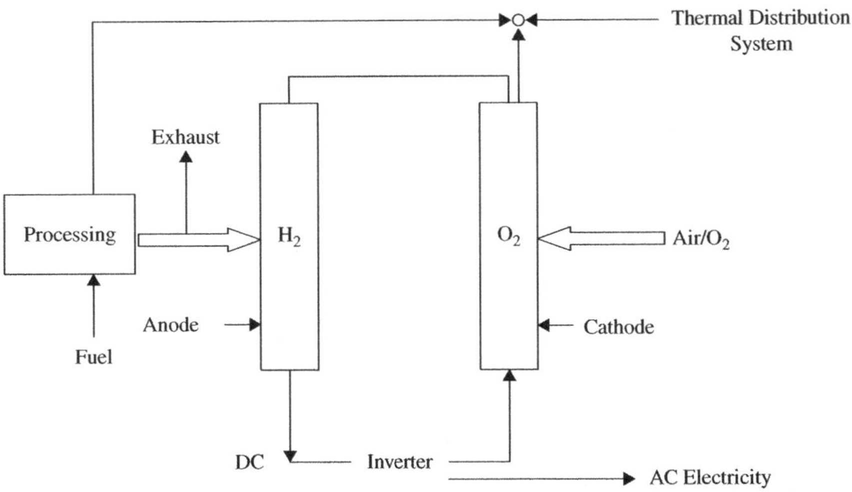

Figure 9.9 Schematic diagram of fuel-cell system.

and botanical ceramic fuel cells (BCFC). Fuel cells produce virtually no emissions of air pollutants or greenhouse gases. However, their costs are significantly higher than those of conventional technologies.

Operation of Fuel Cells

Although fuel cells use different types of fuels, they operate using the same basic principle. A fuel cell consists of two electrodes: an anode and a cathode, separated by an electrolyte, as seen in Figure 9.9. Through the hydrogen catalyst, atoms split into a proton H + and an electron, and the proton passes through the electrolyte to the positive cathode. A DC current results. Using a converter, we can easily generate an AC current. The combined hydrogen and oxygen at the cathode produce water and heat.

Details on the differences between fuel cells are based on materials and manufacturing costs, operating temperature, efficiency, and power-to-volume (weight) ratio. Additional models of fuel cells and their distinguishing features are available in the literature.

The topology of a fuel cell is defined as a stack that consists of the part of the fuel cell that holds the electrodes and the electrolytic material. Hydrogen is extracted from gasoline, propane, and natural gas refineries to operate commercial fuel cells. Emissions from fuel cells are very low, so they have minimal environmental impacts. Their high efficiency leads to lower fuel costs and minimal maintenance due to a lack of moving parts. They have virtually no pollutant emissions, and CO_2 is rather low. Fuel cells are a good fit for green power and premium power. In addition, they provide a moderately high thermal-quantity output and hence are ideal for CHP applications. However, they perform poorly as peaking power due to extremely high capital cost. Developmental research work to refine the effectiveness of fuel cells is receiving greater attention.

Fuel cell efficiency ranges from 40 to 80 percent. Two of the most commonly used fuel cell types are PAFCs, which operate at relatively high temperature and use an external water-cooling system to cool the stack and PEM cells, which operate at a lower temperature than most of the other fuel cells and contain no chemicals such as liquid acids or molten bases that would cause concerns about the materials of construction.

Power and Energy Relationships

$$Power : P = VI \tag{9.26}$$

$$\text{Energy} : E = VI \cdot t \tag{9.27}$$

The energy of chemical input and output, i.e., of H_2, O_2, and H_2O, is defined as a chemical energy given as enthalpy, Helmholtz function, or Gibbs free energy. Fuel-cell energy can also be expressed in terms of calorific value. However, Gibbs free energy (G_f) is the preferred measure of fuel-cell energy, defined as the energy to do external work, neglecting any work done by changes in pressure or volume. Gibbs energy represents the external work involved in moving electrons around an external circuit.

G_f energy is used to represent the zero-energy point, and a change in G_f is given by

$$\Delta G_f = G_{f_{product}} - G_{f_{reactant}}. \tag{9.28}$$

Consider $2H_2 + O_2 \rightarrow 2H_2O$, which is equivalent to the following (after chemical interaction) as

$$H_2 + \frac{1}{2}O_2 \rightarrow H_2O \tag{9.29}$$

where the new product is one mole of H_2O. Reactants are 1 mole of H_2 and 1/2 mole of O_2, which gives a balance of energy

$$\Delta \bar{g}_f = \bar{g}_f \ of \ products - \ \bar{g}_f \ of \ reactant \tag{9.30}$$

i.e.,

$$\Delta \bar{g}_f = (\bar{g}_f)_{H_2O}o - (\bar{g}_f)_{H_2} \frac{1}{2} \left(\bar{g}_f\right) O_2. \tag{9.31}$$

However, the G_f of the function is not constant. It changes with temperature and the state of the liquid or gas.

Continuing the calculation in terms of electrical work, we get 2N electrons around the equivalent fuel-cell circuit. N is the Avogadro number, i.e., the charge on one electron.

$$-2Ne = -2F \tag{9.32}$$

where F is the Faraday constant or charge on 1 mole of electron.

$$\text{Electrical work done} = \text{charge} \times \text{voltage}$$

$$= 2F \cdot E\ joules \tag{9.33}$$

$$\therefore\ \Delta\bar{G}_f = -2F \cdot E \tag{9.34}$$

$$E = -\frac{\Delta\bar{G}_f}{2F} \tag{9.35}$$

This equation gives the fundamental emf = reversible open-circuit voltage of the fuel cell.

9.4.7 Ocean Energy

Ocean energy is another type of renewable energy. The ocean can produce thermal energy from the sun as well as heat and mechanical energy from the tides and waves. Ocean thermal energy has a variety of applications involving electricity generation. It uses a simple conversion from warm surface water or boiled seawater to turn a turbine that activates a generator. The conversion of ocean power to electricity involves heavy use of a mechanical turbine. A dam is usually used to convert sea-power energy to electricity. Active research on ocean energy is under development in the Pacific West of the United States and in Europe.

Supply Characteristics

The energy available from the ocean's surface-wave motion is almost unlimited, but it has proved difficult to capture in any appreciable quantity. Many ingenious systems have been proposed but, except for small installations, very few are generating electricity commercially and most have been thwarted by practical problems.

Some of these proposals are outlined below. Most are still in an experimental phase and many are not scalable into high-capacity systems.

(a) **Oscillating float systems** are one of the simplest and most common solutions. A float is housed inside a cylinder-shaped buoy, which is open at the bottom and moored to the seabed. Inside the cylinder the float moves up and down on the surface of the waves as they pass through the buoys. Various methods have been employed to turn the motion of the float into electrical energy. These include:

- In hydraulic systems, air is compressed in a pneumatic reservoir above the float during its upward movement on the crests of the waves. After the crests have passed, the air expands and forces the float downward into the following troughs of the waves. A hydraulic system then uses the reciprocating movement of the float to pump water through a water turbine, which drives a rotary electrical generator.
- In pneumatic systems, air displaced in the cylinder is used to power an air turbine, which drives the generator.

- Linear generators turn the reciprocating motion of the float directly into electrical power.
- Instead of generating the electricity on board the buoy, some systems pump the hydraulic fluid ashore to power shore-based generators.

(b) Oscillating paddle systems use large paddles moored to the ocean floor to mimic the swaying motion of sea plants in the presence of ocean waves. The paddles are fixed to special hinged joints at the base that use the swaying motion of the paddles to pump water through a turbine generator.

(c) Oscillating snake systems use a series of floating cylindrical sections linked by hinged joints. The floating snake is tethered to the seabed and maintains a position head-on into the waves. The wave-induced motion at the hinges is used to pump high-pressure oil through hydraulic motors via smoothing accumulators. The hydraulic motors in turn drive electrical generators to produce the electrical power.

(d) Overtopping wave systems use reservoirs filled by wave actions where the collected water is released to drive turbine and generator in same way as hydro power plants does. They focus waves onto a tapered ramp, which causes their amplitude to increase. The crests of the waves overtop the ramp and spill into a low dam. Water from the low dam then flows through hydroelectric turbines back into the sea beneath the floating structure.

(e) Lever-based systems have been developed that use long levers mounted onto steel piles or floating platforms. Large floats or buoys are attached to the extremities of the levers, which move up and down with the waves. Movement of the lever arms forces fluid into a central hydraulic accumulator and through to a turbine-generator system. Alternatively, high-pressure water can be pumped ashore to power shore-based generators.

(f) Oscillating water columns are formed within large concrete structures built on the shoreline or on rafts. The structure is open at both the top and the bottom. The lower end is submerged in the sea and an air turbine fills the aperture at the top. The rising and falling of the water column inside the structure moves the air column above it, driving the air through the turbine generator. The turbine has movable vanes that rotate to maintain unidirectional rotation when the movement of the air column reverses. Figure 9.10 shows a diagram of an oscillating water-column system.

(g) Wave capture systems use a narrowing ramp to funnel waves into an elevated reservoir, as shown in Figure 9.11 [8]. Waves entering the funnel over a wide front are concentrated into a narrowing channel, which causes the amplitude of the waves to increase. The increased wave height coupled with the momentum of the water is sufficient to raise a quantity of water up a ramp and into a reservoir situated above the sea level. Water from the reservoir can then be released through a hydroelectric turbine located below the reservoir to generate electricity.

(h) Pressure transducer systems use a submerged, gas-filled tank with rigid sides and base and a flexible, bellows-like top. The gas in the tank compresses

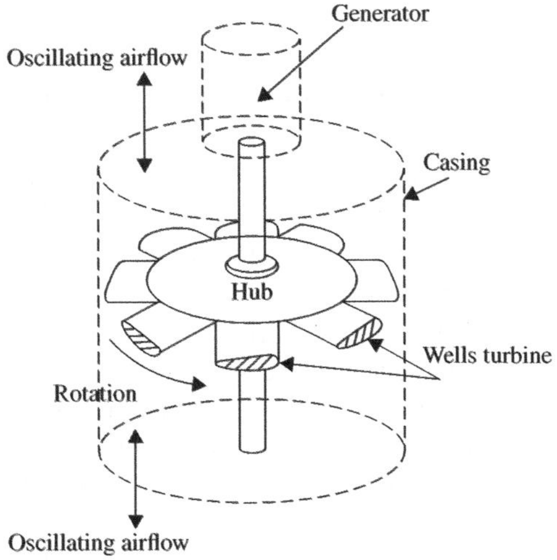

Figure 9.10 Oscillating water-column system.

and expands in response to pressure changes from the waves passing overhead, causing the top to rise and fall. A lever attached to center of the top drives pistons, which pump pressurized water ashore for driving hydraulic generators.

Technical Challenges and Solutions

Formidable technical challenges are involved in designing practical systems for capturing wave energy.

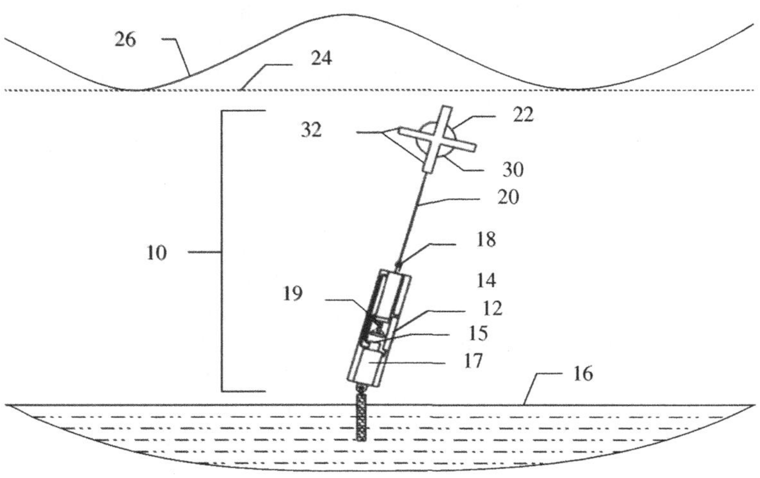

Figure 9.11 Wave-capture system.

- **Variability of sea conditions:** Sea conditions are notoriously variable and the system must be able to cope with a wide range of wave amplitudes and frequencies as well as changes in the directions of currents.
- **Matching the generating equipment with the wave characteristics:** Mechanisms are required to convert the power of the irregular oscillating mechanical forces induced by the waves into electrical power synchronized with the grid. This could involve some expensive power electronics. Typical rotating machines used for power generation operate at a synchronous speed of 1,200 rpm (20 rps) whereas the frequency of waves driving the generator is likely to be between five and ten cycles/second. A mechanical gearing system is needed to match this 200:1 ratio in operating speeds, possibly combined with special purpose, slow-speed generators, incorporating a large number of pole pairs. One way around these problems is to use hydraulic accumulators either in situ or on shore to smooth out the energy delivery to the generator.

A number of solutions have been proposed to solve these technical challenges.

- **Equipment construction:** For reasonably sized systems, very high mechanical forces will be involved in converting the wave energy into mechanical energy for driving the electrical generator.
- **Housing and mooring the equipment**: Substantial housings must be provided to protect the generating equipment from the harsh ocean environment. Holding the installation in place is also particularly difficult in deep water.
- **Energy transmission**: Low-loss armored and insulated cables or high-pressure pipes must be developed for delivering the electrical or hydraulic energy back to the shore.
- **Resistance to storm damage:** Storm damage is a major threat. The frequency of occurrence of waves of any particular amplitude follows a Rayleigh distribution similar to that which applies to wind speeds. Though the frequency of serious storms may be rather small, a wave of ten times the average amplitude may be expected once every fifty years. From the power calculation detailed in the next subsection, the wave power is proportional to the square of the wave amplitude. This means that the installation must be designed to withstand forces one hundred times greater than the normal working level. This adds considerably to the costs.

Available Power

The wave power per unit length of the wave front P_L is given by

$$P_L = \rho g a^2 \lambda / 4T \,(\mathrm{W/m}) \tag{9.36}$$

where ρ is the density of the water (10^3 kg/m^3), a is the wave amplitude (half of the wave height), g is the gravitational constant (10 m/sec^2), λ is the wavelength of the oscillation, and T is the period of the wave.

Example 6

Estimate the power per meter in a wave with amplitude 1.82 m, length 82.15 m, and period 3.5 seconds. Take density of seawater as 1,029 kg/m^3 by

$$
\begin{aligned}
P_L &= \frac{\rho g a^2 \lambda}{4T} \\
&= \frac{1,029 \times 10 \times 1.82^2 \times 82.15}{4 \times 3.5} \\
&= 200\,\text{kW/m}
\end{aligned}
$$

9.4.8 Geothermal Heat Pumps

Another minor source of renewable energy is the geothermal heat pump. This form of power is based on accessing underground steam or hot water from wells several miles into the earth. Pumping hot water to drive conventional steam turbines, which drive the electrical generator to produce electrical power, does the conversion. The water is then recycled back to earth to recharge the reservoir for a continuous energy cycle.

There are several types of geothermal power plants, namely dry steam, flash steam, and binary cycle. Dry-steam plants draw water from the reservoirs of steam, while both flash-steam and binary-cycle plants draw their energy from the recycled hot water reservoir.

Geothermal power is currently under development in the United States, and some reasonable levels of power have been produced in California, Utah, Nevada, and Hawaii. Various applications of geothermal power exist, such as heat pumps, agricultural applications, fishing farms, and food processing. Geothermal projects force significant up-front capital investment for exploration, drilling wells, and capital equipment. Exploration risks and environmental impacts are also considered in geothermal power plant projects.

9.4.9 Microturbines and Stirling Engines

Microturbines are a new generation of gas turbines that are small in size, typically producing between 25 and 500 kW of power. The technology is derived from the auxiliary power systems used in aircraft, diesel engines, turbochargers, and automotive designs. It consists of a compressor, combustor, turbine, and generator, as shown in Figure 9.12.

Description

Incoming air is compressed to about 3 atm of pressure and sent to a heat exchanger called a recuperate, where hot exhaust gases raise the temperature. The heated steam is mixed with fuel and burned with enough energy to drive the turbine, which subsequently powers the electrical generator. The turbine has only one moving part, which drives the generator at 96,000 rpm on the air bearings and hence does not require lubrication and cuts down on operational/maintenance costs. The generator creates AC current but can be rectified for DC output as needed. It can easily be operated for 60 Hz and reverted to a 50 Hz supply.

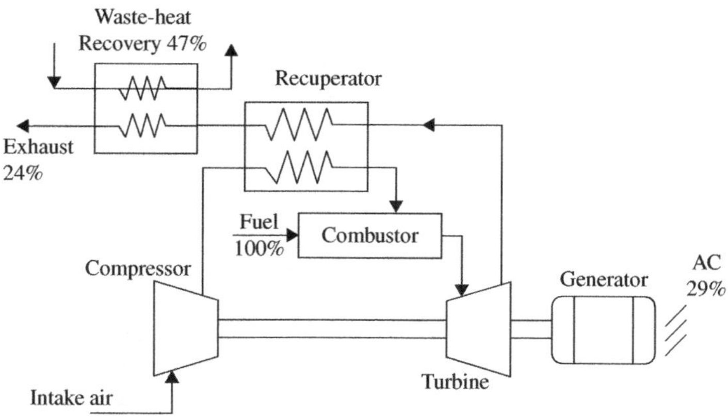

Figure 9.12 Schematic diagram of microturbine.

Microturbines can also easily be started in parallel for increased power output to 30 to 60 kW.

In general, microturbine emissions are comparable with those of large turbines. Manufacturers base NOx levels on field tests and projections. Emission control to achieve an acceptable standard is focused on combustion design and flame control.

Microturbines are moderately applicable for peaking power but can be used as a stand-alone in limited areas. Their inverter-based generators offer a high premium of power quality. If efficiency degrades as temperature increases, it leads to CO_2 emissions, which further degrade its efficiency. It is a moderate fit for CHP applications.

Further development of microturbines to lower costs due to electronics, power conditioning, and grid connection are concerns. Applications of fuel diversity (low BTU) and digester gas are under development. Microturbine efficiency is up to 30 to 40 percent for non-recuperated units and 20 to 30 percent for recuperated units. The available natural gas pressure level mostly impacts microturbine efficiency. Further work that hybridizes microturbines with fuel cells will facilitate the generation of additional electricity of up to 60 percent efficiency.

There are many models of microturbines produced by a number of manufacturers, e.g., Elliot Group, Ingersoll Rand, Honeywell, and Capstone Turbine Corporation. Microturbines are rated in terms of rated power, fuel input, heat rate, efficiency, emissions (NOx, CO_2), turbine rotation, weight, noise, and size.

9.4.10 Stirling Engine

Stirling engines (Figure9.13) [12] are classified as external combustion engines in which energy is supplied to the working fluid inside the engine from a source outside the engine. They are scaled systems with an inert working fluid, either helium or hydrogen. They are generally found in sizes of 1 to 25 kW.

The efficiency of stirling engines is typically less than 30 percent. Potential applications include use as small-scale portable power for battery chargers and as co-generators for electricity and thermal/cooling. Stirling engines have low emissions when natural gas is used.

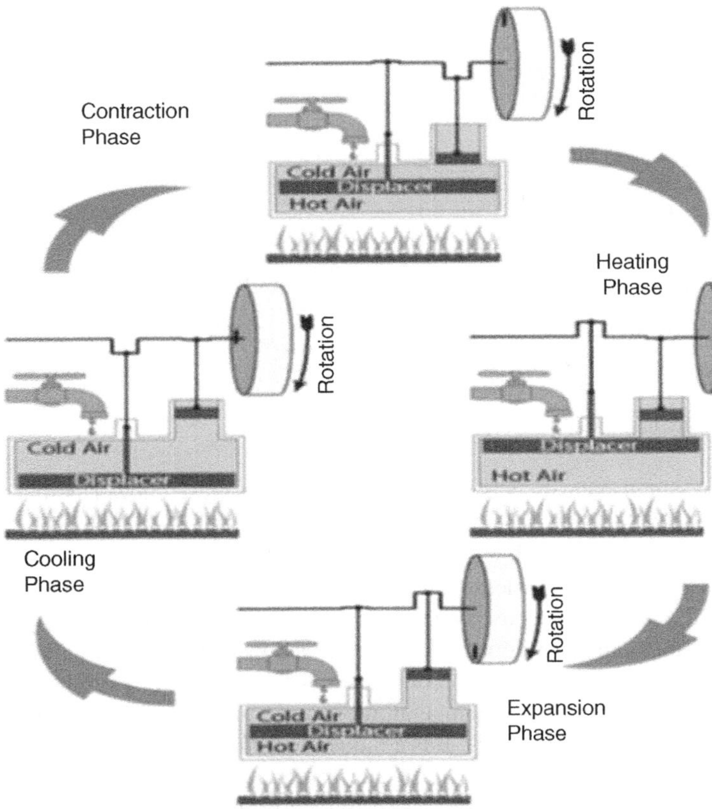

Figure 9.13 Stirling engine.

9.4.11 Summary of Renewable-Energy Classifications

Table 9.1 compares the various types of renewable energy we have reviewed in this section.

With the analysis of the energy needs complete, the next step is an evaluation of the power options available. This raises questions such as:

- Should power be generated on-site, or should power or a renewable-energy certificate (REC) be purchased from outside vendors?
- Which type of green power is suitable?

The answers to these questions would be individual to the user's circumstance and location. On-site generation brings with it the need for an up-front investment but also a long-term reduction in the consumption of conventional energy and increased reliability of the power supply. Limitations to options available to the consumer stem from factors such as the electricity market structure, which varies by state as well as the availability and quality of resources such as solar, wind, or biomass fuel.

Storage technologies are summarized on Table 9.2 as shown below.

TABLE 9.1 Summary of Renewable Energy Resources

Technology	Size Ranges (kW)	Efficiency (%)		Generation Electricity ($/MWh)	Environmental Issues/ Emission Control	Reliability Drawbacks on Fuels Source
Reciprocating diesel	30–5,000	26–43	85–90	7.1–14.2	Controls required for NO_X and CO_X	Yes
Turbines/ microturbines	5–10	20–30	60–75	11.9–18.9	Low impact	Yes
Fuel-cell PEM*	1–250	27–40	40–75	21.9–31.3	Nearly zero emissions	Yes
Photovoltaic (PV)	5–5,000	9–14	8–35	18.0–36.3	Zero direct emissions	No
Wind	5–1,000	20–40	20–26	0.2–28.5	Zero direct emissions	No
Biomass	20,000–50,000	5–10	5–20	0.05–0.09	Indirect emission	No
Geothermal	5,000–10,000	5–15	5–25	0.03–0.05	Low emission, only excess steam	No
Ocean	100–10,000	5–45	5–60	5–7	Zero direct emissions	No
Small hydro	100–1,000	50–55	60–90	0.03–0.25	Zero direct emissions	No

*PEM: Polymer electrolyte membrane

TABLE 9.2 List of Storage Technologies

Technologies	Properties
Solid-state batteries	• nano composite electrolyte • compact with thin-film stack
Thermal storage	• next-generation, non-intermittent, cost-competitive solar-thermal power plant • thermal energy from the sun stored in the form of chemical bonds
Flywheel	• high-energy density and reliability • rotational speeds higher than 100,000 rpm
Pumper hydro storage	• two water reservoirs at different elevations to pump water from the lower to the upper reservoir (charging) • efficiency ranges from 60 to 75 percent • long lifetime and practically unlimited cycle stability
Compressed-air energy storage	• air is compressed using electricity • compressed air is mixed with natural gas, burned and expanded in a modified gas turbine

9.5 RENEWABLE-ENERGY PENETRATION

Penetration refers to the proportion of a power source on a system, expressed as a percentage. There are several ways that this can be calculated, with the different methods yielding different penetrations. It can be calculated as:

- nominal capacity of a power source divided by peak demand
- nominal capacity of a power source divided by total capacity
- average power generated by a power source, divided by the average system demand

The level of penetration of intermittent variable sources is significant for the following reasons:

- Power grids with significant amounts of pumped storage, hydropower, or other peaking power plants—such as natural-gas-fired power plants—are more inherently capable of accommodating fluctuations from intermittent power.
- Isolated, relatively small systems with only a few wind plants may only be stable and economic with a lower fraction of wind energy (e.g., the system now used in Ireland), although mixed wind/diesel systems have been used in isolated communities with success at relatively high penetration levels.

Renewable electricity supply in the 20–50+ percent penetration range has already been implemented in several European systems, albeit in the context of an integrated European grid system. In 2010, four German states, totaling 10 million people, relied on wind power for 43 to 52 percent of their annual electricity needs. Denmark is not far behind, supplying 22 percent of its power from wind in 2010 (26 percent in an average wind year). The Extremadura region of Spain gets up to 25 percent of its electricity from solar, while the whole country meets 16 percent of its demand from wind. From 2005 to 2010, Portugal vaulted from 17 percent to 45 percent renewable electricity.

Discussion of acceptable or unacceptable penetration figures should be treated and used with caution, as the relevance or significance will be highly dependent on local factors, grid structure and management, and existing generation capacity.

9.6 MAXIMUM PENETRATION LIMITS OF RENEWABLE-ENERGY RESOURCES

There is no generally accepted maximum penetration of wind energy that would be feasible in any given grid. Rather, economic efficiency and cost considerations are more likely to dominate as critical factors; technical solutions may allow higher penetration levels to be considered in the future, particularly if cost considerations are secondary.

High-penetration scenarios may be feasible in certain circumstances:

- Retaining the existing power stations can provide power generation for periods of little or no wind generation. The cost of using existing power stations for this purpose may be low since fuel costs dominate the operating costs. The actual cost of paying to keep a power station idle but usable at short notice may be estimated from published spark spreads and dark spreads. As existing traditional plants age, the cost of replacing or refurbishing these facilities will become part of the cost of high-penetration wind energy if they are used only to provide operational reserve.
- Automatic load shedding of large industrial loads and its subsequent automatic reconnection is established technology, used in the US—and in the UK by frequency-service contractors, which switch off and on several GW each month in this way. Reserve-service contractors offer fast-response gas turbines and even faster diesels in the UK, France, and the US to control grid stability.
- In a close to 100 percent wind scenario, surplus wind power can be allowed for by increasing the levels of the existing reserve- and frequency-service schemes and by extending the scheme to domestic-sized loads. Energy can be stored by advancing deferrable domestic loads such as storage heaters, water heaters, fridge motors, or even hydrogen production, and load can be shed by turning such equipment off.
- Additionally, power can be exported to neighboring grids and reimported later. High-voltage DC cables are efficient with 3 percent loss per 1,000km and may be inexpensive in certain circumstances. For example, an 8 GW link from UK to France would cost about £1 billion using high-voltage DC cables. Under such scenarios, the amount of transmission capacity required may be many times higher than currently available.

9.7 CONSTRAINTS TO IMPLEMENTATION OF RENEWABLE ENERGY

Due to high fuel and electricity prices and low supply reliability in developing countries, some renewable-energy technologies are more feasible in some places than others.

Another constraint facing renewable energy is the difficulty of adequate funding for renewable-energy projects, some of which can be very cost intensive (e.g., solar). Most developing countries cannot afford to finance significant renewable-energy projects.

In most developing countries, government policies are not supportive of renewable-energy development. There is therefore a need for policy review to favor renewable-energy development across the world.

Illustrative Problems and Examples

1. Consider a solar module rated 5.3 V, 38 W. If it is used to provide power to an application that needs 30 V and 25 A current, determine
   - (a) total number of solar modules required
   - (b) type of connection needed in terms of series and parallel-connected modules
   - (c) necessary load-resistance value to get the rated power

Solution:

Solar module rating 5.3 V, 38 W

Peak rating = no. of modules × rated power = 30 × 25 = 750 W

(a) 750 W = no. of modules × 38
no. of modules = 19.74 ≈ 20 modules

(b) No. of modules in series = 30 V/5.3 = 5.7 ≈ 6
no. of modules in parallel = 20/4 ~ 4
total number of modules in parallel = 24

(c) $38 = I^2R$, (where P = 38W)

$$R = 0.061\,\Omega$$
$$= 61\,\text{m}\Omega$$

2. Consider a wind turbine with blade radius 7.2m. At the location of installation, wind speed is 15 m/s for 3 hours and 25 m/s for another 3 hours. Determine the amount of energy that can be intercepted by the wind turbine

Solution:

Blade radius r = 7.2m

$V_1 = 15\text{m/s}, t_1 = 3hrs$

$V_2 = 25\text{m/s}, t_2 = 3hrs$

Power extracted from the wind $P = 0.5\rho AV^3 = 0.5\rho\pi r^2V^3$

$\rho = 1.25\text{kg}/m^3$

Energy $E = P \times t = 0.5\rho\pi r^2V^3t$

$E_1 = 0.5 \times 1.25 \times \pi \times 7.2^2 \times 15^3 \times 3 = 1,030.73\ \text{kWh}$

$E_2 = 0.5 \times 1.25 \times \pi \times 7.2^2 \times 25^3 \times 3 = 4,772\ \text{kWh}$

Amount of energy intercepted by the wind turbine = 1,030.73 + 4,772 = 5,802.7 kWh.

3. A wind turbine installed in a desert where the average wind speed is 5.5 m/s is to generate power for a small settlement. If the length of each turbine blade is 2.6 m, efficiency of the turbine-gear system is 0.62 while generator efficiency is 0.72. Determine whether the machine will be able to meet community needs if daily energy demand is 45,000 kWh.

Solution:

Radius $r = 2.6m$

Efficiency of turbine gear system $\eta_{tgs} = 0.62$

Efficiency of generator $\eta_{gen} = 0.72$

$\rho = 1.25\text{kg}/m^3$ (assumed)

Power extracted $P = 0.5\eta_{tgs}\eta_{gen}C_p\rho AV^3$
where C_p is the Betz limit taken to be 0.59
$P = 0.5 \times 0.62 \times 0.72 \times 0.59 \times 1.25 \times \pi \times 7.2^2 \times 15^3 = 581.62\text{W} = 0.581\text{kW}$
Daily energy produced $E = P \times t = 0.581 \times 24 = 13.96\,kWh$

If the community requires a daily energy of 45,000 kWh, then the daily energy produced by wind turbines is less than 45,000 kWh, hence the wind turbine cannot meet the energy needs of the community.

Chapter Summary Renewable-energy systems have attracted a lot of attention in the past decade owing to environmental friendliness and sources that are freely available in nature. In this chapter, we discussed some of the key renewable-energy resources and associated energy-conversion systems. Different evaluation criteria of renewable energy in terms of efficiency, cost, interoperability, etc., were given and serve as a guide for selecting which renewable energy to use in a typical microgrid design. This is important because of the roles of renewable-energy resources in future electric power systems. Because of their easy deployment on a stand-alone basis, renewable-energy resources are readily deployed in a microgrid.

EXERCISES

1. Consider the rooftop of a home measuring 15 × 20 m, which is covered with PV cells to provide the electrical energy required by that home. If the peak sun per day is four hours and the efficiency of a solar cell is 22 percent, find the average daily electric energy converted by the rooftop. If a unit of energy costs $0.1, what is the cost of energy harvested by the home in one year?

2. A circular cell has a diameter of 6cm. If its rating at 25° is 1,200mA and 0.45V in full sunshine, estimate the efficiency.

3. Compare the performance of an onshore and offshore wind turbine in terms of return on investment, assuming that the two turbines are the same in all respects. Consider all possible factors and make appropriate assumptions.

4. Consider a wind turbine that is rated at 100 kW in a 10 m/s wind speed in air at standard conditions. If power output is directly proportional to air density, what is the power output of the wind turbine in a 10 m/s wind speed at an elevation 2,000m above sea level at a temperature of 20°C?

5. Discuss the following subjects with respect to renewable energy:
   (a) sustainability of renewable energy
   (b) hybridization of renewable energy systems
   (c) disadvantages of renewable energy

BIBLIOGRAPHY

[1] "Renewable Energy Constraints and Opportunities: Analysis of London Borough of Redbridge," 2010, www.londonheatmap.org.uk.

[2] T. Gönen, *Electrical Machines with MATLAB*, 2nd ed., Taylor & Francis Group, Boca Raton, FL, 2011.
[3] Electropaedia, "Hydroelectric Power," www.mpoweruk.com/hydro_power.htm.
[4] J. Twiddel and T. Weir, *Renewable Energy Resources*, 2nd ed., Taylor & Francis Group, London, 2006.
[5] Practical Action, "Micro-Hydro Power," http://practicalaction.org/micro-hydro-power.
[6] US Office of Energy Efficiency & Renewable Energy, "Types of Hydropower Plants," http://energy.gov/eere/water/types-hydropower-plants.
[7] US Office of Energy Efficiency & Renewable Energy, "Power Tower System Concentrating Solar Power Basics," August 20, 2013, http://energy.gov/eere/energybasics/articles/concentrating-solar-power-tower-system-basics.
[8] C. B. Carroll, Surface Wave Energy Capture System, US Patent 6,229,225, filed May 5, 1998, and issued May 8, 2001.
[9] http//www.theeconomist.com/news/scienceandtechnology/Modern Stirling Engine.
[10] Sandia National Laboratories, "Floating Oscillating Water Column Reference Model Completed," February 24, 2015, www.energy.sandia.gov/floating-oscillating-water-column-reference-model-completed.
[11] Fichtner Management Consulting, *Hydroelectric Power: A Guide for Developers and Investors*, February 2015.
[12] B. H. Van Arsdell, *Around the World by Stirling Engine*, American Stirling Company, San Diego, CA, 2003.

CHAPTER 10

STORAGE SYSTEMS IN THE SMART GRID

10.1 INTRODUCTION

Remote locations or sites where there is no power supply are motivated by a continuous need for energy to make provisions for energy-storage devices such as rechargeable batteries and lamps—for the same reason that we make provision for a flask when out on a cold day. The human need for energy is extensive and continuous, hence energy is an important subject in a technology-driven world. That is why much attention is directed toward energy production. Energy may not always be available when and where needed and at the required quantity, making energy storage an important component of energy studies [1].

10.2 FORMS OF ENERGY

Energy exists either naturally or in a transformed form as heat, sound, light, kinetic, potential, gravitational, thermonuclear, magnetic, electric, chemical, and nuclear. These can all be categorized broadly as electrical energy, mechanical energy, chemical energy, thermal energy, and electrochemical energy. Today, interconversion of energy from one form to another is not only possible but is being improved upon by the day through continuous research and development efforts all around the world.

10.3 ENERGY STORAGE SYSTEMS

Energy storage is any means employed to store energy and ensure energy security in the future. Some energy storage technologies include:

- flywheel energy storage system
- pumped hydro energy storage system
- battery storage systems

Energy Processing and Smart Grid, First Edition. James A. Momoh

- supercapacitor or ultracapacitor energy storage system
- solar thermal energy storage system
- compressed air energy storage system
- superconducting magnetic energy storage system
- fuel cell energy storage system

The storage systems are discussed as follows:

(a) Flywheel Energy Storage System

A flywheel (Figure 10.1) [4] is a rotating mechanical device used to store rotational energy. Flywheels have a significant moment of inertia, and thus resist changes in rotational speed. The amount of energy stored in a flywheel is proportional to the square of its rotational speed. Energy is transferred to a flywheel by applying torque to it, thereby increasing its rotational speed and hence its stored energy. Conversely, flywheels release stored energy by applying torque to a mechanical load, thereby decreasing their rotational speed.

Mathematical Models of Flywheels

$$Ek = \frac{1}{2}I\omega^2 \tag{10.1}$$

where E_k is the kinetic energy, ω is the angular velocity, and I is the moment of inertia of the mass.

The moment of inertia of a solid cylinder is

$$\mathrm{I} = \frac{1}{2}\mathrm{mr}^2 \tag{10.2}$$

where m denotes the mass and r denotes the radius.

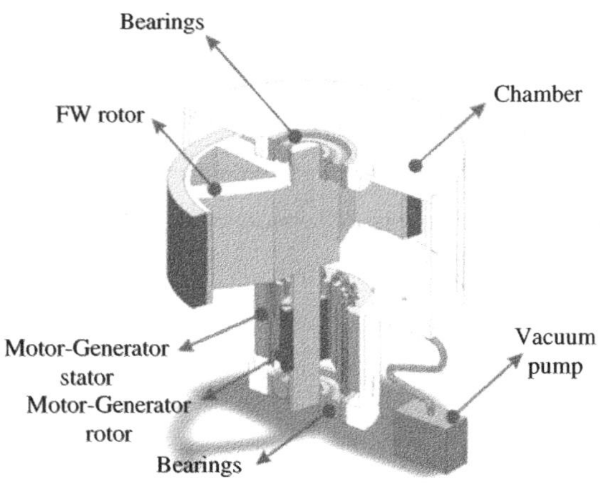

Figure 10.1 Flywheel.

The amount of energy that can safely be stored in the rotor depends on the point at which the rotor will warp or shatter. The hoop stress on the rotor is a major consideration in the design of a flywheel energy storage system.

$$\sigma_t = \rho r^2 \omega^2 \tag{10.3}$$

where σ_t is the tensile stress on the rim of the cylinder, ρ is the density of the cylinder, r is the radius of the cylinder, and ω is the angular velocity of the cylinder.

Applications of Flywheel Energy Storage

A flywheel is capable of providing continuous energy when the main energy source is disconnected. They are also used to store energy over time and then release it at a rate beyond the rate of the main source. Finally, they are used for controlling the orientation of the mechanical system. To control the orientation of mechanical systems, the angular momentum of a flywheel is transferred to a load when energy is transferred to and from the flywheel.

Advantages of Flywheel Energy Storage

One advantage of flywheel storage is that a flywheel cannot be affected by climate change; it can operate at a wide range of temperatures. Flywheel has indefinite working life span unlike lithium ion polymer batteries which operates for limited period. Flywheels are also less damaging to the environment because they is widely made of inert or benign materials, without a negative effect on the environment.

Disadvantages of Flywheel Energy Storage

One limit to the flywheel system is the tensile of the material used to build the rotor. A stronger disc will make it spin faster and store more energy. When the tensile strength of flywheel is exceeded the flywheel will shatter, releasing all of its stored energy once. Hence, it requires strong containment vessels as a safety precaution which increases total mass of the device.

Additional limitations for some flywheel types are energy storage time and energy loss.

Environmental Implications of Flywheel Energy Storage

Flywheels are potentially very environmentally friendly and efficient. In addition, they can be built with harmless materials and without hazardous chemicals. Energy efficiency is important in an energy system where there is still a deficit of renewable-energy sources. Flywheels are green and do not require cooling or ventilation like conventional battery systems, nor do they pose the same environmental hazards as batteries.

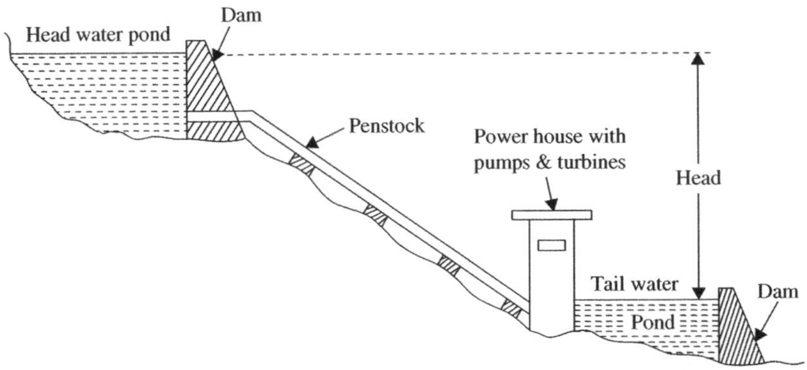

Figure 10.2 Pumped hydro energy storage system.

Flywheel Energy Storage Costs

Compared to batteries, flywheels generally have a high initial expense, but lower installation and maintenance costs.

(b) Pumped Hydro Energy Storage System

Hydro pump storage techniques (Figure 10.2 [8]) are mainly used for load balancing in a power system. These systems store energy by pumping water uphill to a reservoir at a higher elevation from a reservoir at the lower elevation.

When the demand for electricity is low, pumped hydro storage facilities store energy by pumping water from a lower reservoir to an upper reservoir. During periods of high electrical demand, the water is released back to the lower reservoir and turns a turbine, generating electricity. This principle is the same as a hydroelectric station. The water is released through the penstock to the blades of a turbine, which is connected to a generator to produce electricity. This type of technique has an overall efficiency of 75 to 85 percent.

Model and Formulas of Pumped Hydro Energy Storage

$$Q = n \,.\, Vstroke \,.\, \mathfrak{N}vol \tag{10.4}$$

where Q is the low in cubic meter second $\left[\frac{m3}{S}\right]$, n is the revolution per second $\left[\frac{rev}{S}\right]$, V_{stroke} is the swept volume in cubic meters [m^3], η_{vol} is the volumetric efficiency.

$$\text{Power } P = {nV_{stroke}\Delta p}/{\eta_{mech\ hydr}}$$

where P is the power in watts (Nm/s), n is the revs per second, V_{stroke} is the swept volume in m^3, Δp is the pressure difference over pump in N/m^2, and $\eta_{mech\ hydr}$ is the mechanical/hydraulic efficiency.

TABLE 10.1 Hydro Pump Capacity

Station	Country	Capacity (MW)
Bath County Pumped Storage Station	United States	3,003
Guangdong Pumped Storage Power Station	China	2,400
Okutaragi Hydroelectric Power Station	Japan	1,932
Ludington Pumped Storage Power Plant	United States	1,872
Tianhuangping Pumped Storage Power Station	China	1,836

Advantages of Pumped Hydro Energy Storage

Among the advantages of the hydro pump are that it can be used to store energy through pumped storage, the generation of electricity to meet demand can easily be adjusted, fuel is not burned so there is minimal pollution, water to run the power plant is provided free by nature, and with tidal power it can produce energy for up to ten hours a day.

Disadvantages of Pumped Hydro Energy Storage

It is important to mention that the hydro pump has disadvantages, including the large reservoirs required for the operation of hydroelectric power stations, which result in submersion of extensive areas upstream of dams, destroying biologically rich and productive lowland and riverine valley forests, marshlands, and grasslands. The loss of land is often exacerbated by habitat fragmentation of surrounding areas caused by the reservoir.

Size of Pumped Hydro Energy Storage Plants

Data on the five largest operational pumped-storage plants are shown in Table 10.1.

(c) Battery Storage Systems

A battery is a device that converts chemical energy into electrical energy. This is done by means of an electrochemical oxidation-reduction reaction of its active materials. This process involves the transfer of electrons from one material to another through an electric circuit. An oxidation-reduction reaction is defined as a reaction in which electrons are transferred. Oxidation means loss of electrons. Reduction is the process of accepting electrons.

The basic electrochemical unit is the "cell." A battery of any number of cells is used depending on the desired output voltage. In modern usage, a battery may refer to just one cell or a group of cells.

Types of Batteries

There are two types of batteries:

- primary batteries
- secondary or storage batteries

Primary batteries can provide only one continuous or intermittent discharge. Bringing together individual chemical components and assembling the battery in a charged state forms a primary battery. In the process of discharge these components are irreversibly changed. Electrical energy is obtained from chemical energy. Primary batteries cannot be recharged; they are used as a source of DC power for everyday items such as flashlights and transistor radios.

A secondary or storage battery is made of several chemical and elemental materials. These materials change during charging and discharging and this change is reversible. After the battery has discharged, it is brought back to a charged state by causing the current to flow back through the battery in the opposite direction. The electrodes are thus returned to approximately their original state.

Secondary batteries are used as a source of DC power when the battery is the main source of power and many discharge and charge cycles are required, such as in electrical vehicles, mine locomotives, or submarines, or when standby power is required, such as in telephone exchanges and emergency lighting. They are often used to supply large, time-repetitive power requirements such as car and airplane batteries. They are also used for load leveling of an electric power supply network.

Disposable batteries are intended to be used once and rechargeable batteries can be used multiple time. Batteries come in multiple sizes ranging from the small-size batteries used in wristwatches to large batteries used for power system applications such as backup power for data centers.

Batteries are made up of cells. A cell is composed of a positive pole and a negative pole immersed in a chemical medium where an exchange of charged particles takes place when the two poles are bridged through a load, e.g., a bulb. This exchange of charged particles is referred to as *electric current flow*. Various kinds of batteries available are shown in Figure 10.3.

Batteries in Series and Parallel Construction

Cells or batteries may be connected in a series or in parallel, shown in Figure 10.4. When connected in a series, the potential differences between the poles is the summation of the individual battery potential differences. Thus, if n batteries are connected in a series, the overall potential difference between the terminals is given by

$$V_T = V_1 + V_2 + V_3 + \cdots + V_n \text{ (in series)}. \tag{10.5}$$

If the potential difference of each of the batteries is equal,

$$V_T = nV. \tag{10.6}$$

For n batteries of equal potential difference connected in parallel, the overall potential difference is the potential difference across any one of the batteries, that is

$$V_T = V. \tag{10.7}$$

Note that it is unconventional to connect batteries of different potential differences together in parallel. If this is done, the battery with the lower potential difference drains down the battery with the higher potential until an equilibrium potential V_e is reached.

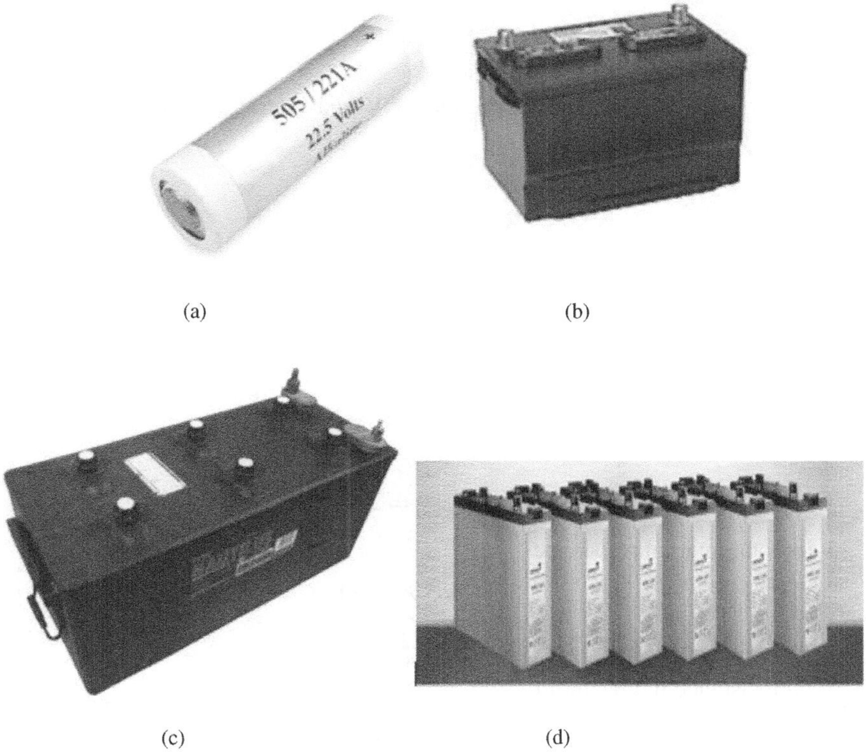

Figure 10.3 Different kinds of batteries. (a) Dry cell. (b) Automotive battery. (c) Deep-cycle battery. (d) Bank of deep-cycle batteries.

Battery Internal Resistance

Secondary cells possess internal resistance, and there is usually some voltage drop across this internal resistance. The practical approach to reducing the battery internal resistance is to increase the size of the cell plates. The limitation to this approach is overall battery weight and cost. Hence an optimal plate size is selected and the plates

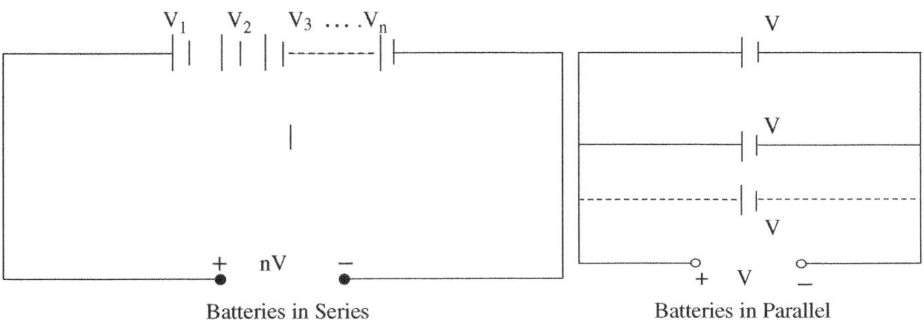

Figure 10.4 Batteries in series and parallel.

are connected in parallel, which amounts to connecting resistance in parallel with the overall effect of reduced internal resistance.

Example 1

Five batteries, each with potential difference 5 V and internal resistance 0.8Ω are connected in

(a) series

(b) parallel

with a load resistance of 10Ω. Calculate the current flowing through the load in each case.

Solution:

(a) For a series connection of batteries

$$V_T = nV = 5 \times 5$$
$$= 25V$$

with total internal resistance

$$R_T = 5 \times 0.8\,\Omega$$
$$= 4\,\Omega$$

total resistance in the circuit

$$(4 + 10)\,\Omega$$
$$= 14\,\Omega$$

and current flowing through the load

$$I = \frac{V}{R} = \frac{25}{14}$$
$$= 1.79A.$$

(b) For a parallel connection

$$V_T = V = 5V$$

$$R_T = n\frac{1}{R_i} + \frac{1}{R_L}$$
$$= 5\left(\frac{1}{0.8}\right) + \frac{1}{10}$$
$$= 6.35\,\Omega$$

with current flowing through the load

$$I = \frac{V}{R} = \frac{5}{6.35}$$
$$= 0.79\text{A}.$$

If the load resistance were connected in series with the parallel connection of batteries, then the overall circuit resistance is

$$= 5\left(\frac{1}{0.8}\right) + 10$$
$$= 16.25\,\Omega.$$

Hence current flowing through the load

$$I = \frac{V}{R} = \frac{5}{16.25}$$
$$= 0.31\text{A}.$$

Example 2

A battery with terminal voltage 12.5 V and internal resistance 1.2Ω is charged with a 96 V DC source. If a charging current of 3.5A is desired, calculate:

- **(a)** size of the series resistance to be connected
- **(b)** how much energy is expended in charging up the battery if it is charged up in eight hours
- **(c)** charging cost if a unit of electricity is $0.15

Solution:

(a) Net charging voltage = (96−12.5) = 83.5 V

is series resistance is R Ω

total resistance = (R + 1.2) Ω

and charging current $I = \frac{V}{R}$

$$3.5 = \frac{83.5}{R + 1.2}$$
$$= 22.66\,\Omega.$$

(b) Energy expended for charging

$$E = VIt$$
$$= 83.5 \times 3.5 \times 8$$
$$= 2,338\ \text{Wh}$$
$$= 2.338\ \text{kWh}$$

(c) Energy cost

$$= 2.338 \times 0.15$$
$$= \$0.35\ \text{cents}$$

Battery Capacity and Specifications

The capacity of a battery is the product of the current in amperes and the time during which it can supply power until the voltage falls to a particular value V_o, usually specified by the battery manufacturer. Usually, V_o is 1.8 V/cell. Battery capacity is measured in ampere-hours (AH). Batteries are usually specified in terms of capacity and terminal voltage. Thus, a battery rated as 200AH, 12 V can supply 200A continuously for one hour, 100A continuously for two hours, 50A continuously for four hours, and so on.

Battery Efficiency

Battery efficiency is defined as the ratio of the discharge AH to the charging AH:

$$\eta_b = \frac{\text{Discharging Amp} - \text{Hour}}{\text{Charging Amp} - \text{Hour}}$$

$$\eta_b = \frac{E_d}{E_c} \tag{10.8}$$

Battery State of Charge and Depth of Charge

State of charge (SOC) and depth of charge (DOC) are variables that can describe a battery's charge condition. The main difference is that the SOC describes the percentage of remaining charge relative to the nominal capacity, while the DOC represents that relative to the actual capacity under a specific discharge current. These two variables can be obtained by calculating the charge consumed and the battery capacity:

$$SOC = 1 - \frac{Q_e}{C} \tag{10.9}$$

$$DOC = 1 - \frac{Q_e}{C_I} \tag{10.10}$$

where Q_e, C, and C_I are charges consumed by load, battery nominal capacity, and actual capacity under discharge current, respectively.

Depth of Discharge

Depth of discharge (DOD) is the percentage of a storage device's useful capacity that has been depleted. For instance, a 90 percent DOD implies that 10 percent of the battery's energy remains.

Self-Discharge Time

Self-discharge is the time required for a fully charged, non-interconnected battery to reach a DOD. The relationship between self-discharge time and DOD is rarely linear so self-discharge times must be measured and compared at a uniform DOD.

Cycle Life

A battery's cycle life is the number of consecutive charge/discharge cycles a storage installation can undergo while maintaining the installation's other specifications within certain limited ranges. Cycle-life specifications are made against a chosen

DOD depending on the application of the battery. Cycle life is a measure of how long the battery will last before it needs to be replaced.

Battery-Specific Power

Specific power, in units of power per mass, is perhaps the most important parameter measuring electrical energy storage requirements for hybrid electric vehicles. Because hybrids normally depend on the electrical energy stored onboard to provide the power for accelerations and hill climbs, the higher the power for less mass the better. Hence, high-specific-power batteries are critical to the success of hybrid electric vehicles.

Battery-Specific Costs

Battery-specific costs refers to the cost per unit of energy (kWh) of battery and are related to the economic viability of the battery. Because lead-acid batteries have been around the longest and are the most fully developed battery technology, they have the lowest specific cost at around $125/kWh currently with projections to decrease to $75 with bipolar batteries. Currently, nickel–cadmium and nickel–metal hydride batteries are four to five times more expensive than lead-acid batteries.

Battery Model

In recent years, much of the development of electric energy storage systems have focused on battery storage devices. Presently, a wide range of batteries is commercially available with others in design stages. The most commercially developed type is the lead-acid battery. The first commercially available battery was the flooded lead-acid. However, the valve-regulated lead-acid (VRLA) battery, which is the latest commercially available option, requires low maintenance, is spill and leak proof, and is relatively compact. The development of battery performance models has been under active investigation. Of particular interest here is the kinetic battery model, which consists of three parts:

- capacity model
- voltage model
- lifetime model

The capacity model assumes a first-order chemical rate process; the voltage model is based on the adaptation of the battery energy storage test model in combination with capacity estimates from the capacity part of the model; and the lifetime model initially uses the assumption that the number of cycles a battery can tolerate is a function only of the depth of discharge of the cycles, with each cycle starting with a fully charged battery.

The capacity model is given by:

$$q_{max}(I) = \frac{q_{max,0}kcT}{1 - e^{-kT} + c(kT - 1 + e^{-kT}} \tag{10.11}$$

where $q_{max,0}$ is the maximum capacity (at infinitesimal current), k is the rate constant, and C is the ratio of available charge capacity to total capacity.

The voltage model is given by

$$E = E_o + AX + \frac{CX}{(D - X)} \tag{10.12}$$

where E_o is the fully charged/discharged internal battery voltage (after the initial transient); A is the parameter reflecting the initial linear variation of internal battery voltage with state of charge (A is reckoned as positive during charging and negative during discharging); C is the parameter reflecting the decrease/increase of battery voltage when battery is progressively discharged/charged; D is a parameter reflecting the decrease/increase of battery voltage when the battery is progressively discharged/charged, and is normally approximately equal to the maximum capacity; and X is the normalized maximum capacity at the given current.

The normalized maximum capacity X in charging is defined in terms of the charge in the battery by:

$$X = \frac{q}{q_{\max}(I)}. \tag{10.13}$$

While in discharging it is

$$X = \frac{q_{\max}(I) - q}{q_{\max}(I)}. \tag{10.14}$$

The voltage model reflects the observation that terminal voltage depends on:

- state of the battery (charging or discharging)
- state of charge of the battery
- internal resistance of the battery
- magnitude of the charging and discharging current

Lifetime Model

The lifetime model uses a double exponential curve fit of data to characterize cycle failures versus cycle depth. The equation used is of the form

$$C_F = a_1 + a_2 e^{-a_3 R} + a_4 e^{-a_5 R} \tag{10.15}$$

where C_F is the cycle of failure, a_1 is the fitting constant, and R is the range of cycle (fractional depth of discharge normalized using $q_{max,0}$).

Benefits of Battery Storage

Storage batteries have many advantages, depending on the type of batteries. For example, low-internal-resistance batteries (around 0.001ohm) can rapidly deliver high initial currents, which can be used for starting large devices such as car engines. Another benefit is the ability to recharge batteries; it is convenient for users to have energy stored in a battery later use but also be able to recharge the battery multiple times. Rechargeable batteries are also beneficial for the environment—rechargeable

TABLE 10.2 Costs of Different Battery Technologies

Battery Type	Cost ($ per Wh)	Wh/kg	Joules/kg	Wh/liter
Lead-acid	$0.17	41	146,000	100
Alkaline long-life	$.019	110	400,000	320
Carbon-zinc	$0.13	36	130,000	92
NiMH	$0.99	95	340,000	300
NiCad	$1.50	39	140,000	140
Lithium-ion	$0.47	128	460,000	230

batteries will decrease the millions of batteries used and discarded every year and save the environment from the chemical products used in batteries.

Environmental Concerns

One of the main environmental concerns is the toxic metals and hazardous chemicals used in batteries. Many countries are now recycling or using proper disposal to prevent dangerous elements such as lead, mercury, and cadmium from penetrating the environment.

Costs of Battery Storage

The costs of battery storage depend on the types of batteries, as shown in Table 10.2.

Battery Storage and Renewable Energy

Battery storage is used as a complement to renewable-energy sources (such as solar PV) to provide power when renewable energy is not available. For example, in a solar power system maintenance-free batteries store solar energy to be used as needed. Grid power systems are another area of renewable energy where batteries are used to store energy for peak hours, especially at night.

A system configuration for a solar PV battery storage system is shown in Figure 10.5. It includes a maximum power tracker, battery subsystem, power-conditioning system, switchgear, and structural/mechanical system for transmission of power. The PV array consists of fixed PV modules that use large-area, solid-state semiconductor devices to convert ultraviolet rays from the sun into DC power. The power-conditioning subsystem includes an inverter system, which converts DC power to AC power, and a step-up transformer, which raises the AC voltage to system-transmission-level voltage. While transmission of power is taking place, the battery storage system is simultaneously being charged up. At sundown or during periods of heavy cloud cover, when solar insolation is drastically reduced and power supply from the system drops below an acceptable threshold, the battery automatically switches on to supply stored DC power, which goes through the same process of conversion to AC, transformation, and transmission. This way, the battery storage system is able to ensure power reliability.

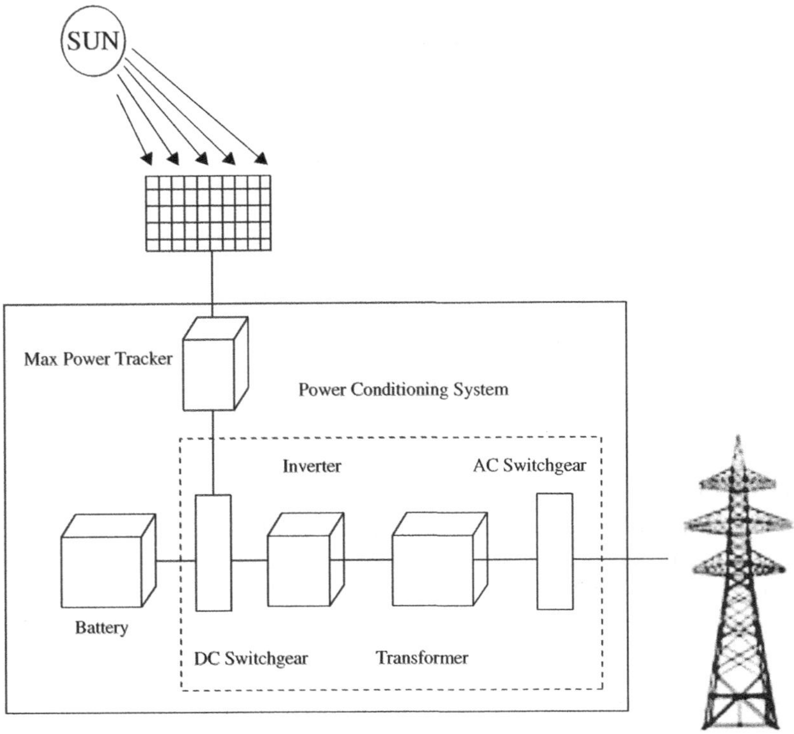

Figure 10.5 Battery storage application in a solar PV system.

(d) Supercapacitor Energy Storage System

Supercapacitors (also known as ultracapacitors), Figure 10.6, are special types of capacitors that can store at least one hundred times more energy per unit volume than an ordinary electrolytic capacitor. They have an advantage over batteries because of a faster and simpler charging process and faster release of stored energy. Thus, they have found application in situations where a short time burst of energy is required.

Figure 10.6 A supercapacitor.

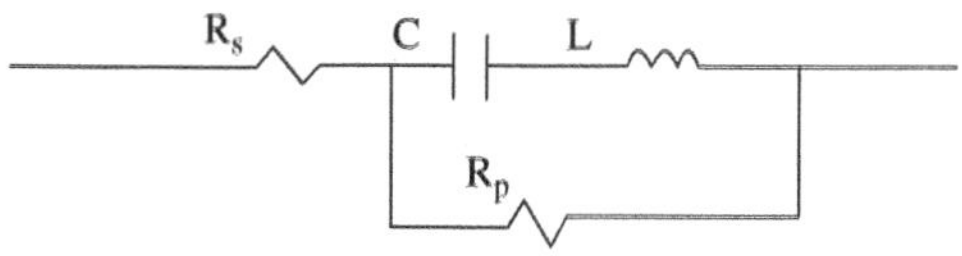

Figure 10.7 Typical equivalent model of an electromechanical double-layer capacitor.

The secret of their high capacitance and energy storage ability is the large size of their plates and the shorter distance between them.

Advantages of Supercapacitor Energy Storage

Supercapacitors have:

- an unlimited life cycle, hence can be charged and discharged continuously an unlimited number of times
- a simple, fast, and reliable charging system
- no problem with overcharge as encountered in batteries
- excellent thermal charge and discharge performance
- ability to deliver high power to drive high-current load demands because of low resistance

Supercapacitors are, however, limited by high cost and high self-discharge; hence their full energy spectrum is not always available for utilization.

Applications of Supercapacitor Energy Storage

Supercapacitors have found applications in many areas, including:

- Providing needed energy storage for firming up a renewable-energy system's output, thereby providing grid stability. This application is very useful, especially in dealing with load frequency control and bus voltage stability issues in microgrids.
- Serving as a good energy back-up system for data centers and other sensitive facilities between outage and startup of an uninterruptible power system.
- In lifting operations where sharp burst of power is needed, supercapacitors can prove very useful.

Supercapacitor Energy Storage Model

Electromechanical double-layer capacitors (EDLCs) or supercapacitors are electromechanical capacitors that usually have high energy density compared to common capacitors. For the same size as conventional capacitors, EDLCs have a capacitance of several farads, an improvement of about two to three orders of magnitude in capacitance but usually at a lower working voltage. Consequently, EDLCs have very high densities [2]. A typical model is shown in Figure 10.7.

R_s is the series resistance due to the electronic and ionic conductor equivalent ohmic resistance; R_p is the parallel resistance, which is responsible for electrical losses that generate internal heating; and L is the parasitic inductance due to its geometry.

The capacitance is composed of a constant C_0 and linear voltage dependent part,

$$C_v = k_v V. \tag{10.16}$$

where k_v is a coefficient that depends on technology. The total capacitance at the voltage U is given by

$$C = C_o + C_v \tag{10.17}$$

$$I(t) = (C_o + 2k_v V)\frac{dV}{dt}. \tag{10.18}$$

It can be shown that the stored energy is given by

$$E(V) = \left(C_o + \frac{4}{3}kV\right)\frac{V^2}{2} \tag{10.19}$$

(e) Solar Thermal Electric Storage System

Thermal energy can be stored as sensible heat, as latent heat, or as the heat of a thermochemical reaction. Sensible heat Q is stored in a material of mass m and specific heat C_p by raising the temperature of the storage material and is expressed as

$$Q = mC_p\Delta T. \tag{10.20}$$

The most common heat-storage materials are water, organic oils, rocks, ceramics, and molten salts. Water has the highest specific heat value of 4,190 J/kg°C.

(f) Compressed Air Energy Storage System

Compressed air systems use electric power to run compressors that push air into a tank at very high pressure, and then use the air under pressure in a piston or turbine to generate power on demand. These systems store compressed air in geologic structures such as aquifers or abandoned salt mines or other mines. Generally, compressed air systems have fixed cost that are too high for DG applications. The smallest practical size of such storage is of the order of 10MW power and 100 MWh energy. Compressed air systems have high power and energy density in comparison with pumped hydro.

(g) Superconducting Magnetic Energy Storage

Superconducting magnetic energy storage (SMES) systems store energy in a superconducting magnetic coil immersed in a very cold liquid such as liquid helium, contained in a highly insulated thermal bottle. Superconducting magnetic coils have zero electrical resistance, so electric current being circulated will not diminish over time. The coils themselves are not very large, approximately two feet in diameter and one foot high. Power is stored in an SMES system by circulating a DC electric current

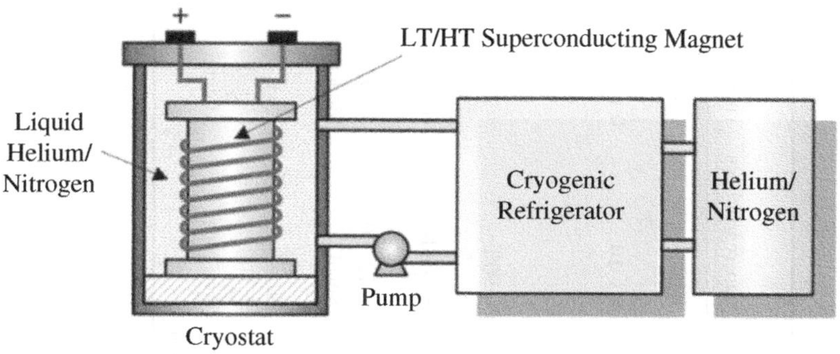

Figure 10.8 Illustration of an SMES system.

in the coil magnetically. It is withdrawn through reversal of the process. A typical SMES unit stores about 250 kWh of energy in a space the size of a large refrigerator weighing approximately 700 pounds. The capacity of an SMES ranges from 0.3 to 3.0MW. Figure 10.8 [7] shows a superconducting magnetic energy storage system that supplies electricity in case of sudden drops in voltage from lightning strikes and other natural phenomenon.

Advantages of SMES

The advantages of SMES are efficiency, robustness, reliability, and noiselessness in operation. Its power conditioning is of the highest quality.

Disadvantages of SMES

The disadvantages of SMES are cooling requirements and high cost of energy per unit, in the range of $2,500 to $3,800/kWh of storage energy. This is twenty times that of lead-acid batteries. Other disadvantages include sensitivity to temperature and very high magnetic field, on the order of 9Tesla or more, which may be a problem for communication equipment. Table 10.3 shows characteristics of energy storage technologies.

10.4 COST BENEFITS OF STORAGE

Cost–benefit analysis (CBA) is a tool for analyzing the relation between the costs and benefits of a project and comparing similar projects on the basis of those results. CBA is a formal technique for balancing the benefits that a project produces against the costs of producing that benefit. It may be viewed as an art consisting of a series of techniques for decision-making. CBA is also a systematic, quantitative method of assessing the life-cycle costs and benefits of competing alternative approaches and determining which alternative is best.

The standard criterion for justifying a project is that the benefits exceed the costs over the life cycle of the project. The competing alternative with the greatest

TABLE 10.3 Characteristics of Energy Storage Technologies [8]

Characteristic	Pumped Hydropower	CAES	Flywheels	Batteries	SMES	Supercapacitors	Thermal
MJ Energy Range	$1.8 \times 10^6 - 36 \times 10^6$	$18 \times 10^4 - 18 \times 10^6$	1–1,800	1,800–180,000	$1,800 - 5.4 \times 10^6$	1.10	1–100
Power Range MWe	100–1,000	50–1,000	1.10	Lead acid 60–180; nickel– metal hydride –370; Li–ion 400–600	10–100	0.1–10	0.1–10
Cycle Overall Efficiency	64–80%	60–70%	~ 90%	~ 75%	~ 95%	~ 90%	~ 80 – 90%
Charge/Discharge Time	Hours	Hours	Minutes	Hours	Minutes to hours	Seconds	Hours
Life Cycle	≥ 10.000	≥ 10.000	≤ 10.000	≤ 20.000	≥ 10.000	> 10.000	> 10.000
Footprint/Unit Size	Large if above ground	Moderate if under ground	Small	Small	Large	Small	Moderate
Siting Ease	Difficult	Difficult to moderate	N/A	N/A	Unknown	N/A	Easy
Maturity	Mature	Early development	Early development	Lead acid mature, others under development	Early R&D stage, under development	Available	Mature

net benefit (benefits minus costs) should be selected. In considering a CBA for storage system, it should be noted that:

- The cost of energy storage depends on the specific technology involved.
- The benefits of energy storage depend on the service it provides and the local energy market or utility electric tariffs.

However, energy storage added to an electric power system at any level is capable of:

- improving the grid's operation
- reducing overall electricity costs, which is good for all players in the energy sector
- improving electric energy reliability
- integration of renewable-energy systems such as solar and wind
- allowing excess energy to be stored for dispatch at periods of low or zero generation

10.4.1 Battery Energy Storage System Case Study

A battery energy storage system is a mature storage technology that enjoys wide applications in industrial, commercial, and domestic day-to-day needs. An average individual is constantly using at least one battery at any moment in time through a wristwatch, cell phone, wearable device, iPod, computer, or other electronic device that has become part of our daily existence. More energy-intensive applications include electric vehicles and home or commercial energy applications. In recent years, the deployment of battery storage systems within the power grid have received increased attention and implementation. It is important to ascertain the cost benefits of this technology, especially at utility scale, and particularly for renewable-energy penetration. The following cost components are associated with deployment of battery energy storage systems.

- **Battery costs** in large-scale applications, especially at the grid level, are a very substantial fraction of the cost of a storage-deployment project. Batteries for any application may be acquired new or refurbished. Obviously, new costs more than refurbished.
- **Power electronics inverters** for grid connection are in addition to the cost of batteries. The cost of the inverters includes all the wiring for the connections of battery packs and/or modules (for series/parallel connections for desired voltage and current levels) and the inverter's control system (digital signal processor for inverter control) and sensors.
- **Operation and maintenance costs** include periodic and unscheduled maintenance, repair, or replacement of failed or damaged battery packs and maintenance of the power electronic converter. The annual cost, estimated based on a similar system, is divided by the power rating to obtain the dollars per kilowatt

per year value. The annual operation and maintenance costs do not include the recharging costs of the batteries as this is application dependent.

- **Shipping/transportation costs** depend on the classification of batteries as either new or used. Utility-scale battery energy storage systems are quite substantial in weight and consequently are a major cost factor.
- **Hazardous materials or hazardous waste costs** may be incurred if used batteries must be transported from a dealer to a refurbishing factory by truck; the driver may be an employee of the refurbishing company, and the variable costs include gasoline and maintenance. The costs of the driver and truck will add only a few cents per kilowatt hour to the fixed cost.

The total system cost for this analysis encompasses the price of batteries, transportation, power electronics, and operation and maintenance. Clearly, the price of batteries depends on the supply, demand, and the price of new batteries.

10.4.2 Battery Energy Storage System Benefits

- **Electric supply capacity**: While sharp permanent increases in demand require utility companies to permanently add capacity, additional capacity is not necessarily the solution to small increases in demand. Because of this, some utility companies rent generation capacity in the wholesale market. Energy storage alternatives can be used in lieu of adding or renting additional capacity. The financial value of the benefit can be estimated by using either of two options (i.e., adding or renting capacity).
- **Load following**: The marginal cost of generation is primarily fuel and maintenance. This cost can be avoided or reduced if energy storage is used instead of generation to provide load-following service. The capacity-related costs (i.e., costs incurred to add generation capacity) depend on the need, which can vary depending on the region and the time of year. Additional generation capacity could be from hydroelectric to combined cycle-generation capacity.
- **Area regulation**: For the calculation of energy storage benefits of area regulation, the storage benefit will be the same as conventional, generation-based regulation and reflect the prevailing price for the service in dollars per megawatt. Unlike generation used for area regulation, efficient storage can provide 2 kW for each 1 kW of rated output because it can provide for regulation while charging and discharging.
- **Electric supply reserve capacity**: The electric supply reserve capacity benefit is somewhat small because generation-based reserves are inexpensive. The estimate of benefit is based on the price paid for reserves and the number of hours per year during which storage provides reserves.
- **Voltage support**: During outages the voltage-support application power rating may vary from 1MW to 10MW, assuming a distributed deployment to locally serve the voltage support. The required time durations for voltage support may vary from fifteen minutes to one hour considering the time needed for system stabilization and orderly load shedding. For the cost calculations for the

voltage-support application, a power rating of 1MW and thirty minutes of discharge duration (500 kWh storage capacity) are usually the basis for analysis.

- **Transmission support**: Energy storage can complement the load-carrying capacity of transmission systems, avoid transmission outages, and defer the need to add transmission capacity (purchase of additional transmission and distribution (T&D) equipment and/or rent to participate in wholesale electric marketplace). The estimated benefit is the cost of the most likely alternative if storage was not deployed. For example, if capacitors are proposed as the standard solution, then the energy storage offset is the cost of capacitors.
- **Transmission congestion**: Transmission congestion occurs when there is insufficient energy to meet the demands of all customers. The power rating for the transmission-congestion-relief application varies from 1MW to 100MW depending on the severity of the congestion. Based on the work in transmission congestion management, 20MW peak power was selected for transmission-congestion relief. The discharge duration for this application may vary from three to six hours based on the peak demand hours as congestion typically occurs in peak demand hours. Therefore, four hours of discharge capability would be reasonable for relieving the congestion, resulting in 80 MWh total energy capacity. Cost of benefit may be computed by comparing these costs with the possible revenue loss due to an outage.

10.4.3 Computation of CBA

In computing a cost–benefit ratio (CBR), the financial value of all costs and benefits must be evaluated. The undiscounted CBR $= \sum_{i=1}^{n} \frac{C_i}{B_i}$.

If the concept of time value of money is factored in, CBR may be computed from

$$\textit{Cost–Benefit Ratio} = \frac{\sum_{i=1}^{n} C_i\left(1+r_i\right)^{-t}}{\sum_{i=1}^{n} B_i\left(1+r_i\right)^{-t} - C_i\left(1+r_i\right)^{-t}} \tag{10.21}$$

where, C_i is the overall cost associated with the ith option, B_i is the overall monetized benefits associated with the ith option, r_i is the interest rate attached to the ith option, and n is the time span.

10.4.4 Generalized Procedure for CBA Implementation

The general procedure for performing CBA is as follows:

1. Determine/define objectives.
2. Document current process and estimate future requirements.
3. Make any relevant CBA assumptions (e.g., the discount rate for the benefit stream may be set equal to that of the cost stream for some projects).
4. Define or break down the decision process into its elements.

5. Calculate, research, or estimate the cost and benefits associated with each element.
6. Identify all tangible and intangible costs, benefits, and respective distribution weights.
7. Aggregate all real costs and benefits based on their priority and social or distributional weights.
8. Present worth or discount the costs and benefits if necessary.
9. Compute the CBR.
10. Rank the "elements" into a hierarchy that reflects the impact of their potential success/failure on the entire process.
11. Assign weighting values to each element, ranked on their instrument value.
12. Estimate the likelihood of success or failure of each element.
13. Compute the risk associated with each element or perform sensitivity analysis.
14. Evaluate alternatives and make final recommendations.

CBA is significant in assessing the benefits and costs of storage technologies and systems related to the integration of distributed energy resources, such as solar, wind, and microgrids, into the planning and operations of the power system. This helps in evaluation and sizing of the suitable energy storage technology for the microgrid when considering optimal operation.

10.5 CHAPTER SUMMARY

In this chapter, we presented and discussed energy storage. Electric power is generated and consumed in real time. However, to manage certain contingencies, energy storage devices store excess energy during peak energy production. Stored energy may then be released during periods of energy shortage. Different types of storage systems and their characteristics and models were presented. The rationale for studying storage systems is based on the fact that some key renewable-energy resources do not produce energy around the clock. It is therefore important to make provision for periods when energy production is low or unavailable. In addition, storage devices were shown to play other critical roles in microgrids besides backup duties. Different components of storage technologies were given to help distinguish their capabilities and limitations.

BIBLIOGRAPHY

[1] Maxwell Technologies, "Ultracapacitor Overview," maxwell.com/products/ultracapacitors.
[2] A. B. Cultura and Z. M. Salameh, "Modeling, Evaluation and Simulation of a Supercapacitor Module for Energy Storage Application," International Conference on Computer Information Systems and Industrial Applications, January 2015.
[3] H. Binder, T. Cronin, J. Manwell, et al., *Lifetime Modelling of Lead Acid Batteries*, Riso National Laboratory Report, April 2005.

[4] N. Pfleger, T. Bauer, D. Martin, et al., "Thermal Energy Storage: Overview and Specific Insight into Nitrate Salts for Sensible and Latent Heat Storage," *Beilstein J. Nanotech.* **6** (2015).

[5] C. K. Narula, R. Martinez, O. Onar, et al., *Final Report: Economic Analysis of Deploying Used Batteries in Power Systems*, Oak Ridge National Laboratory, June 2011.

[6] J. Gonçalves de Oliveira, "Power Control Systems in a Flywheel-Based All-Electric Driveline," PhD Diss., Uppsala University, 2011.

[7] Fichtner Management Consulting, *Hydroelectric Power: A Guide for Developers and Investors*, February 2015.

[8] J. W. Tester, E. M. Drake, M. W. Golay, M. J. Driscoll, and W. A. Peters, "Sustainable Energy - Choosing Among Options," MIT Press, Cambridge, MA, 2005.

[9] Energy Storage Sense, "Superconducting Magnetic Energy Storage (SMES)," http://energystoragesense.com/superconducting-magnetic-energy-storage-smes

[10] Pumped Storage Hydroelectricity, https://en.wikipedia.org/wiki/pumped_storage_hydroelectricity, March 2018.

CHAPTER 11

POWER ELECTRONICS

11.1 INTRODUCTION

Power electronics is the application of solid-state electronics to the control and conversion of electric power.

The first high-power electronic devices were mercury-arc valves. In modern systems, power conversion is performed with semiconductor switching devices such as diodes, thyristors, and transistors, pioneered by R. D. Middlebrook and others beginning in the 1950s. An AC/DC converter (rectifier) is the most typical power electronics device found in many consumer electronic devices, e.g., television sets, personal computers, battery chargers. The power range is typically from tens of watts to several hundred watts. In industry, a common application is the variable speed drive (VSD) that is used to control an induction motor. The power range of VSDs starts from a few hundred watts and end at tens of megawatts. Power electronics started with the development of the mercury-arc rectifier. It was used to convert AC into DC.

Power electronics are important in applications to power systems. Power electronic devices include high-power semiconductor devices used in rapid switching operations in protection devices like relays and circuit breakers. In the generation system of a power system, they are components of protection devices used in protecting generators and transformers. In transmission systems, they are key components of flexible alternating current transmission system (FACTS) devices used in increasing the power-transfer capability of the transmission network, increasing the loading capability of the lines to their thermal capabilities, providing direct control of power flow over assigned transmission line routes, and facilitating loss reduction. In distribution systems, they are components of devices used for power-actor correction and in protection of transformers. In renewable-energy sources when integrating with the grid or stand-alone, converters are needed for both integration and applications such as the use of DC to AC converters in the conversion of renewable DC sources such as PV to AC sources in microgrid and smart-grid systems. They are also the choice of technology for AC/DC/AC conversion needed in different loads. Power electronic devices are used in the decentralization of renewable-energy resources in microgrids and smart grids.

Energy Processing and Smart Grid, First Edition. James A. Momoh

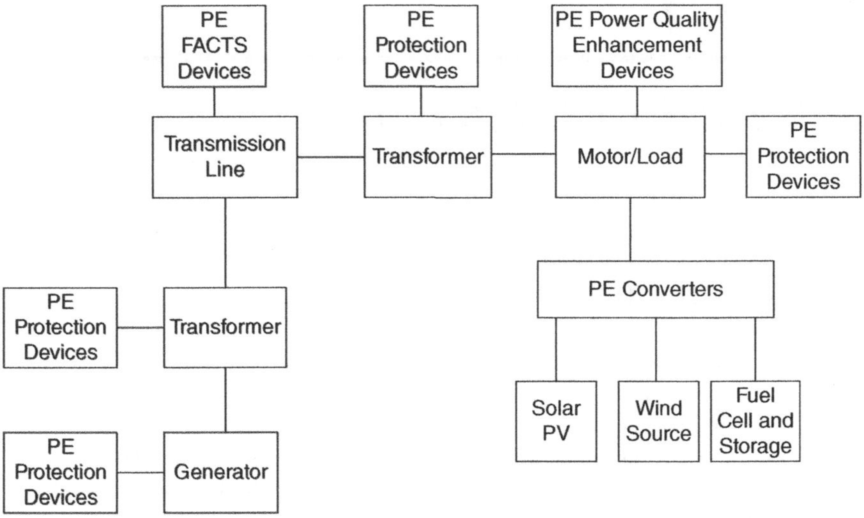

Figure 11.1 Architecture of power system with power electronic devices.

This chapter discusses the elements of power electronics and provides a comparative study of power electronic devices. Different power electronic devices, such as power diodes and power transistors, are discussed with respect to their principles of operation and characteristics and a comprehensive performance analysis. Different applications of power electronics in utilities, smart grid, and ship systems are discussed.

11.2 POWER SYSTEMS WITH POWER ELECTRONICS ARCHITECTURE

Figure 11.1 shows the architecture of a power system with renewable-energy sources. The power electronic devices are applied in protection, transmission lines capacity enhancement, power quality enhancement, and power conversion. The protection devices include circuit breakers and relays, fuses, and isolators for the protection of generators, transformers, and other devices or machines. FACTS devices that have thyristors and other power electronic devices as key components include static synchronous compensators (STATCOMs), static var compensators (SVC), and unified power flow controllers (UPFCs) used to enhance or boost the power-transmission capability of the transmission lines. Also, an application of power electronic device is the use of thyristor-controlled reactors (TCRs) and thyristor-switched capacitors (TSCs), which help power-factor improvement of the distribution system. Lastly, renewable-energy sources can be integrated into the network using converters (AC to DC, DC to DC, DC to AC, and AC to AC) for various applications.

11.3 ELEMENTS OF POWER ELECTRONICS

Power electronics and converters utilizing them made a head start when the first silicon-controlled rectifier (SCR) device was proposed by Bell Laboratories and commercially produced by General Electric in the earlier 1950 s. Mercury-arc rectifiers were well in use by that time, and the robust and compact SCR started replacing it in rectifiers and cyclo-converters. The necessity arose of extending the application of SCRs beyond the line-commutated mode of action, which called for external measures to circumvent its turn-off incapability via its control terminals. Various turn-off schemes were proposed and their classifications suggested but it became increasingly obvious that a device with turn-off capability was desirable to permit wider application.

The power metal oxide semiconductor field-effect transistor (MOSFET) became successful commercially in the late 1970 s. This device represented the first successful marriage between modern integrated circuit and discrete power semiconductor manufacturing technologies. Its voltage-drive capability—giving it again a higher gain, ease of paralleling, and, most importantly, much higher operating frequencies reaching up to a few MHz—saw it replace the bipolar at the sub-10 kW range, mainly for switch mode power supply applications. Extension of very-large-scale integration manufacturing facilities for the MOSFET reduced its price vis-à-vis the bipolar. Different versions of gate turn-off thyristor, were proposed by various manufacturers. Requirements for an extremely high turn-off control current via the gate and the comparatively high cost of the device restricted its application only to inverters rated above a few hundred kva. Brief comparison of power semiconductor devices commonly used for energy processing are given in Table 11.1.

11.4 POWER SEMICONDUCTOR DEVICES

11.4.1 Diodes

Diodes are made from semiconductor material, silicon or germanium, in which half is positively doped and is known as the *p region* and the other half is negatively doped and referred to as the *n region*. The regions are separated at the p–n junction by a depletion region. The p region is the anode while the n region the cathode. Figure 11.2 shows the schematic representation of a diode.

Typically, the anode (A) and cathode (C) are identifiable on a diode, depending on the physical configurations. A band or some other feature usually identifies the cathode.

Diode Operation

(a) Forward biasing: To bias a diode, a DC voltage is applied across it. Forward bias is when current is allowed to flow through the p–n junction. A DC-voltage source is connected by conductive material (contacts and wires) across a diode to produce forward bias. This external-bias voltage is designated as V_{BIAS}. A

TABLE 11.1 Comparison of Power Semiconductor Devices for Energy Processing

Semiconductor Device	Description	Symbol(s)	Applications
Diodes	Electrical device that allows current to move through it in one direction with far greater ease than in the other	PN diode  Zener Diode 	• limit circuits • clamp applications • rectifier circuits • Zener diodes, used in voltage regulation
Bipolar Junction Transistor (BJT)	Active semiconductor device formed by two P–N junctions whose function is amplification of an electric current; uses both electron and hole charge carriers	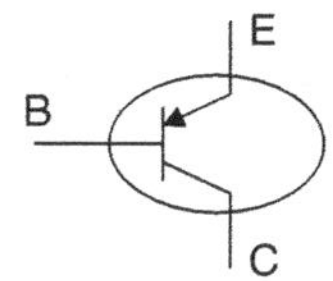 PNP type 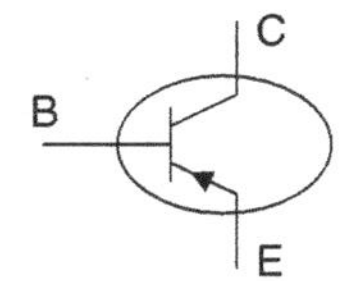 NPN type	• high-speed digital logic • amplifiers • temperature sensors • logarithmic converters
Field Effect Transistor (FET)	Unipolar devices that operate only with one type of charge carrier; voltage controlled, where the voltage between two of the terminals (gate and source) controls the current through the device	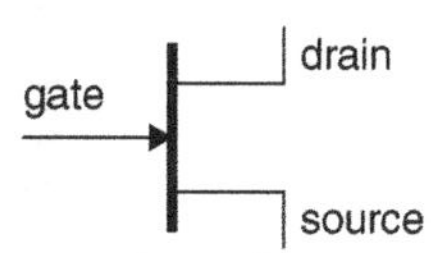 n-channel 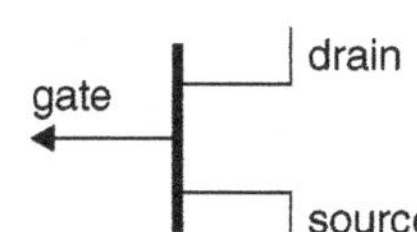 p-channel	• amplifiers (for example, due to large input resistance and low output resistance, is effective as a buffer in common-drain (source follower) configuration)
Insulated Gate Bipolar Transistor (IGBT)	Combines features of BJTs and MOSFETs; has high input impedance like MOSFETs and low on-state conduction losses like BJTs	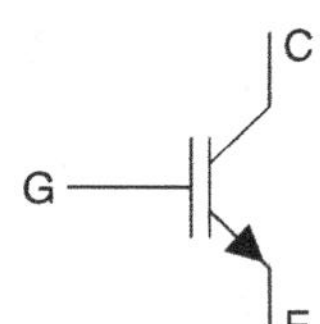	• switching internal combustion engine ignition coils, where fast switching and voltage blocking capabilities are important • medium-power applications such as DC and AC motor drives, power supplies, solid-state relays, and contactors

Metal Oxide Semiconductor Field Effect Transistor (MOFET)	Voltage-controlled devices with very high input impedance; used for switching and amplifying electronic signals in electronic devices; can operated in a circuit in two configurations: depletion type or enhancement type	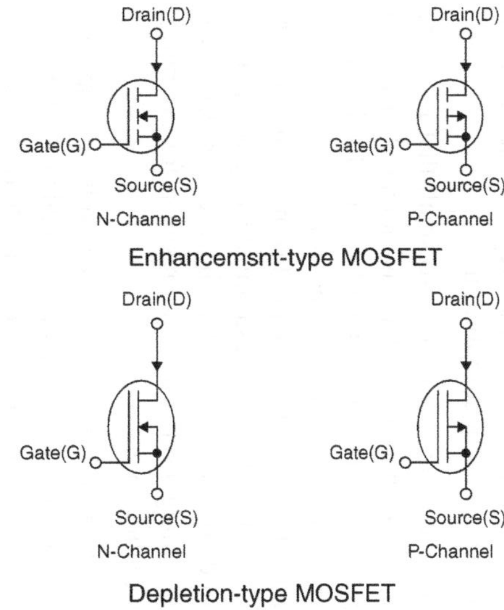	• digital-integrated circuits, such as microprocessors • calculators, memories, and logic complementary metal–oxide–semiconductor gates • analog switches • amplifiers • applications of power electronics and switch-mode power supplies • oscillators in radio systems • automobile sound systems and in sound reinforcement systems
Thyristor	Bistable semiconductor device made of three or more junctions that can be switched from the off state to the on state and vice versa	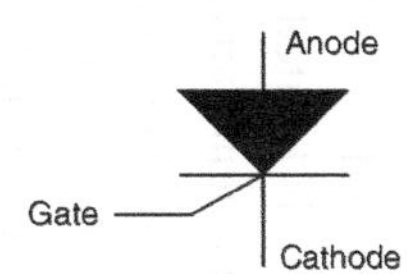	• controls power to many kinds of loads: a simple lamp, power supplies, voltage regulators, and industrial motors

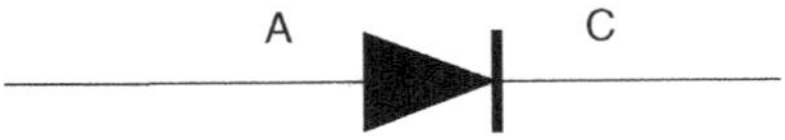

Figure 11.2 Schematic symbol of a diode.

resistor is used in the arrangement to limit the forward current to a value that will not damage the diode. The negative side of the voltage source is connected to the n region of the diode and the positive side is connected to the p region. This is one requirement for forward bias. A second requirement is that the bias voltage, V_{BIAS}, must be greater than the barrier potential.

(b) Reverse biasing: Reverse bias is the condition that essentially inhibits current through the diode. A DC-voltage source is connected across a diode to produce reverse bias. This external bias voltage is known as V_{BIAS} just as in forward bias. The positive side of the DC voltage V_{BIAS} is connected to the n region of the diode and the negative side is connected to the p region. The depletion region is much wider than in forward bias or equilibrium.

(c) Reverse breakdown: Normally the reverse current is so small that it can be neglected. However, if the external reverse-bias voltage is increased to a value called the *breakdown voltage*, the reverse current will drastically increase. This multiplication of conduction electrons is termed the *avalanche effect*, and reverse current can increase dramatically if steps are not taken to limit the current. When the reverse current is not limited, the resulting heating will permanently damage the diode. Most diodes are not operated in reverse breakdown, but if the current is limited (by adding a series-limiting resistor, for example), there is no permanent damage to the diode.

(d) V–I characteristic for forward bias: When a forward-bias voltage is applied across a diode, current flows to the circuit. This current is called the forward current and is designated I_F. V–I characteristic for forward bias shows what happens when a forward-bias voltage is increased positively. A resistor is used to limit the forward current to prevent overheating and damage to the diode.

With 0 V across the diode, there is no forward current. As the forward-bias voltage is gradually increased, the forward current and the voltage across the diode gradually increase. A portion of the forward-bias voltage is dropped across the limiting resistor. When the forward-bias voltage is increased to a value where the voltage across the diode reaches approximately the barrier potential, the forward current begins to increase rapidly, as illustrated in Figure 11.3.

With further increase in the forward-bias voltage, the current continues to increase very rapidly, but the voltage across the diode increases only gradually. This small increase in the diode voltage above the barrier potential is due to the voltage drop across the internal dynamic resistance of the semiconductive material.

(e) V–I characteristic for reverse bias: When a reverse-bias voltage is applied across a diode, there is only an extremely small reverse current (I_R) through

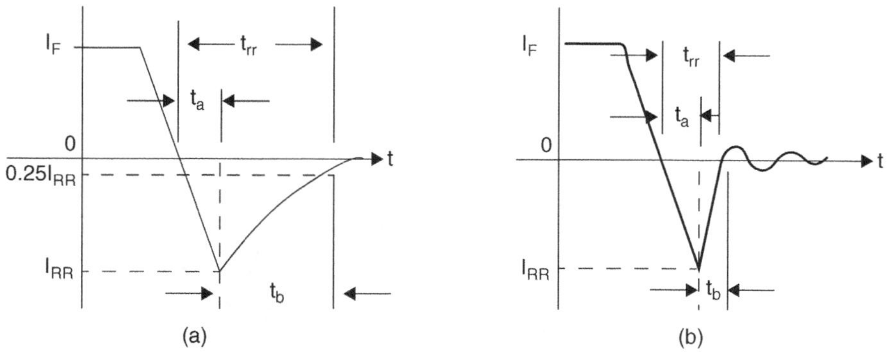

Figure 11.3 (a) Soft reverse-recovery characteristics. (b) Abrupt reverse-recovery characteristics.

the p–n junction. With 0 V across the diode, there is no reverse current. As you gradually increase the reverse-bias voltage, there is a very small reverse current and the voltage across the diode increases. When the applied bias voltage is increased to a value where the reverse voltage across the diode (V_R) reaches the breakdown value (V_{BR}), the reverse current begins to increase rapidly. A continuous increase in the bias voltage causes the current to continue increasing very rapidly, but the voltage across the diode increases very little above V_{BR}. Breakdown, with exceptions, is not a normal mode of operation for most p–n junction devices.

(f) Reverse recovery characteristics: Once a diode is in a forward-conduction mode and its forward current is reduced to zero (due to the natural behavior of the diode circuit or by applying a reverse voltage), the diode continues to conduct due to minority carriers that remain stored in the p–n junction and the bulk semiconductor material. The minority carriers require a certain time to recombine with opposite charges to be neutralized. This is called reverse recovery time of the diode (see Figure 11.3(a) and (b) [5]) and it is given by

$$t_{rr} = t_a + t_b \tag{11.1}$$

where t_a is due to charge storage in the depletion region and represents the time between zero crossing and the peak reverse current I_{rr} and t_b is due to charge storage in the bulk semiconductor material. The ratio $\frac{t_b}{t_a}$ is known as a softness factor (SF).

The peak reverse current can be expressed in reverse di/dt as

$$I_{RR} = t_a \frac{di}{dt} \tag{11.2}$$

where $t_a = t_2 - t_1$ is the period from t_1 to t_2 (as seen in Figures 11.2(a) and 11.3(b)) and $\frac{di}{dt}$ is the rate of change of current.

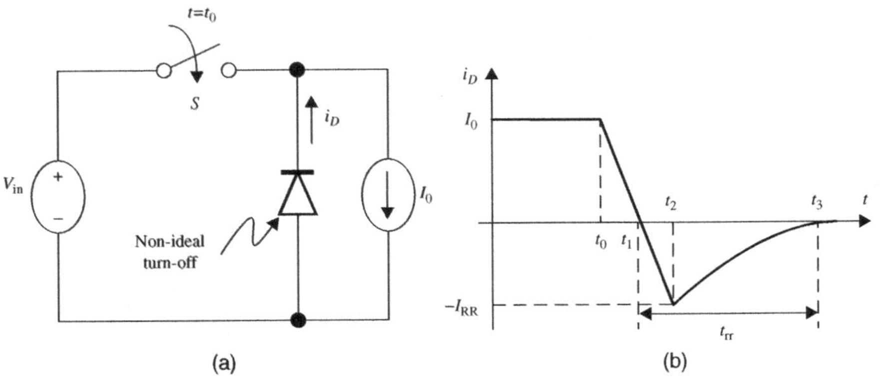

Figure 11.4 Typical diode-switching characteristics. (a) Switching circuit with S closed at $t = t_0$. (b) Diode current.

As seen in Figure 11.4 [5], when the switch is open, the diode is forward biased by current source I_0, which makes it output a maximum current of I_0. When the switch closes at $t = t_0$, the diode is forced to start turning off and so the current starts reducing from t_0 to t_1. At this interval, the diode current is positive, and the forward-diode voltage is small and ${}^{di}/_{dt}$ is constant. At t_1, diode current is zero and it is expected that the diode will turn off, but that is not the case. The diode still conducts until t_2 because of the excess minority carriers that have to be removed from the diode's p–n junction for its reverse voltage to begin to rise. The point at which it gets to t_2 is the I_{RR}, which is a function of ${}^{di}/_{dt}$ and the t_1 to t_2 interval.

From interval t_2 to t_3, the diode current rises exponentially to zero at t_3, which is caused by the charge stored in the bulk of the semiconductor material. At t_3, the diode is fully switched off.

For most practical purposes, a diode can be regarded as an ideal switch, whose characteristics are shown in Figure 11.5(b). The V–I characteristics shown in Figure 11.5(a) can be expressed by the Shockley diode equation, given as

$$I_D = I_s\left(e^{V_D/nV_t} - 1\right) \tag{11.3}$$

where I_D is the current through the diode, V_D is the diode voltage with anode positive with respect to cathode V, I_S is the leakage (or reverse-saturation) current, V_T is a constant called *thermal voltage* given by

$$V_T = \frac{kT}{q} \tag{11.4}$$

k is Boltzmann's constant: 1.3806×10^{-23} J/K, T is the absolute temperature in Kelvin (K = 273 + °C), and q is the electron charge: 1.6022×10^{-19} Coulomb (C).

Basic Diode Waveform Illustrations

From Figure 11.6(a) [5] and assuming the diode is ideal, during the positive half cycle of the input waveform, the diode acts in *reverse-bias mode*, so no current flows

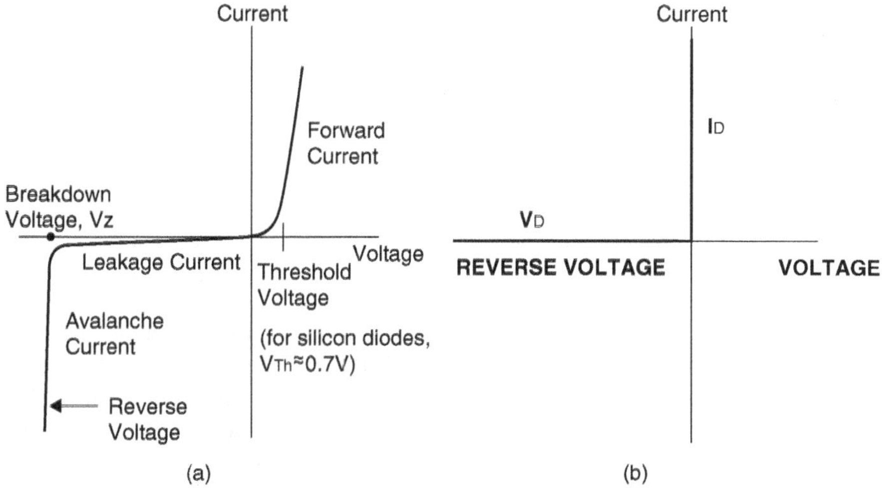

Figure 11.5 V–I characteristics of a diode. (a) Practical characteristics. (b) Ideal characteristics.

through it. It acts as an open circuit, meaning the full voltage is impressed on the diode and therefore the voltage dropped on the resistor is zero as seen in the output waveform. During the negative half cycle of the input waveform, the diode acts in the forward-bias, mode making current flow through the resistor. Therefore, there is voltage (full voltage) dropped on the resistor, as seen in the output waveform in Figure 11.6(a). Because there is conduction in the negative cycle, the output voltages appear in the negative cycle.

From Figure 11.6(b) [5] and assuming the diode is ideal, during the positive half cycle of the input waveform, the diode acts in *forward-bias mode*, making current flow

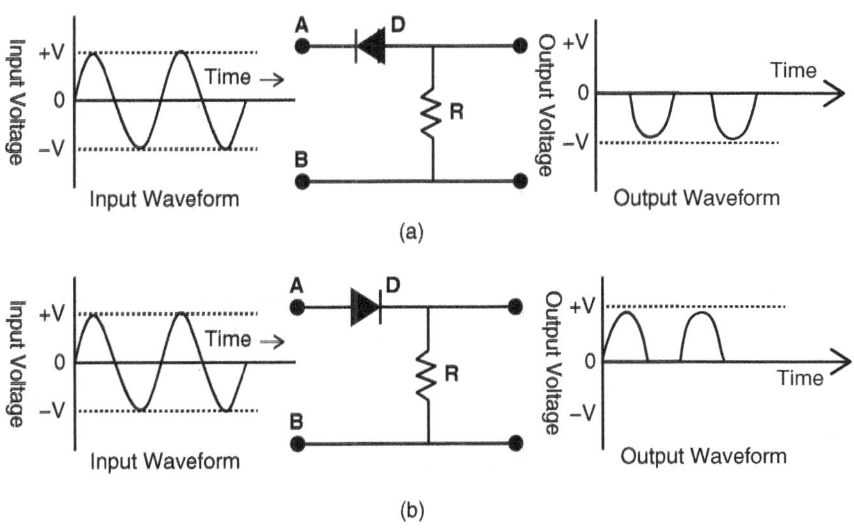

Figure 11.6 Diode waveform illustrations. (a) Positive clipper. (b) Negative clipper.

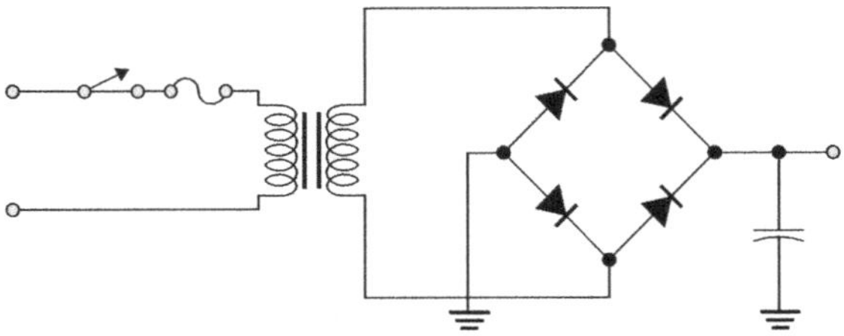

Figure 11.7 Diode as a rectifier.

through the resistor. Therefore, there is voltage (full voltage) dropped on the resistor as seen in the output waveform in Figure 11.6(b). During the negative half-cycle of the input waveform, the diode acts in reverse bias mode and so no current flows through it. It acts as an open circuit, meaning the full voltage is impressed on the diode and therefore the voltage dropped on the resistor is zero as seen in the output waveform. Because there is conduction in the positive cycle, the output voltages appear in the positive cycle.

Example 1
The forward-voltage drop of a power diode is $V_D = 1.2\,V$ at $I_D = 350\,A$. Assuming that $n = 3$ and $V_T = 25.8\,mV$, find the saturation current I_S.

Solution:
From Equation 11.3

$$I_D = I_s\left(e^{V_D/nV_t} - 1\right)$$

$$350 = I_s\left(e^{1.2/3 \times 25.8} - 1\right)$$

$$I_S = 22,400.45\ A$$

Power Diodes
Silicon-power diodes are the successors of selenium rectifiers, having significantly improved forward characteristics and voltage ratings. They are classified mainly by their turn-off (dynamic) characteristics, as seen in Figure 11.7. The minority carriers in the diodes require finite time - t_{rr} (reverse recovery time) to recombine with opposite charges and neutralize. Large values of $Q_{rr} = (Q_2 + Q_2)$, the charge to be dissipated as a negative current when the diode turns off, and $t_{rr} = t_2 - t_0$), the time it takes to regain its blocking features, impose strong current stresses on the controlled device in series. Also, a "snappy" type of recovery of the diode effects high $^{di}/_{dt}$ voltages on all associated power devices in the converter because of load or stray inductances present in the network. There are broadly three types of diodes used in power electronic applications:

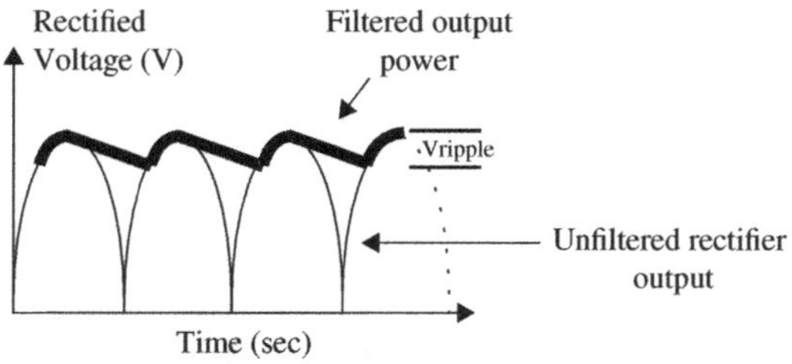

Figure 11.8 Resultant output waveform.

- **Line-frequency diodes**: These PN diodes with general-purpose, rectifier-type applications, are available at the highest voltage (~5 kV) and current ratings (~5kA) and have excellent over-current (surge rating about six times average current rating) and surge-voltage withstanding capability. They have relatively large Q_{rr} and t_{rr} specifications.
- **Fast-recovery diodes**: Fast-recovery diffused diodes and fast-recovery epitaxial diodes have significantly lower Q_{rr} and t_{rr} (~1.0 sec). They are available at high powers and are mainly used in association with fast-controlled devices, such as freewheeling or DC–DC choppers and rectifier applications. They are used in DC–DC and DC–AC converter circuits where the speed of recovery is of critical importance. Fast-recovery diodes also find application in induction heating, uninterruptible power systems, and traction.
- **Schottky diodes**: These are the fastest rectifiers, being majority carrier devices without any Q_{rr}. However, they are available with voltage ratings up to one hundred volts only, though current ratings may be high. Their conduction voltages specifications are excellent (~0.2 V). The freedom from minority-carrier recovery permits reduced snubber requirements. Schottky diodes face no competition in low-voltage, self-powered microsystem applications and in instrumentation.

Applications of Diodes

- **Rectifier circuits**: Because of their ability to conduct currents in one direction and block currents in the other direction, diodes are used in circuits called *rectifiers* that convert AC voltage into DC voltage. Rectifiers are found in all DC power supplies that operate from an AC voltage source. A power supply is an essential part of each electronic system from the simplest to the most complex. Figure 11.7 and Figure 11.8 shows the circuit diagram for diode as a rectifier and its respective output curve.
- **Clamping applications**: Sometimes it is necessary to limit the range of signal (for instance not to exceed certain voltage limits and not destroy a device). The circuit shown in Figure 11.9 accomplishes this. The diode prevents the output from exceeding 5.6 V, with no effect on voltages smaller than this, including negative voltages. The only limitation is that the input must not be so negative that the reverse-breakdown voltage is exceeded. Diode clamps are the standard

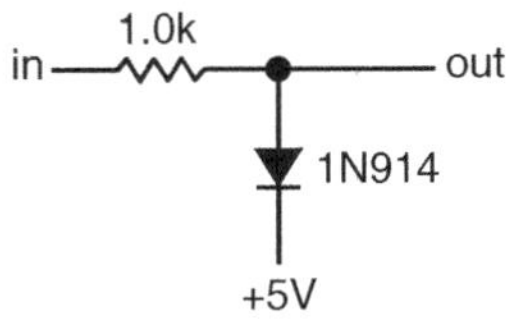

Figure 11.9 Clamping configuration.

equipment on all inputs in the complementary metal oxide semiconductor family of digital logic. Without them, the delicate input circuits are easily destroyed by static electricity.

- **Limiter circuits**: Diode limiters Figure 11.10 are often used as input protection for high-gain amplifiers. For large-voltage amplifiers, the input must always be near zero voltage, otherwise the output is in a state of saturation. For an op amp with a gain of 1,000, the amplifier operates with supply voltage ± 15 V; sometimes it can be ± 12 V to ± 18 V. It will never give output voltage larger than the supply voltage, i.e., ± 15 V, so the input signal ± 15 mV (± 15 V/1,000) or bigger will saturate the output.

11.4.2 Bipolar Junction Transistors

Bipolar junction transistors (BJTs) are constructed with three doped semiconductor regions separated by two p–n junctions. The three regions are called the *emitter*, *base*, and *collector*. Physical representations of the two types of BJTs are shown in Figures 11.11 and 11.12. One type consists of two n regions separated by a p region (npn), and the other consists of two p regions separated by an n region (pnp). *Bipolar* refers to the use of both holes and electrons as current carriers in the transistor structure. In order for a BJT to operate properly as an amplifier, the two p–n junctions must be correctly biased with external DC voltages. In this section, we mainly use the npn transistor for illustration. The operation of the pnp is the same as for the npn except that the roles of the electrons and holes, bias-voltage polarities, and current directions are all reversed.

Biasing

The emitter current is slightly greater than the collector current because of the small base current that splits off from the total current injected into the base region from the emitter.

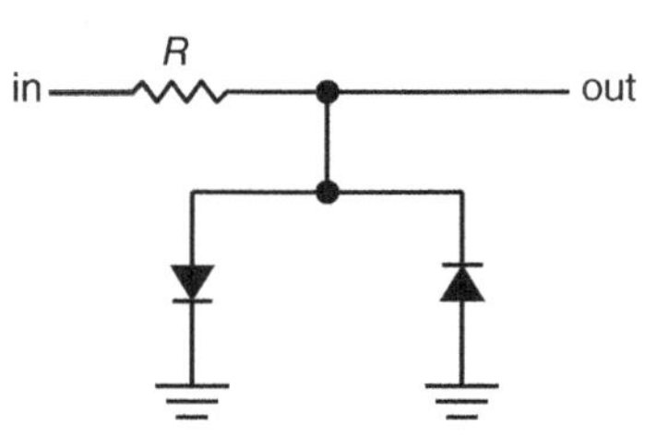

Figure 11.10 Diode-limiter circuit.

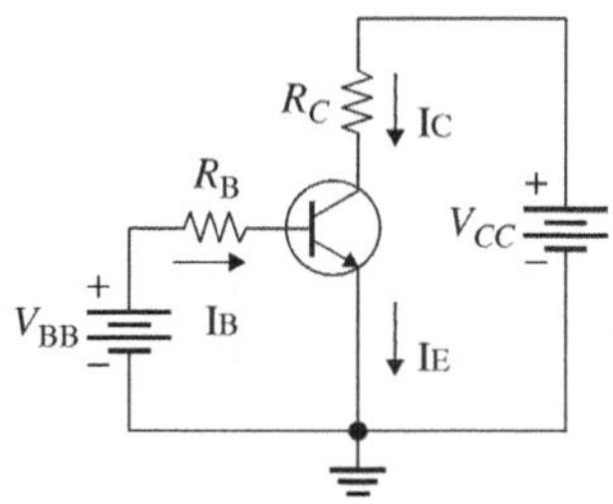

Figure 11.11 npn transistor.

When a transistor is connected to DC-bias voltages, as shown in Figures 11.11 and 11.12 for both npn and pnp types, V_{BB} forward biases the base-emitter junction and V_{CC} reverse biases the base-collector junction. V_{CC} is normally taken directly from the power-supply output and V_{BB} (which is smaller) can be produced with a voltage divider.

Circuit Analysis

Before analyzing the BJT characteristics, it is important to talk about its modes of operation (Figure 11.13):

- Common base (CB) mode has current gain but no voltage gain.
- Common emitter (CE) mode has voltage gain but no current gain.
- Common collector (CC) mode has both current and voltage gain.

CB Characteristics

- **Input characteristics**: Consider a pnp transistor where the input current is the emitter current (designated as I_E) and the input voltage is the collector-base voltage (designated as V_{CB}). Since the emitter-base junction is forward biased, input current I_E increases for a fixed emitter-base voltage V_{EB} when V_{CB} increases. Figure 11.14 shows that the graph of I_E versus V_{EB} is akin to the forward-bias characteristics of a p–n diode.
- **Output characteristics**: The output characteristics show the relation between output voltage and output current. I_C is the output current and collector-base voltage and the emitter current I_E is the input current and works as the parameters. Figure 11.15 shows the output characteristics for a pnp transistor in CB

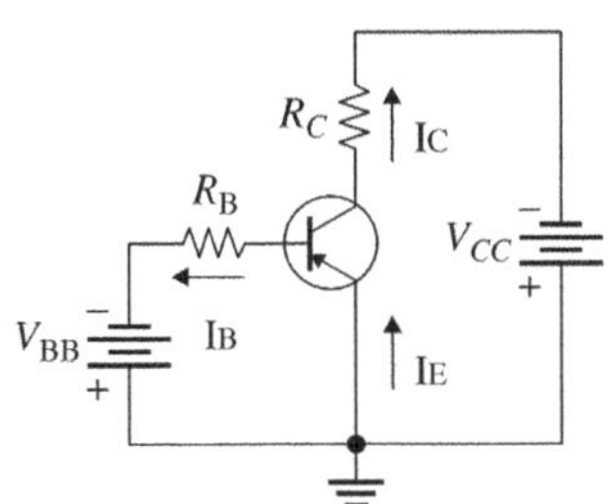

Figure 11.12 pnp transistor.

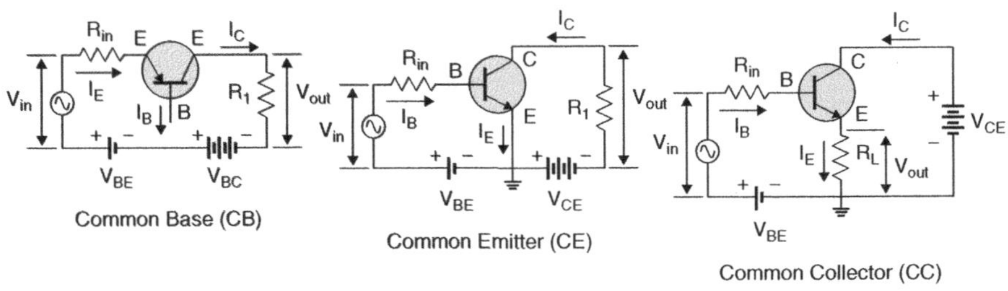

Figure 11.13 Circuit operation modes of the BJT.

mode. As we know for pnp transistors, I_E and V_{EB} are positive and I_C, I_B, V_{CB} are negative. These are the curve-active, saturation, and cutoff regions. The active region is the region where the transistor operates normally. Here the emitter junction is reverse biased. Now the saturation region is the region where both the emitter and collector junctions are forward biased. Finally, the cutoff region is where both the emitter and collector junctions are reverse biased.

CE Characteristics

- **Input characteristics**: I_B (base current) is the input current; V_{BE} (base-emitter voltage) is the input voltage for CE mode. So, the input characteristics for CE will be the relation between I_B and V_{BE} with V_{CE} as parameter. The typical CE input characteristics are similar to that of a forward-biased p–n diode. But as V_{CB} increases, the base width decreases. The characteristics are shown in Figure 11.16.
- **Output characteristics**: Output characteristics for CE mode are the curve or graph between collector current (I_C) and collector-emitter voltage (V_{CE}) when the base current I_B is the parameter. The characteristics are shown in Figure 11.17.

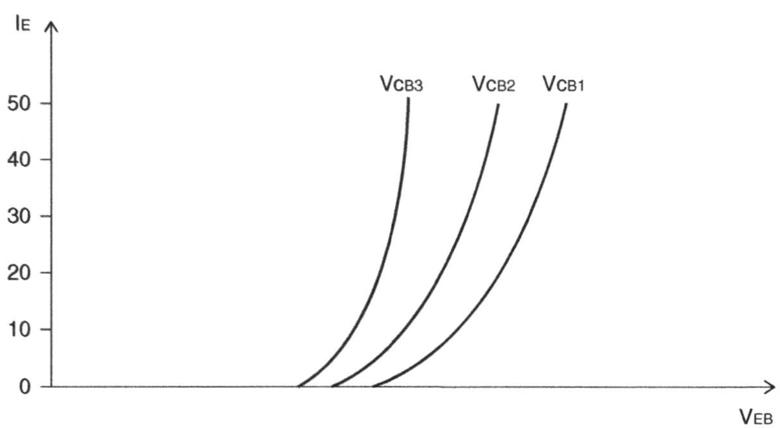

Figure 11.14 CB input characteristics.

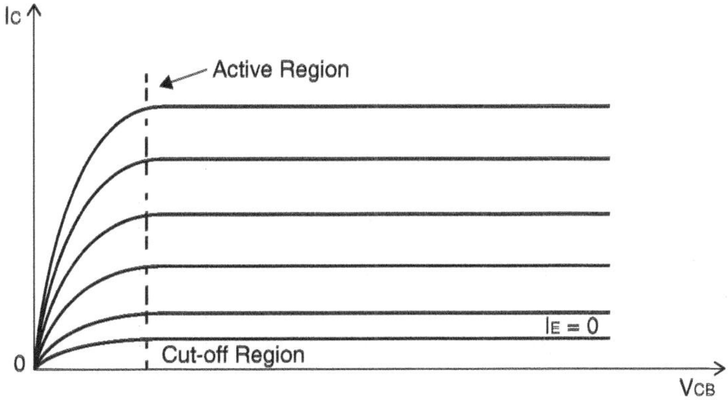

Figure 11.15 CB output characteristics.

Like the output characteristics of a common-base transistor, CE mode has also three regions: active, cutoff, and saturation. The active region has a collector region reverse biased and the emitter junction forward biased. For the cutoff region the emitter junction is slightly reverse biased and the collector current is not totally cut off. And finally, for the saturation region both the collector and the emitter junction are forward biased.

The current gain of a transistor is the ratio of the collector current (I_C) to the base current (I_B) and is designated as beta (β).

Typical values of β range from less than 20 to 200 or higher. The ratio of the DC collector current (I_C) to the DC emitter current (I_E) is alpha (α). The alpha is a less-used parameter than beta in transistor circuits.

Typically, values of α range from 0.95 to 0.99 or greater, but α is always less than one. This is because I_C is always slightly less than I_E by the amount of I_B. For example, if $I_E = 100A$ and $I_B = 1$ A, then $I_C = 99A$ and $\alpha = 0.99$.

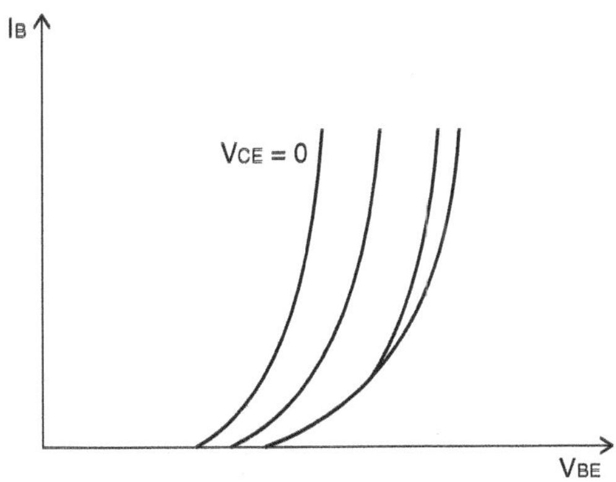

Figure 11.16 CE input characteristics.

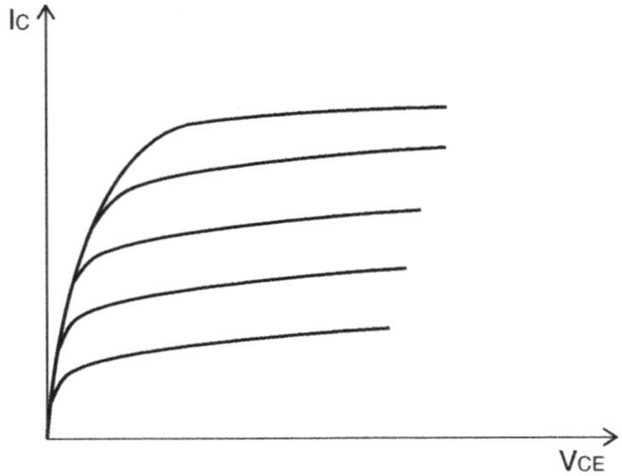

Figure 11.17 CE output characteristics.

CC Characteristics

- **Input characteristics**: The input characteristics of a CC configuration are quite different from the CB and CE configurations because the input voltage V_{BC} is largely determined by V_{EC} level. Here,

$$V_{EC} = V_{EB} + V_{BC}$$

$$V_{EB} = V_{EC} - V_{BC}.$$

Input characteristics are obtained between the input current I_B and the input voltage V_{CB} at constant output voltage V_{EC}. Keep the output voltage V_{EC} constant at different levels, vary the input voltage VBC for different points, and record the IB values for each point. Figure 11.18 shows the graph of IB versus VCB.

- **Output characteristics**: The output characteristics shows the relation between output voltage VEC and output current IE at constant input current IB. In the

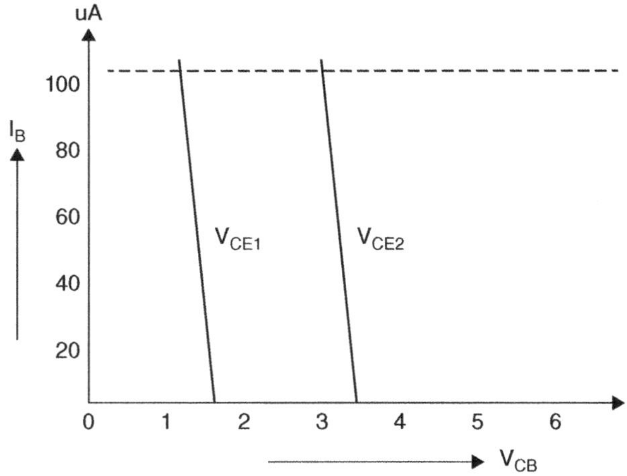

Figure 11.18 CB input characteristics.

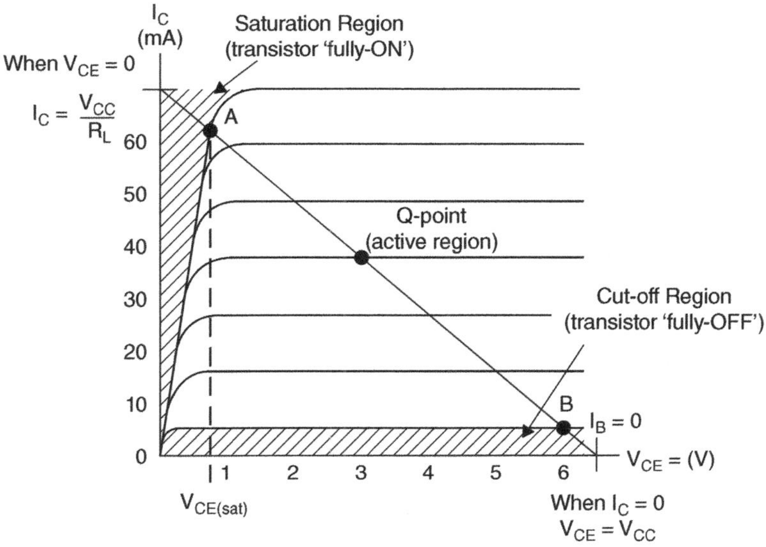

Figure 11.19 CC output characteristics.

operation of CC circuit, if the base current is zero then the emitter current also becomes zero. As a result, no current flows through the transistor. Figure 11.19 shows the output characteristics for a transistor in CC mode.

Saturation and Cutoff

Assume a transistor is connected in a CB configuration, V_{BB} is set to produce a certain value of I_B, and V_{CC} is zero. For this condition, both the base-emitter junction and the base-collector junction are forward biased, while the emitter and the collector are at 0 V. The base current is through the base-emitter junction because of the low impedance path to ground and therefore I_C is zero. When both junctions are forward biased, the transistor is in the saturation region of its operation. Saturation is the state of a BJT in which the collector current has reached a maximum and is independent of the base current.

As V_{CC} is increased, V_{CE} increases as the collector current increases. I_C increases as V_{CC} is increased due to the forward-biased base-collector junction.

Ideally, when V_{CE} exceeds a stipulated base voltage, the base-collector junction becomes reverse biased and the transistor goes into the active or linear region of its operation. Once the base-collector junction is reverse biased, I_C levels off and remains essentially constant for a given value of I_B as V_{CE} continues to increase. Actually, I_C increases very slightly as V_{CE} increases due to widening of the base-collector depletion region. This results in fewer holes for recombination in the base region, which effectively causes a slight increase in β. When V_{CE} reaches a sufficiently high voltage, the reverse-biased base-collector junction goes into breakdown and the collector current increases quickly. A transistor should never be operated in this breakdown region.

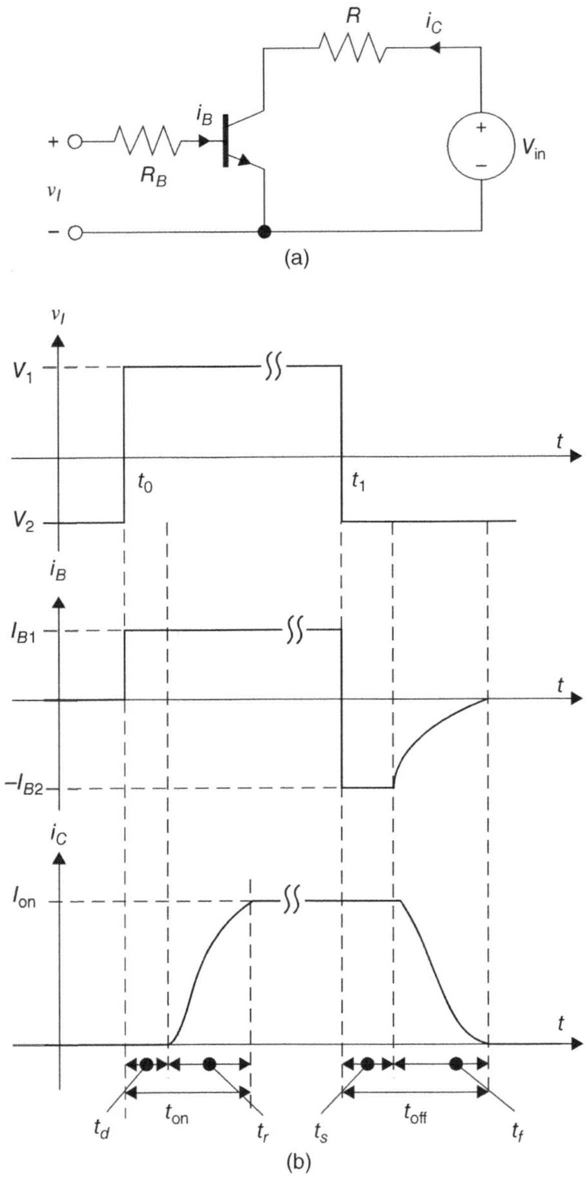

Figure 11.20 Switching characteristics of BJT. (a) Circuit. (b) Switching waveform

When I_B is equal to zero, the transistor is in the cutoff region, although there is a very small collector leakage current. Cutoff is the nonconducting state of a transistor.

Figure 11.20 [26] shows the circuit diagram of a simple inverter circuit with BJT application of CE configuration. The voltage v_1 is the base-driving voltage, which has V_1 and V_2 as the positive polarity for driving base current into the base, and negative polarity for discharging the base current. At t = t_0, v_1 applies V_1, the collector current starts to flow after a delay time t_d, and it flows exponentially until it gets to its peak—this period is the rise time t_r. The combination of t_d and t_r give t_{on}. When a negative-base voltage is applied, the BJT is turned off but it does not do so immediately, as the current remains at maximum because of t_s, which is the storage time (caused by the delay in the removal of the stored saturation charge). After this, the current exponentially reduces to zero. The combination of t_s and t_f gives t_{off}.

Example 2

The reverse-saturation current of an npn transistor in a CB configuration is 135 A. If the emitter current is 200 A and collector current is 184 A, calculate:

(a) current gain

(b) base current

Solution:

(a) Given

$$I_{SAT} = 135 \text{ A}$$
$$I_E = 200 \text{ A}$$
$$I_C = 184 \text{ A}$$
$$I_c = \alpha I_E + I_{SAT}$$

Therefore, the current gain is given by

$$\alpha = \frac{I_C - I_{SAT}}{I_E} = \frac{184 - 135}{200} = 0.245 \text{ A.}$$

(b) The base current is given as

$$I_B = I_E - I_C = 200 - 184 = 16 \text{ A}$$

Example 3

For a certain transistor in a CC circuit, $I_C = 5.5$ A, $I_B = 50$ A, and saturation current is given as $I_{SAT} = 5$ A. Determine the values of

(a) α, β, and emitter current I_E

(b) new level of base current I_B required to make $I_C = 10$ A

Solution:

(a) The emitter current is given by $I_E = I_B + I_C$

$$5.5 \text{ A} + 50 \text{ A} = 55.5 \text{ A}$$

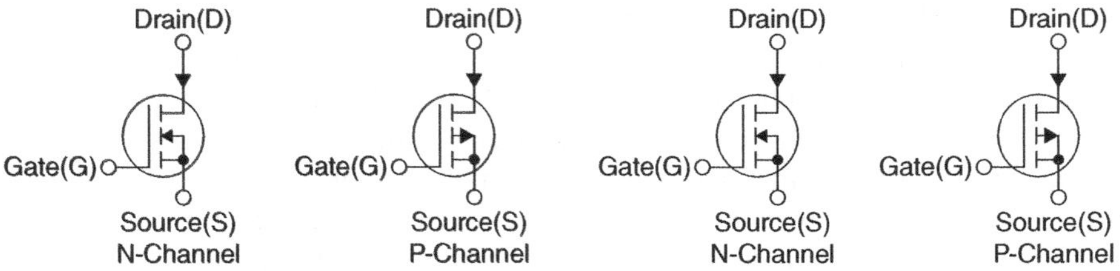

Figure 11.21 Circuit symbols of enhancement-type and depletion-type MOSFETs. (a) Enhancement-type MOSFET (b) Depletion-type MOSFET.

Also, the collector current is given by

$$I_c = \alpha I_E + I_{SAT}$$

therefore

$$\alpha = \frac{I_C - I_{SAT}}{I_E} = \frac{5.5 - 5}{55.5} = 0.99$$

$$\beta = \frac{\alpha}{1-\alpha} = \frac{0.99}{1-0.99} = 100.$$

(b) $I_C = \beta I_B + (1+\beta)I_{SAT}$ $\quad 5.5 = 100 I_B + (1+100)5$

$$I_B = 94.95\ \text{A}.$$

11.4.3 MOSFET

The MOSFET is a semiconductor device used for switching and amplifying electronic signals in electronic devices. It is one type of FET transistor. In these transistors, the gate terminal is electrically insulated from the current-carrying channel so that it is also called an insulated gate FET (IG-FET). Due to the insulation between gate and source terminals the input resistance of MOSFET may be very high, such as in megaohms (MΩ). The MOSFET is the core of integrated circuit and it can be designed and fabricated in a single chip because of these very small sizes. The MOSFET is a four-terminal device with source (S), gate (G), drain (D), and body (B) terminals. The body of the MOSFET is frequently connected to the source terminal, making it a three-terminal device like a field-effect transistor. The MOSFET is by far the most common transistor and can be used in both analog and digital circuits. MOSFET can operate in a circuit in two configurations (Figure 11.21):

- depletion type
- enhancement type

Depletion-Type MOSFET

Depletion-type MOSFETs, normally known as *switched-on devices*, are usually closed when there is no-bias voltage at the gate terminal. If the gate voltage increases in the positive, then the channel width increases in depletion mode. As a result, the drain current I_D through the channel increases. If the applied gate voltage is more negative, then the channel width is less and MOSFET may enter into the cutoff region. The depletion-mode MOSFET is a rarely used type of transistor in electronic circuits. When there is no voltage on the gate, the channel shows its maximum conductance. As the voltage on the gate is either positive or negative, the channel conductivity decreases.

Enhancement-Type MOSFET

Enhancement-type MOSFETs are a commonly used type of transistor, equivalent to a normally open switch because it does not conduct when the gate voltage is zero. If the positive voltage ($+V_{GS}$) is applied to the N-channel gate terminal, then the channel conducts and the drain current flows through the channel. If this bias voltage increases to more positive then channel width and drain current through the channel increase. But if the bias voltage is zero or negative ($-V_{GS}$), then the transistor may switch off and the channel is in a nonconductive state. So now we can say that the gate voltage of the enhancement-type MOSFET enhances the channel.

As shown in Figure 11.21, within depletion-type and enhancement-type MOSFETs are:

- **P-channel MOSFETs**: MOSFET that has P channel region between source and gate is called P-channel MOSFET. The drain and source are in a heavily doped p + region and the body or substrate is n-type. The flow of current is positively charged holes. When negative-gate voltage is applied, the electrons present under the oxide layer are pushed downward into the substrate with a repulsive force. The deflection region is populated by the bound positive charges, which are associated with the donor atoms. The negative-gate voltage also attracts holes from the p + source and drain region into the channel region.
- **N-channel MOSFETs**: MOSFET that has n channel region between source and gate is called N-channel MOSFET. The drain and source are in a heavily doped n + region and the substrate or body is P-type. The current flow is as a result of negatively charged electrons. At application of the positive-gate voltage, the holes present under the oxide layer are pushed downward into the substrate with a repulsive force. The deflection region is populated by the bound negative charges, which are associated with the acceptor atoms. The electron-reach channel is formed. The positive voltage also attracts electrons from the n + source and drain regions into the channel.

Steady-State Characteristics

MOSFETS are voltage-controlled devices with very high input impedances. The gate draws a very small leakage current, on the order of nano-amperes. The current gain, which is the ratio of drain current I_D, to input gate current I_G, is typically on the order of 10^9. However, the current gain is not an important parameter. The trans-conductance, which is the ratio of drain current to gate voltage, defines the transfer characteristics and is a very important parameter. Transfer characteristics of MOSFETs are shown in Figure 11.22 [5].

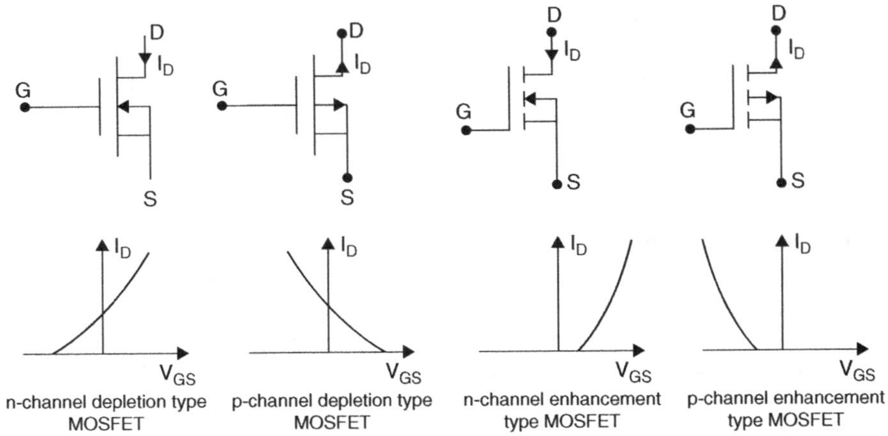

Figure 11.22 Transfer characteristics of MOSFETs.

Due to high-drain current and low-drain voltage, the power MOSFETs are operated in the linear region for switching actions. In the saturation region, the drain current remains almost constant for any increase in the value of V_{DS} and the transistors are used in this region for voltage amplification. It should be noted that saturation has the opposite meaning to that for bipolar transistors. The steady-state model is the same for both depletion-type and enhancement-type MOSFETs. The trans-conductance g_m is defined as

$$g_m = \frac{\Delta I_D}{\Delta V_{GS}}. \tag{11.5}$$

The output resistance is defined as

$$R_{DS} = \frac{\Delta V_{DS}}{\Delta I_D}. \tag{11.6}$$

Example 4

Two MOSFETs connected in parallel carry a total current of $I_T = 20$ A. The drain-to-source voltage of MOSFET M_1 is $V_{DS1} = 2.5$ V and that of MOSFET M_2 is $V_{DS2} = 3$ V. Determine the drain current of each transistor and the difference in current sharing if the current-sharing series resistances are (a) $R_{S1} = 0.3\Omega$ and $R_{S2} = 0.2\Omega$.

Solution:

$$I_{D1} + I_{D2} = I_T \tag{11.7}$$

$$V_{DS1} + I_{D1}R_{S1} = V_{DS2} + I_{D2}R_{s2} = R_{s2}\left(I_T - I_{D1}\right) \tag{11.8}$$

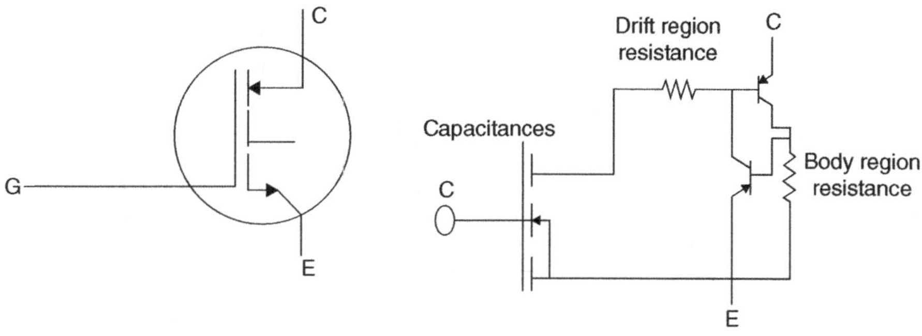

Figure 11.23 Circuit symbol and equivalent circuit for an IGBT.

therefore

$$I_{D1} = \frac{V_{DS2} - V_{DS1} + I_T R_{S2}}{R_{s1} + R_{s2}} \tag{11.9}$$

$$\frac{3 - 2.5 + 20 \times 0.2}{0.3 + 0.2} = 9\ \text{A}$$

$$I_{D2} = 20 - 9 = 11\ \text{A}$$

$$I_{D1} = \frac{3 - 2.5 + 20 \times 0.5}{0.5 + 0.5} = 10.5\ \text{A}$$

$$I_{D2} = 20 - 10.5 = 9.5\ \text{A}$$

Application of MOSFETs

MOSFET devices have a variable number of applications, including linear-amplifier stages, analog switches, basic chopper amplifiers, and switched-capacitor circuits.

11.4.4 Insulated Gate Bipolar Transistor

An insulated gate bipolar transistor (IGBT) combines the advantage of BJTs and MOSFETs, as shown in Figure 11.23 [3]. An IGBT has high input impedance, like MOSFETs, and low on-state conduction losses, like BJTs. But there is no second breakdown problem like BJTs. By design and structure, the equivalent drain-to-source resistance R_{DS} is controlled to behave like that of a BJT. An IGBT is made of four alternate pnpn layers and could latch like a thyristor given the necessary condition that $\alpha_{npn} + \alpha_{pnp} > 1$.

Steady-State Characteristics of IGBTs

The VI characteristics of an n-channel IGBT are shown in Figure 11.24. They appear qualitatively like those of a logic-level BJT except that the controlling parameter is not a base current but the gate-emitter voltage.

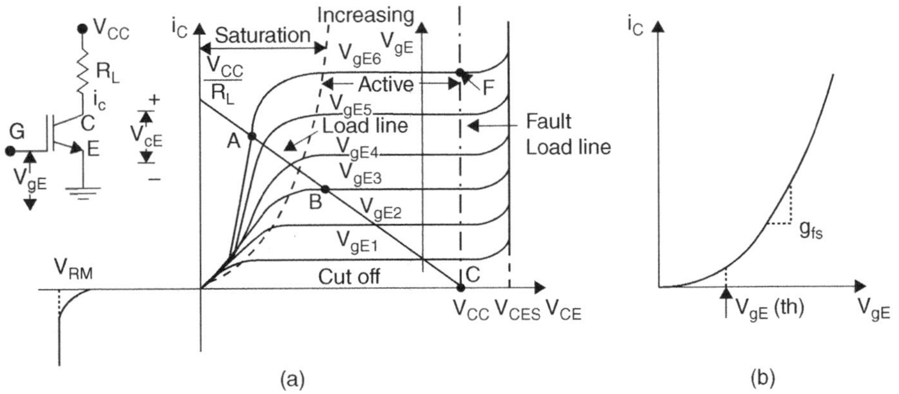

Figure 11.24 Static characteristics of an IGBT. (a) Output characteristics (b) Transfer characteristics.

An IGBT is a voltage-controlled device similar to a power MOSFET. It has lower switching and conducting losses while sharing many of the appealing features of power MOSFETs, such as ease of gate drive, peak current, capability, and ruggedness. An IGBT is inherently faster than a BJT. However, the switching speed of IGBTs is inferior to that of MOSFETs. The current rating of a single IGBT can be up to 400 A, 1,200 V, and the switching frequency can be up to 20 kHz.

A general comparison between BJTs, MOSFETs and IGBTs is given in Table 11.2.

11.4.5 Thyristors

A thyristor is a bistable semiconductor device made of three or more junctions that can be switched from the off state to the on state and vice versa. Electronically controlled switches use thyristors in industrial, military, aerospace, commercial, and consumer applications. Thyristors can control power to many kinds of loads, such as a simple lamps, power supplies, voltage regulators, and industrial motors. Circuits using AC or DC can use thyristors for power control.

TABLE 11.2 Properties of BJTs, MOSFETs, and IGBTs

Device Characteristic	Power BJT	Power MOSFET	IGBT
Voltage Rating	High < 1kV	High < 1kV	Very high > 1kV
Current Rating	High < 500A	Low < 200A	High > 500A
Input Drive	Current, h_{FE} 20–200	Voltage, V_{GS} 3–10V	Voltage, V_{GE} 4–8V
Input Impedance	Low	High	High
Output Impedance	Low	Medium	Low
Switching Speed	Slow (uS)	Fast (nS)	Medium
Cost	Low	Medium	High

The most commonly used thyristors for power control are:

- silicon-controlled rectifier (SCR), a reverse blocking/unidirectional triode, gate-triggered, three-terminal thyristor that has positive anode to cathode voltages and exhibits a reverse blocking state for negative anode to cathode voltages
- triode for alternating current (TRIAC), a bidirectional triode, gate-triggered, three-terminal thyristor that switches for either positive anode to cathode voltages or negative anode to cathode voltages

SCR Characteristics

An SCR conducts when the anode to cathode is forward biased and a gate-trigger signal is applied. After an SCR is turned on, it remains on as long as it has a minimum anode current. This minimum anode current that maintains conduction is the holding current. Holding current (I_H) is the minimum current required to keep an SCR conducting.

Figures 11.25 and 11.26 [5] show the voltage/current characteristics of an SCR with no gate current applied and with a forward- and reverse-blocking voltage rating. When the forward-blocking voltage is reached, the SCR turns itself on. The anode-to-cathode voltage decreases, and the anode current increases. Reverse-blocking voltage is the maximum reverse voltage that may be applied without turning on an SCR. Forward-blocking voltage is the maximum forward voltage that may be applied without turning on an SCR.

Internally, the SCR is similar to two cross-connected transistors.

As seen in Figure 11.27, a gate-trigger signal turns on transistor Q2. Q2 then forward biases the Q1 base-emitter junction, and both transistors turn on, providing a current path from the anode to the cathode. Both transistors are then latched on until the anode current is removed. SCR can change AC into DC and at the same time can control the amount of power fed to the load. Thus SCR combines the features of a rectifier and a transistor.

An SCR has two states, i.e., either it does not conduct or it conducts heavily. There is no state in between. Therefore, SCR behaves like a switch. There are two ways to turn on the SCR. The first method is to keep the gate open and make the supply voltage equal to the break-over voltage. The second method is to operate SCR with supply voltage less than break-over voltage and then turn it on by means of a small voltage (typically 1.5 V, 30 mA) applied to the gate. Applying small positive voltage to the gate is the normal way to close an SCR because the break-over voltage is usually much greater than supply voltage. To open the SCR (i.e., to make it nonconducting), reduce the supply voltage to zero.

Breakover voltage is defined as the minimum forward voltage, when the gate is open, at which the SCR is turned on. This implies that if the breakover voltage of an SCR is 187 V, SCR remains turned off as long as the supply voltage is less than 187 V.

Holding current is defined as the maximum anode current, when the gate is open, at which SCR is turned off. It is the anode current when SCR turn-off is affected.

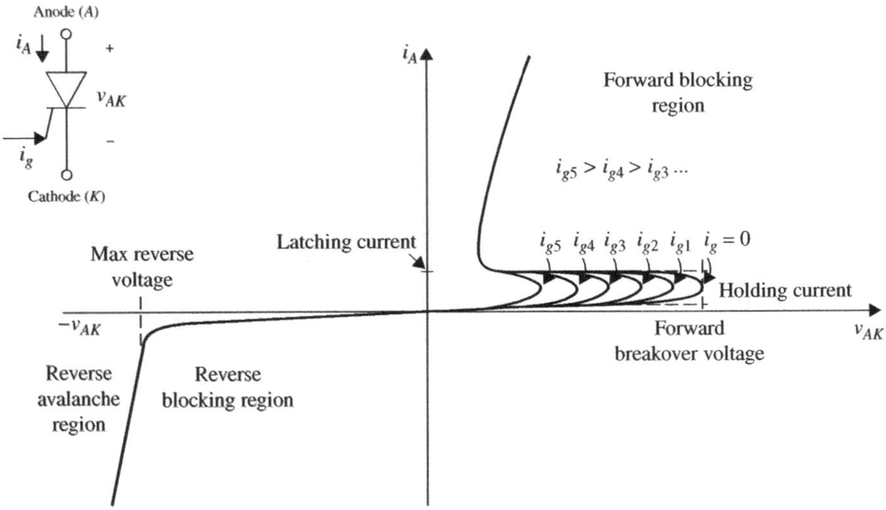

(a)

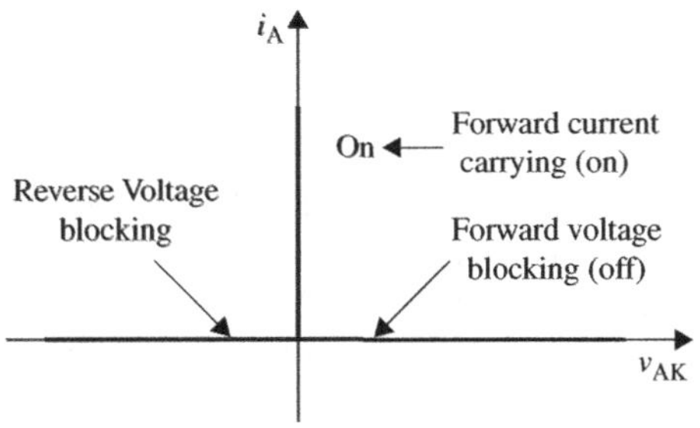

(b)

Figure 11.25 Switching characteristics of SCR. (a) I–V characteristics and symbolic representation (b) Ideal-switching characteristics.

SCR Turn-On and Turn-Off

To facilitate SCR turn-on, the gate voltage is increased to a minimum value to initiate triggering. This minimum value of gate voltage at which SCR turn-on is achieved is called the *trigger voltage*. The resulting gate current is called *trigger current*.

- DC-gate bias: Here, a positive, gate-triggering voltage is applied to the gate to turn on the SCR. A resistance is shunt connected in the circuit to facilitate noise suppression and improve the turn-on time. The primary factor on which the turn-on time depends is the magnitude of the gate-current AC. A higher gate current implies shorter turn-on time.

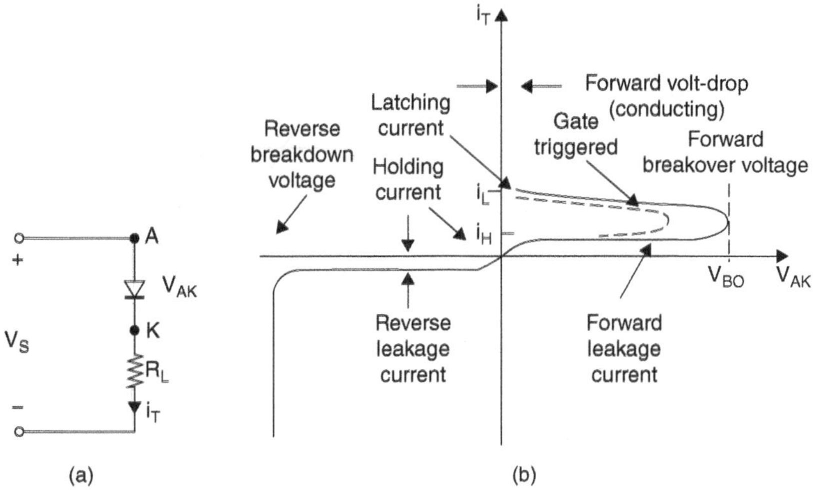

Figure 11.26 SCR circuit and characteristics. (a) Circuit. (b) v-i characteristics.

- AC trigger method: During the positive half-cycle of the AC gate current, the SCR is turned on.

SCR turn-off is more complex than turn-on. This is because the gate loses control once turn-on is achieved. There are two methods of thyristor turn-off:

- anode current reduction, when the anode current is reduced below the holding current and the SCR is turned off
- forced commutation, which discharges a capacitor in parallel with an SCR

SCR Half-Wave, Full-Wave, and Full-Wave-Bridge Waveforms

Figure 11.28 (a) shows a half-wave SCR with resistor R representing the load. Figure 11.28(b) shows the output voltage waveform and the trigger-pulse waveform. When the input voltage is in the positive half cycle, the thyristor is forward biased.

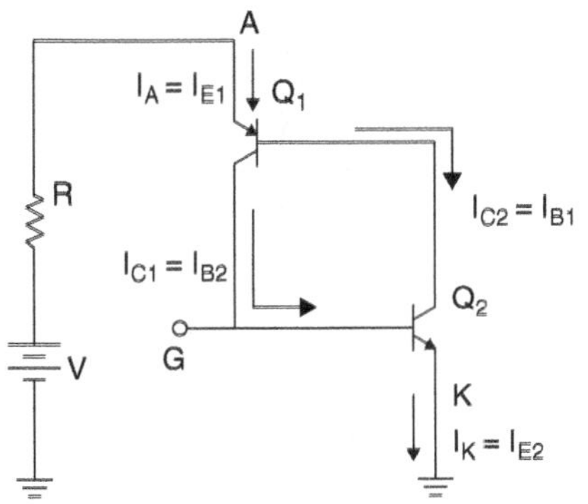

Figure 11.27 SCR equivalent circuit.

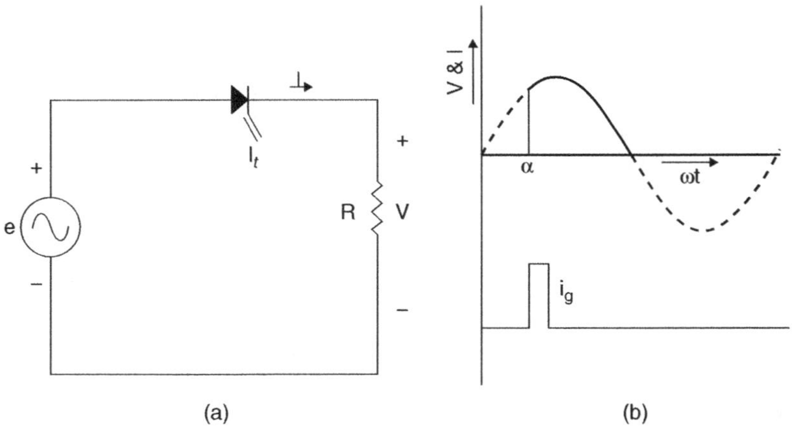

Figure 11.28 SCR half-wave rectifier and waveform. (a) SCR half-wave rectifier. (b) Waveform and trigger pulse.

The SCR will only start conducting after the gate signal α has been applied, which will conduct from α to π, the output voltage as seen in Figure 11.28(b). At $\omega t = \alpha$, the SCR will be naturally commutated since, at that point, the output voltage and current are zero. In the negative half cycle, the output will be zero since the SCR will be in reverse-bias mode. In the next positive half cycle, the SCR will conduct if the gate has been fired, and this continues periodically.

Figure 11.29(a) shows a full-wave SCR with a load and Figure 11.29(b) shows the output voltage waveform. When the input voltage is in the positive half cycle, the thyristor SCR1 is forward biased while SCR2 is reverse biased. The SCR1 will only start conducting after the gate signal α has been applied, which will conduct from α

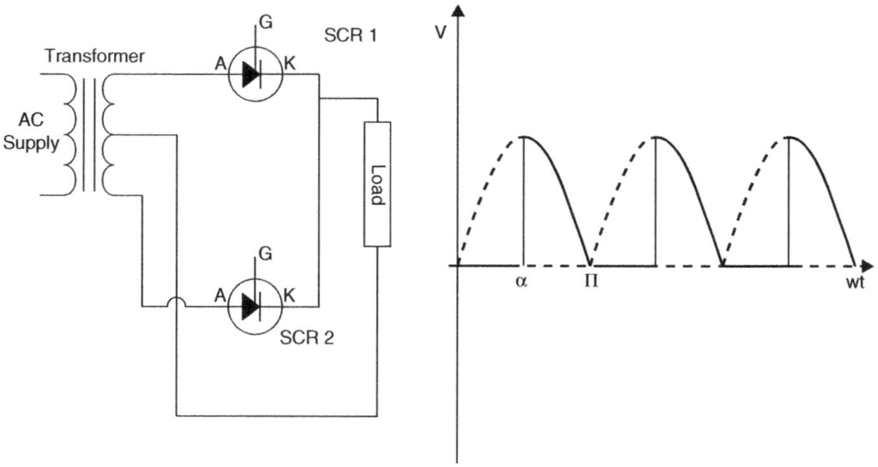

Figure 11.29 SCR full-wave rectifier and waveform.

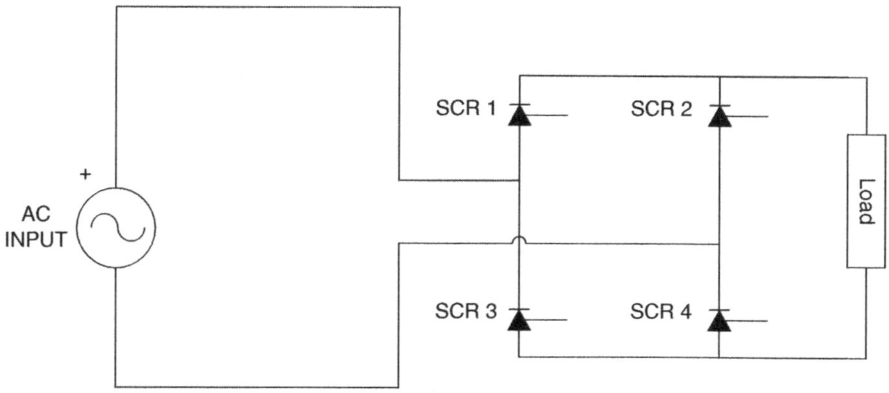

Figure 11.30 SCR full-wave bridge rectifier.

to π, the output voltage as seen in Figure 11.29(b), in which SCR1 will allow load current flow through. In the negative half cycle, the thyristor SCR2 is forward biased while SCR1 is reverse biased. The SCR2 will only start conducting after the gate signal α has been applied, which will conduct from α to π, the output voltage as seen in Figure 11.29b, in which SCR2 will allow load current flow through. In the next positive half cycle, the process continues periodically as previously.

As shown in Figure 11.30, during the positive half cycle of the input-voltage waveform, thyristors SCR1 and SCR4 are forward biased and thus conducting while SCR2 and SCR3 are in reverse-bias mode. During the negative half cycle of the input-voltage waveform, thyristors SCR2 and SCR3 are forward biased and thus conducting, while those of SCR1 and SCR4 are reverse biased. The output waveform is like that of the full-wave rectifier shown in Figure 11.29(b).

Application of Thyristors

Power thyristors are used in applications where high voltages and currents are present. They are generally used in order to control AC. Power thyristors can also be used as the control elements for phase-fired controllers. They are commonly used in inverter circuits for converting direct power to alternating power of specified frequency. These are also used in converters to convert an alternating power into alternating power of different amplitude and frequency.

11.4.6 TRIAC

A TRIAC is a three-terminal, bidirectional thyristor that, unlike SCR, is capable of conducting in either direction between main terminals. A TRIAC can be triggered in the positive or negative half cycle of the AC voltage source by applying a positive or negative gate signal, respectively. Similar to SCR, TRIAC requires minimum current to remain in conduction. This minimum current is the holding current. The i–v characteristics and ideal-switching characteristics with symbolic representation of TRIAC are shown in Figure 11.31.

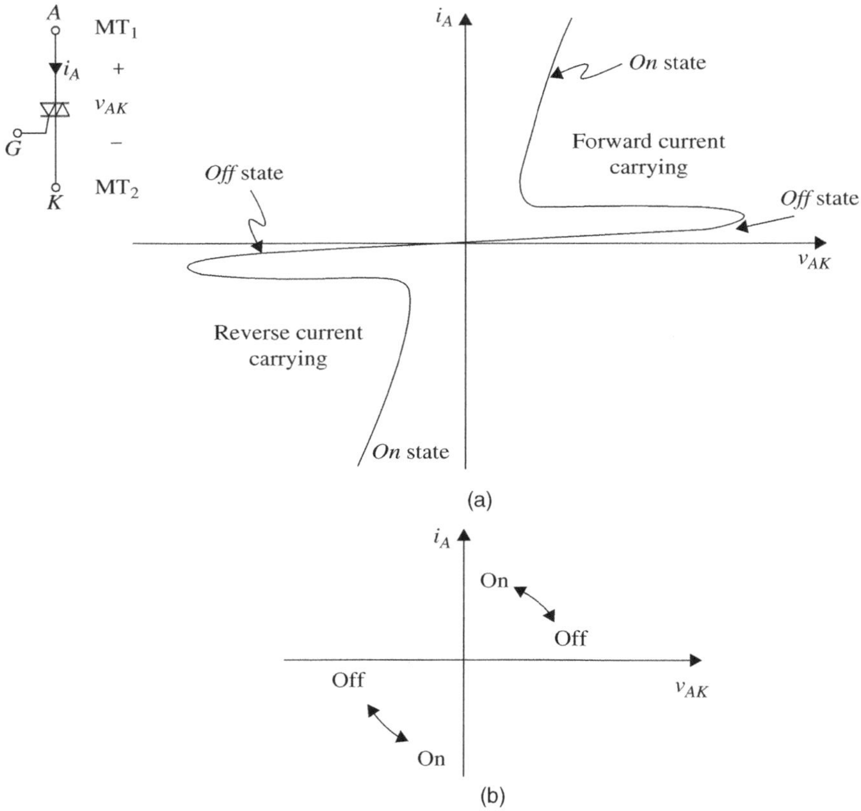

Figure 11.31 Switching characteristics of a TRIAC. (a) I–V characteristics and symbolic representation (b) Ideal-switching characteristics.

11.5 APPLICATIONS OF POWER ELECTRONICS DEVICES TO MACHINE CONTROL

Choppers

A chopper is a static-power electronic device that converts fixed DC voltage or power to variable DC voltage or power. It is an on/off switch that produces a chopped voltage across a load when connecting the load to and disconnecting the load from a DC source. The two types of choppers are the step-up chopper and step-down chopper.

Figure 11.32 shows a step-down chopper with resistive load. The thyristor in the circuit acts as a switch. The supply voltage appears across the load when thyristor is on and the voltage across the load will be zero when the thyristor is off. The output voltage and current waveforms are shown in Figure 11.33.

V_{dc} = average value of output or load voltage

I_{dc} = average value of output or load current

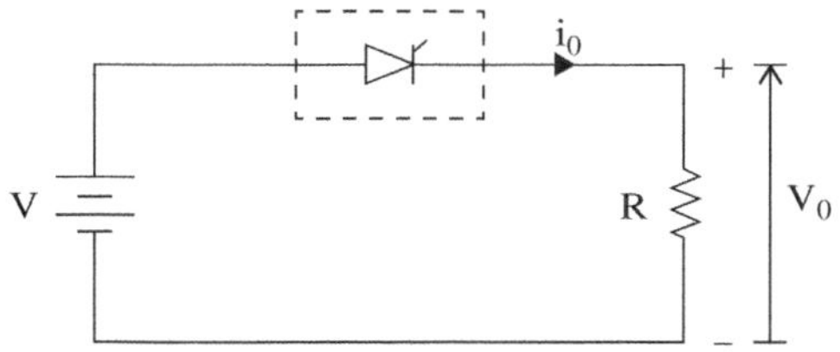

Figure 11.32 Step-down chopper with resistive load.

t_{ON} = time interval for which SCR conducts

t_{OFF} = time interval for which SCR is off

$T = t_{ON} + t_{OFF}$ = period of switching or chopping period

$f = \frac{1}{T}$ = frequency of chopper switching or chopping frequency.

Average output voltage

$$V_{dc} = V\left(\frac{t_{ON}}{t_{ON} + t_{OFF}}\right) \quad (11.10)$$

$$V_{dc} = V\left(\frac{t_{ON}}{T}\right) = V.d \quad (11.11)$$

$$\left(\frac{t_{ON}}{t}\right) = d = \text{dutycycle} \quad (11.12)$$

Average output current

$$I_{dc} = \frac{V_{dc}}{R} \quad (11.13)$$

$$I_{dc} = \frac{V}{R}\left(\frac{t_{ON}}{T}\right) = \frac{V}{R}d \quad (11.14)$$

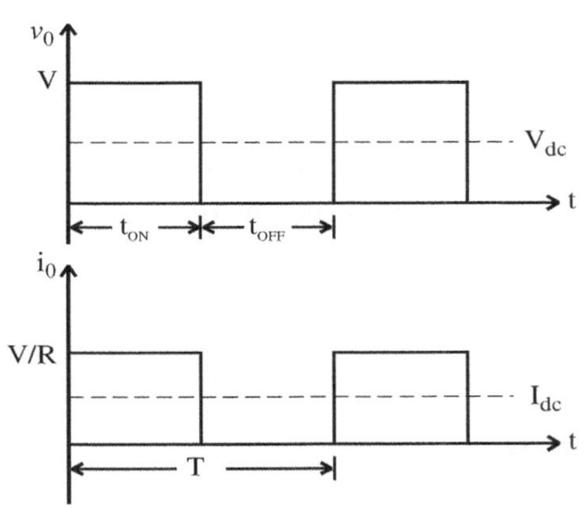

Figure 11.33 Step-down choppers: output voltage and current waveforms.

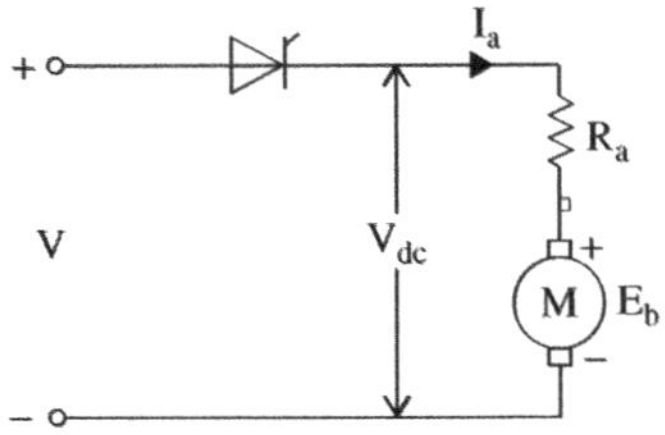

Figure 11.34 DC Motor Chopper Circuit.

Example 5

A chopper circuit is operating on time-ratio control at a frequency of 2.5 kHz on a 415 V supply. If the load voltage is 300 V, calculate the conduction period of the thyristor in each cycle.

Solution:

$$V = 415\,V,\ V_{dc} = 350\,V,\ f = 2.5\,kHz$$

Chopping period

$$T = \frac{1}{f}$$

$$T = \frac{1}{2.5 \times 10^3} = 0.4\,msec$$

Output voltage

$$V_{dc} = V\left(\frac{t_{ON}}{T}\right)$$

Conduction period of thyristor

$$t_{ON} = \frac{T \times V_{dc}}{V}$$

$$t_{ON} = \frac{0.4 \times 10^{-3} \times 300}{415}$$

$$t_{ON} = 0.289\,\text{msec}$$

Example 6

A 250 V, DC motor fed by a chopper runs at 1,200 rpm with a duty ratio of 0.9, Figure 11.34. What must be the on time of the chopper if the motor must run at 800 rpm? The chopper operates at 120 Hz.

Solution:

Speed of motor $N_1 = 1{,}200$ rpm
Duty ratio $d_1 = 0.9$, f = 120 Hz

We know that back emf of motor E_b is given by

$$E_b = \frac{\emptyset ZNP}{60A}$$

where N is the speed in rpm, ϕ is the flux/pole in Wb, Z is the number of armature conductors, P is the number of poles, and A is the number of parallel paths.

Therefore

$$E_b \propto \emptyset N$$
$$\propto N \text{ if flux } \emptyset \text{ is constant}$$
$$E_b = V_{dc} - I_a R_a$$

where I_a is the armature current and R_a is the armature resistance. Since R_a is not given, $I_a R_a$ drop is neglected.

Therefore

$$E_{b_1} = V_{dc_1} = 250\text{ V}$$
$$V_{dc_1} = d_1 V.$$

Supply,

$$V = \frac{V_{dc_1}}{d_1}$$
$$V = \frac{250}{0.9} = 277.78\text{ V}$$
$$E_{b_1} \propto N_1.$$

This implies that 250 ∝ 1,200, As speed changes, d also changes.

$$\text{Given } N_2 = 800\text{ rpm } E_{b_2} = ?$$
$$E_{b_2} \propto N_2$$

Taking the ratio of E_{b_1} to E_{b_2}, we have that

$$\frac{250}{E_{b_2}} = \frac{1,200}{800}$$
$$E_{b_2} = \frac{800 \times 250}{1,200} = 166.67\text{ V}.$$

But

$$E_{b_2} = V_{dc_2} = d_2 V$$
$$d_2 = \frac{V_{dc_2}}{V} = \frac{166.67}{277.78} = 0.6$$

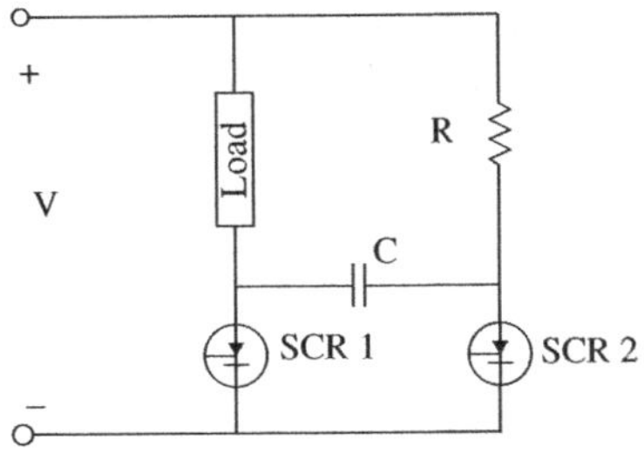

Figure 11.35 Solid-state DC circuit breaker.

Chopping frequency f = 120 Hz

$$T = \frac{1}{f} = \frac{1}{120} = 0.00833\,\text{sec}$$

$$T = 8.3\,\text{milliseconds}$$

$$d_2 = \frac{t_{ON}}{T}$$

On time of chopper

$$t_{ON} = d_2 T$$

$$t_{ON} = 0.6 \times 8.3 \times 10^{-3}$$

$$t_{ON} = 4.98\,milliseconds$$

11.6 APPLICATIONS OF POWER ELECTRONICS DEVICES TO POWER SYSTEM DEVICES

Circuit Breaker

Figure 11.35 [27] shows a DC circuit breaker. It is made up of two thyristors SCR1 and SCR2. The circuit breaker is closed when thyristor SCR1 is conducting and this makes the full DC voltage appear across both the load and the capacitor. Thyristor SCR2 is turned on when the circuit breaker is to be open. Therefore, capacitor C discharges through SCR1 and SCR2, reversing the current through SCR1 and leading SCR1 to be turned off. This operation makes the circuit breaker trip off. Simultaneously, capacitor C will be charged in the reverse direction, causing the current through SCR2 to be turned off. This makes this breaker ideal for high-speed applications.

Solid-State Relays

A solid-state relay (SSR) is an electronic switching device that switches on or off when a small external voltage is applied across its control terminals. An SSR (Figure 11.36 [27]) shows that the control signal from a control circuit turns the thyristor on or off, which in turn controls the main circuit (by actuating the circuit breaker to open, protecting equipment and consumers from harm).

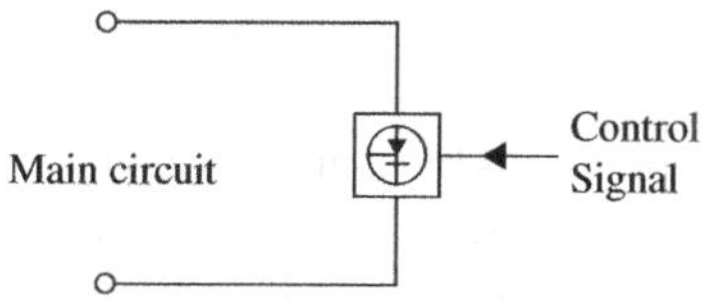

Figure 11.36 Solid-state relay.

11.7 APPLICATIONS OF POWER ELECTRONICS TO UTILITY, AEROSPACE, AND SHIPPING

Applications of power electronics range in size from a switched-mode power supply in an AC adapter, battery chargers, and fluorescent-lamp ballasts, through variable-frequency drives and DC-motor drives used to operate pumps, fans, and manufacturing machinery, and up to gigawatt-scale, high-voltage, DC-power-transmission systems used to interconnect electrical grids. Power electronic systems are found in virtually every electronic device.

Motor drives are found in pumps, blowers, and mill drives for textile, paper, cement, and similar facilities. Drives may be used for power conversion and for motion control [19]. For AC motors, applications include variable-frequency drives, motor soft starters, and excitation systems [20].

In hybrid electric vehicles, power electronics are used in two formats: series hybrid and parallel hybrid. The difference is the relationship of the electric motor to the internal combustion engine. Devices used in electric vehicles consist mostly of DC–DC converters for battery charging and DC–AC converters to power the propulsion motor. Electric trains use power electronic devices to obtain power, as well as for vector control using pulse-width modulation (PWM) rectifiers. The trains obtain their power from power lines. Another new use for power electronics is in elevator systems. These systems may use thyristors, inverters, permanent magnet motors, or various hybrid systems that incorporate PWM systems and standard motors [21].

Smart Grid

A smart grid is a modernized electrical grid that uses information and communications technology to gather and act on information, such as information about the behaviors of suppliers and consumers, in an automated fashion to improve the efficiency, reliability, economics, and sustainability of the production and distribution of electricity [22], [23].

Electric power generated by wind turbines and hydroelectric turbines by using induction generators can cause variances in the frequency at which power is generated. Power electronic devices are utilized in these systems to convert the generated AC voltages into high-voltage direct current (HVDC). The HVDC power can be more easily converted into three-phase power that is coherent with the power associated to the existing power grid. Through these devices, the power delivered by these systems is cleaner and has a higher associated power factor. Wind-power systems' optimum torque is obtained either through a gearbox or direct-drive technologies that can reduce the size of the power electronics device [24].

Electric power can be generated through photovoltaic cells by using power electronic devices. The produced power is usually then transformed by solar inverters. Inverters are divided into three different types: central, module-integrated, and string. Central converters can be connected either in parallel or in series on the DC side of the system. For photovoltaic "farms," a single central converter is used for the entire system. Module-integrated converters are connected in series on either the DC or AC side. Normally several modules are used within a photovoltaic system, since the system requires these converters on both DC and AC terminals. A string converter is used in a system that utilizes photovoltaic cells that are facing different directions. It is used to convert the power generated to each string, or line, in which the photovoltaic cells are interacting [24].

Grid Voltage Regulations

Power electronics can be used to help utilities adapt to the rapid increase in distributed residential/commercial solar-power generation. Germany and parts of Hawaii, California, and New Jersey require costly studies to be conducted before approving new solar installations. Relatively small-scale ground- or pole-mounted devices create the potential for a distributed-control infrastructure to monitor and manage the flow of power. Traditional electromechanical systems, such as capacitor banks or voltage regulators at substations, can take minutes to adjust voltage and can be distant from the solar installations where the problems originate. If voltage on a neighborhood circuit goes too high, it can endanger utility crews and cause damage to both utility and customer equipment. Further, a grid fault causes photovoltaic generators to shut down immediately, spiking demand for grid power. Smart grid-based regulators are more controllable than far more numerous consumer devices.

In another approach, a group of sixteen western utilities called the Western Electric Industry Leaders called for mandatory use of "smart inverters." These devices convert DC to household AC and can also help with power quality. Such devices could eliminate the need for expensive utility-equipment upgrades at a much lower total cost [25].

11.8 FACTS

FACTS have been evolving to a mature technology with high power rating. It has widespread applications and has become a top-rate, reliable technology, based on power electronics. The main purpose of these systems is to supply the network as quickly as possible with inductive or capacitive-reactive power that is adapted to its particular requirements, while also improving transmission quality and the efficiency of the power transmission system.

With the progression and development in power electronics applications comes not only improved performance of AC systems but also feasibility for long-distance applications. FACTS can also help solve technical problems in interconnected power systems. FACTS are available in:

- parallel connection: static var compensators (SVCs) or static synchronous compensators (STATCOM)
- series connection: fixed-series compensation (FSC) or thyristor-controlled/protected series compensation (TCSC/TPSC)

11.8.1 Parallel Compensation

Parallel compensation is any type of reactive power compensation, employing either switched or controlled units that are connected in parallel to the transmission network at a power system node.

Mechanically Switched Capacitors/Reactors (MSCs/MSRs)

Most economical reactive power compensation devices are mechanically switched devices. Mechanically switched capacitors are a simple but low-speed solution for voltage control and network stabilization under heavy load conditions. Their utilization has almost no effect on the short-circuit power but it increases the voltage at the point of connection. Mechanically switched reactors have exactly the opposite effect and are therefore preferable for achieving stabilization under low-load conditions. An advanced form of mechanically switched capacitor is the MSCDN, an MSC with an additional damping circuit for avoidance of system resonances.

Static var Compensators (SVCs)

Static var compensators are a fast and reliable means of controlling voltage lines and system nodes. Switching or controlling reactive power elements connected to the secondary side of the transformer changes the reactive power. Each capacitor bank is switched on and off by a thyristor valve. When system voltage is low, the SVC supplies capacitive reactive power and raises the network voltage. When system voltage is high, the SVC generates inductive reactive power and reduces the system voltage.

Static var compensators perform tasks such as:

- improvement in voltage quality
- dynamic reactive power control
- increase in system stability
- damping of power oscillations
- increase in power transfer capability
- unbalance control (option)

The design and configuration of an SVC, including the size of the installation, operating conditions, and losses, depend on the system condition (weak or strong), the system configuration (meshed or radial), and the tasks to be performed.

11.8.2 Series Compensation

Series compensation is defined as insertion of a reactive power element into transmission lines. The most common application is the fixed-series capacitor.

Thyristor-controlled series capacitors thyristor-protected series capacitors may also be installed.

Fixed-Series Capacitor

The simple and most cost-effective type of series compensation is provided by a fixed-series capacitor (FSC). FSCs comprise the actual capacitor banks, and for protection purposes, parallel arresters (metal oxide varistors or MOVs), spark gaps and a bypass switch for isolation purposes. Among the advantages of an FSC:

- increase in transmission capacity
- reduction in transmission angle

Thyristor-Controlled Series Capacitor

Reactive power compensation by means of a thyristor-controlled series capacitor (TCSC) can be adapted to a wide range of operating conditions. It is also possible to control the current and thus the load flow in parallel transmission lines, which simultaneously improves system stability. Further applications for TCSCs include power-oscillation damping and mitigation of sub-synchronous resonances, a crucial issue in case of large thermal generators.

Additional benefits of thyristor-controlled series compensation:

- damping of power oscillations (POD)
- load-flow control
- mitigation of sub-synchronous resonances
- increase in system stability

Thyristor-Protected Series Capacitor

When high-power thyristors are used, there is no need to install conventional spark gaps or surge arresters. Due to the very short cooling-down times of the special thyristor valves, thyristor-protected series capacitors (TPSCs) can be quickly returned to service after a line fault, allowing the transmission lines to be utilized to their maximum capacity. TPSCs are the first choice whenever transmission lines must be returned to maximum carrying capacity as quickly as possible after a failure.

Illustrative Problems and Examples

1. Evaluate the following for a CB transistor:

   (a) If it has $\alpha = 0.96$ and $I_E = 2\,\text{mA}$, find its current through the collector I_C and the base current I_B.

   (b) If it has $\beta = 20\text{A}$ and $I_E = 2\,\text{mA}$. Compute the value of α, I_C and I_E.

Solution:

(a) $I_C = \alpha I_E = 0.96 \times 2\,\text{mA} = 1.92\,\text{mA}$

$I_B = I_E - I_C = 2\,\text{mA} - 1.92\,\text{mA}$

$I_B = 0.08\,\text{mA}$

(b) $\beta = 20$, $\alpha = \frac{\beta}{1+\beta} = \frac{20}{21} = 0.9524.$

$I_C = \alpha I_E = 0.9524 \times 2 = 1.905\ \text{mA}$

2. During a power electronics laboratory session, a student records the following measurements of a transistor: $I_B = 5.202\ \text{mA}$, $I_C = 50\ m\ \text{A}$, $I_{CBO} = 2\ \text{mA}$. Using these measurements as a base, calculate:

(a) α, β, and I_E

(b) new value of I_B required to make $I_C = 10\ \text{mA}$

Solution:

$$I_{Sat} = I_{CBO} = 2\ \text{mA}$$

$$I_C = \alpha I_E + I_{Sat}$$

$$I_E = \frac{(I_C - I_{Sat})}{\alpha}$$

$$I_E = I_B + I_C = 5.202 + 50 = 55.202\ \text{mA}$$

$$\alpha = \frac{(I_C - I_{Sat})}{I_E} = \frac{50 - 2}{55.202} = 0.87$$

$$\beta = \frac{\alpha}{1-\alpha} = \frac{0.87}{1-0.87} = 6.665$$

To make $I_C = 10\ mA$

$$I_B = I_E - I_C = 55.202–10$$

$$I_B = 45.201\ mA$$

3. Answer the following:

(a) What are the different operation regions of the SCR?

(b) Define latching current and holding current.

(c) Explain why thyristors are referred to as charge controlled device.

(d) What is the relation between the gate signal and forward break-over voltage (VBO)?

(e) List the losses that occur in thyristors during operation.

(f) What are the advantages of speed control using a thyristor.

(g) What happens when a capacitor is connected to a generator load?

(h) Briefly explain why capacitors work on AC only.

4. An npn silicon transistor is connected as a CE amplifier with load R_L and source resistance R_S. The h parameters are: $h_{ie} = 1.1\text{k}$, $h_{re} = 2.5 \times 10 - 4$, $h_{fe} = 50$, and $h_{oe} = 25\mu\text{S}$. If $R_L = 10\text{k}$ and $R_S = 1\text{k}$, find the various gains and input and output impedances. Derive the relations.

5. For npn transistor as shown in Figure 11.11, calculate the exact value of the emitter current and V_{CE} using the values of $R_C = 5k\Omega$, $V_{CC} = 12V$ and $I_B = 9 \times 10^{-5}A$. Assume the transistor to be silicon and $\beta = 100$.
6. An SCR has a break-over voltage of 400 V, trigger current of 10 mA, and holding current of 10 mA. What do you infer from this? What will happen if the gate current is made 15 mA ?
7. An SCR in a circuit is subjected to a 50 A surge that lasts for 12 ms. Determine whether this surge will destroy the device, given that the circuit-fusing rating is 90 A^2s.
8. An SCR has a circuit-fusing rating of 50 A^2s. The device is being used in a circuit where it could be subjected to a 100 A surge. Determine the maximum allowable duration of such a surge.
9. A 220 Ω resistor is connected in series with the gate of an SCR. The gate current required to fire the SCR is 7 mA. What is the input voltage (Vin) required to fire the SCR?
10. What is the difference between an SCR and an ordinary rectifier?
    - (a) Explain clearly why SCRs are typically turned on by gate currents.
    - (b) Why can an SCR not be used as a bidirectional switch?
    - (c) How does SCR control the power fed to the load?
    - (d) Why are SCRs usually used in AC applications?
    - (e) Turning off the SCR is usually more problematic than turning it on. This is because once the device is on, the gate no longer controls it. With the aid of diagrams, explain two turn-off methods.
    - (f) Why is SCR turned on by high-frequency radiation?
11. SCRs are usually found useful in switching and control applications. This is because of their ability to control large currents in a load through a slight gate current. Explain explicitly the application of an SCR:
    - (a) as a static contactor
    - (b) for power control
    - (c) for speed control of a DC shunt motor

11.9 CHAPTER SUMMARY

Power electronics are an important subject of research in electronic and electrical engineering, which deals with the design, control, computation, and integration of nonlinear, time-varying, energy-processing electronic systems with fast dynamics. In this chapter, the elements of power electronics were treated, along with technical details of electronic components. The fundamental issues of modeling, computation steps, and applications of different electronic system/services were discussed. To help the reader, a summary of different types of electronic devices and applications to space, ship, and utility systems were given.

BIBLIOGRAPHY

[1] M. T. Thompson, "Notes 01: Introduction to Power Electronics," *Thompson Consulting Inc.*, 2007, www.thompsonrd.com/NOTES%2001%20INTRODUCTION%20TO%20POWER%20ELECTRONICS.pdf.

[2] IIT Kharagpur, "Power Semiconductor Devices," March 25, 2012, www.nptel.ac.in/courses/Webcourse-contents/IIT%20Kharagpur/Power%20Electronics/New_index1.html.

[3] M. H. Rashid, *Power Electronics Handbook: Devices, Circuits, and Applications*, Academic Press, Burlington, MA, 2007.

[4] D. W. Hart, *Power Electronics*, McGraw-Hill Education, New York, 2010.

[5] N. Mohan, T. Undeland, and W. P. Robbins, *Power Electronics: Converters, Applications, and Design*, John Wiley & Sons, Hoboken, NJ, 2003.

[6] B. K. Bose, "Evaluation of Modern Power Semiconductor Devices and Future Trends of Converters," *IEEE T. Ind. Appl.* 28, 2 (Mar/Apr 1992).

[7] T. L. Floyd, *Electronic Devices: Conventional Current Version*, 9th ed., Prentice Hall, Boston, 2012, 31–95, 385–412.

[8] M. H. Rashid, *Power Electronics Handbook: Devices, Circuits, and Applications*, Academic Press, Burlington, MA, 2007.

[9] A. M. Trzynadlowski, *Introduction to Modern Power Electronics*, John Wiley & Sons, Hoboken, NJ, 2010.

[10] M. H. Rashid, *Power Electronics Handbook: Devices, Circuits, and Applications*, Academic Press, Burlington, MA, 2007.

[11] T. L. Skvarenina, *The Power Electronics Handbook*, CRC Press, Boca Raton, FL, 2001.

[12] F. L. Luo, H. Ye, and M. Rashid, *Digital Power Electronics and Applications*, Elsevier, Amsterdam, 2005.

[13] B. Ozpineci and L. M. Tolbert, "Cycloconverters," University of Tennessee–Knoxville, nd.

[14] C. Klumpner, "Power Electronics 2," University of Nottingham, March 2012.

[15] V. Vodovozov and R. Jansikene, "Power Electronic Converters," Tallinn:TUT, 2006.

[16] S. Kim, S.-K. Sul, and T. A. Lipo, "AC/AC Power Conversion Based on Matrix Converter Topology with Unidirectional Switches," *IEEE T. Ind. Appl.* **36**, 1 (Jan/Feb 2000).

[17] T. Wijekoon, C. Klumpner, P. Zanchetta, et al., "Implementation of a Hybrid AC–AC Direct Converter with Unity Power Voltage Transfer Ratio," *IEEE T. Power Electro.* **23**, 4 (July 2008).

[18] S. Khader, A. Hadad, and A. A. Abu-aisheh, "The Application of PSIM and MATLAB/Simulink Courses in Power Electronics Courses," *IEEE Global Engineering Education Conference*, April 2011.

[19] B. K. Bose, "Power Electronics and Motion Control: Technology Status and Recent Trends," *IEEE T. Ind. Appl.* **29**, 5 (Sep/Oct 1993).

[20] F. Iov, M. Ciobotaru, D. Sera, et al., "Power Electronics and Control of Renewable Energy Systems," IEEE International Conferences on PEDS, November 2007.

[21] M. Yano and Rgounei Uchida, "History of Power Electronics in Japan," IEEE Transactions on Fundamentals and Materials, 2001.

[22] D. J. Hammerstrom, et al. *Pacific Northwest GridWise^{TM} Testbed Demonstration Projects, Part I. Olympic Peninsula Project*, Pacific Northwest National Laboratory, October 2007, www.pnnl.gov/main/publications/external/technical_reports/PNNL-17167.pdf.

[23] Freshersworld.com, "Electrical Questions from Power Electronics," http://placement.freshersworld.com/electrical questions from power electronics/33121966351.

[24] J. M. Carrasco, L. G. Franquelo, J. T. Bialasiewicz, et al., "Power-Electronic Systems for the Grid Integration of Renewable Energy Sources: A Survey," in *IEEE T. Ind. Electron.* **53**, 4 (June 2006): 1002–1016.

[25] Jeff S. John, "Western Utilities Call for Smart Solar Inverters," https://www.greentechmedia.com/articles/read/western_utilities_call_for_smart_solar_inverter

[26] M. Lamonica, "Power Electronics Smooth Solar Transition," *MIT Technology Review*, January 21, 2014, www.technologyreview.com/s/523386/power-electronics-smooth-solar-transition.

[27] I. Batarseh and A. Harb, "Power Electronics," Springer International Publishing, December 2017.

[28] S. A. Nasar, *Electric Energy Conversion and Transmission*, Macmillian, New York, 1985.

CHAPTER 12

CONVERTERS AND INVERTERS

12.1 INTRODUCTION

In order to operate a device properly, it needs to be supplied with the right voltage level. Converters and inverters are needed to achieve this. The main difference between converters and inverters is what they do to the voltage. An inverter changes DC voltage into AC voltage and either increases or decreases it into the appropriate level. A converter changes the voltage level but does not change its type; so an AC voltage would still be AC and a DC voltage would still be DC. Many appliances are designed to work with AC, while small-scale power generators often produce DC.

12.2 DEFINITIONS

A **DC–DC converter** is an electronic circuitry that converts DC from one power level to another. It is widely used in regulated, switch-mode DC power supplies and in DC motor drive applications, Figure 12.1.

Typically, the input to these converters is an unregulated DC voltage, which is obtained by rectifying the line voltage. Therefore, it will fluctuate due to changes in the line-voltage magnitude. Switch-mode DC–DC converters are used to convert the unregulated DC input into a controlled DC output at a desired voltage level.

DC–AC converters are known as inverters. The function of an inverter is to change a DC input voltage to a symmetrical AC output voltage of desired magnitude and frequency. A variable output voltage can be obtained by varying the input DC voltage and the gain of the inverter, which is normally accomplished by pulse-width modulation control within the inverter. Inverters are widely used in industrial applications (e.g., variable-speed AC motor drives, induction heating, standby power supplies, uninterruptible power supplies).

An **AC–DC converter** (also called a *rectifier*) is an electrical device that converts AC, which periodically reverses direction, to DC, which flows in only one direction. The process is known as *rectification*.

Energy Processing and Smart Grid, First Edition. James A. Momoh

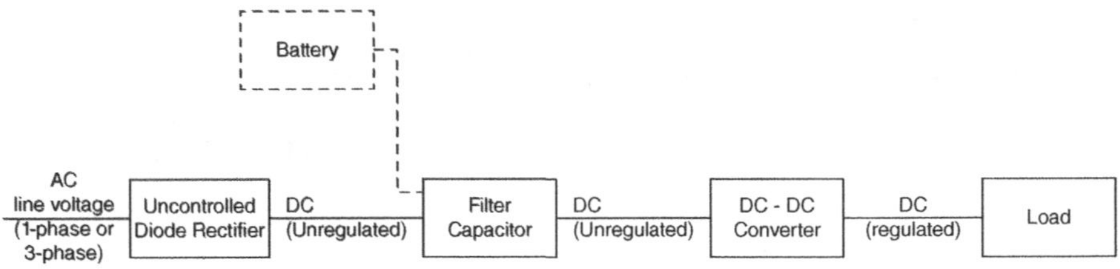

Figure 12.1 A DC–DC converter system.

12.3 DC–DC CONVERTERS

Consider the basic DC–DC converter shown in Figure 12.2(a). The average value of the output voltage V_o depends on t_{on} and t_{off}. The output voltage is controlled by controlling the switch on-and-off durations t_{on} and t_{off}. One method of controlling the output voltage is to employ switching at a constant frequency (hence a constant switching time period $T_s = t_{on} + t_{off}$.) and adjusting the on duration of the switch to control the average output voltage. The other control method is where both the switching frequency and hence the time period and the on duration of the switch are varied. This method is used only in DC–DC converters utilizing force commutated thyristors.

DC–DC converters can be classified as

- non-isolated DC–DC converters
- isolated DC–DC converters

The main difference is that isolated DC–DC converters incorporate electrical isolation achieved by their in-built transformers. Transformer design helps with noise filtering and converting high-ratio voltage efficiently (e.g., very high/low voltage to very low/high voltage). However, the structures of an isolated DC–DC converter are more complicated and have more components.

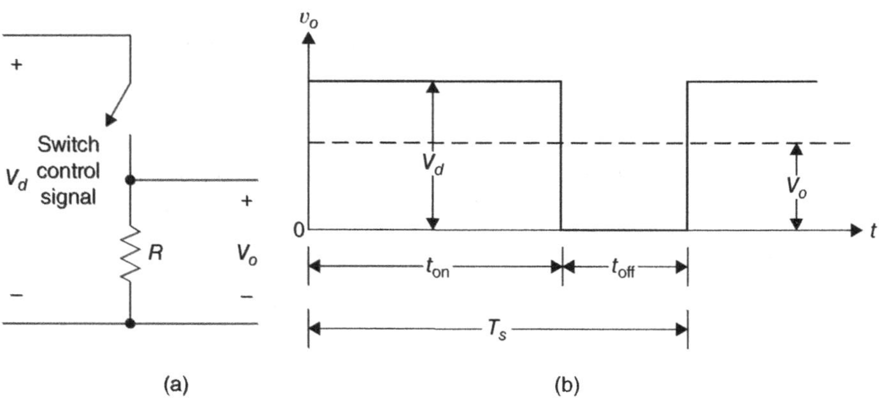

Figure 12.2 Basic DC–DC conversion.

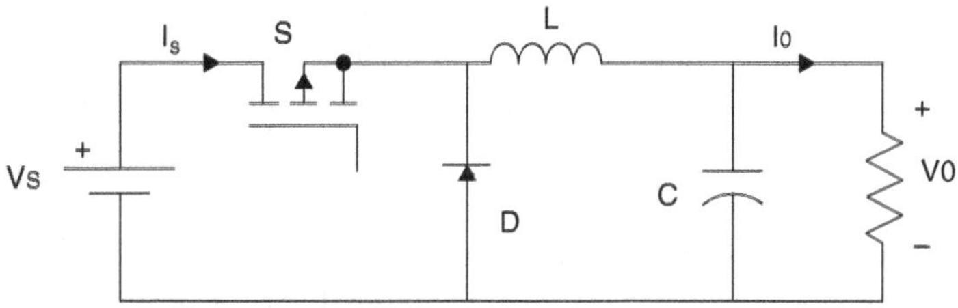

Figure 12.3 Step-down DC–DC converter.

The topologies of non-isolated DC–DC converters are as follows:

- buck converter (step-down voltage only)
- boost converter (step-up voltage only)
- buck–boost converter (inverse voltage, and step-up/step-down voltage)
- Ćuk converter (inverse voltage, and step-up/step-down voltage)

12.3.1 Modeling and Analysis of Buck, Boost, Buck/Boost, and Ćuk Converters

Step-Down (Buck) Converter

The buck converter produces a lower average output voltage than the DC input voltage V_S.

Its main application is in regulated DC power supplies and motor speed control. The basic circuit shown in Figure 12.3 constitutes a step-down converter for a purely resistive load, instantaneous input voltage V_S, and a purely resistive load. The average output voltage can be calculated in terms of the switch-duty ratio:

$$V_O = \frac{1}{T_s}\int_0^{T_s} V_0(t)\,dt = \frac{1}{T_s}\left(\int_0^{t_{on}} V_S\,dt + \int_{t_{on}}^{T_s} 0\,dt\right) = \frac{t_{on}}{T_s}V_S = DV_S. \quad (12.1)$$

The switch-duty ratio is defined as the ratio of the on duration to the switching time period, be expressed as

$$D = \frac{t_{on}}{T_S} = \frac{V_{control}}{\widehat{V}_{st}} \quad (12.2)$$

where T_s is the constant switching time period, equal to $t_{on} + t_{off}$.

Substituting for D in Equation 12.1 gives

$$V_o = \frac{V_S}{\widehat{V}_{st}}V_{control} = kv_{control} \quad (12.3)$$

where

$$k = \frac{V_S}{\widehat{V}_{st}} = constant. \quad (12.4)$$

By varying the duty ratio of the switch, the output voltage can be controlled. It is important to note that the average output voltage V_O varies linearly with the control voltage, as is the case in linear amplifiers. In actual practice, the circuit in Figure 12.3 has two drawbacks:

- In practice, the load would be inductive. Even with a resistive load, there would always be certain associated stray inductance. This means that the switch has to absorb or dissipate the inductive energy and therefore it may be destroyed.
- The output voltage fluctuates between zero and V_S, which is not acceptable in most applications.

The problem of stored inductive energy is overcome by using a diode, as shown in Figure 12.3, and the output voltage fluctuations are very much diminished by using a low-pass filter, consisting of an inductor and capacitor.

During the interval when the switch is on, the diode becomes reverse biased and the input provides energy to the load as well as to the inductor. During the interval when the switch is off, the inductor current flows through the diode, transferring some of its stored energy to the load.

In steady state, the integral of the inductor voltage over one time period must be zero. In the diode-inductor-capacitor loop of Figure 12.3

$$\int_0^{T_s} V_L\,dt = \int_0^{t_{on}} V_L\,dt + \int_{t_{on}}^{T_s} V_L\,dt = 0. \tag{12.5}$$

This implies that

$$\frac{t_{on}}{T_S} = \frac{V_O}{V_S} = D. \tag{12.6}$$

In the continuous-conduction mode, the voltage output varies linearly with the duty ratio of the switch for the given input voltage. It does not depend on any circuit parameter.

Neglecting power losses associated with all circuit elements, the input power P_S equals the output power P_O.

$$P_d = P_O \tag{12.7}$$

$$V_S I_S = V_O I_O \tag{12.8}$$

$$\frac{I_O}{I_S} = \frac{V_S}{V_O} = \frac{1}{D} \tag{12.9}$$

Therefore, in the continuous-conduction mode, the step-down converter is equivalent to a DC transformer, where the turns ratio of this equivalent transformer can be continuously controlled electronically in Equation 12.9.

Step-Up (Boost) Converter

Figure 12.4 [2] shows a step-up converter configuration. Its main application is in regulated DC power supplies and the regenerative braking of DC motors. As the name implies, the output voltage is always greater than the input voltage. When the switch

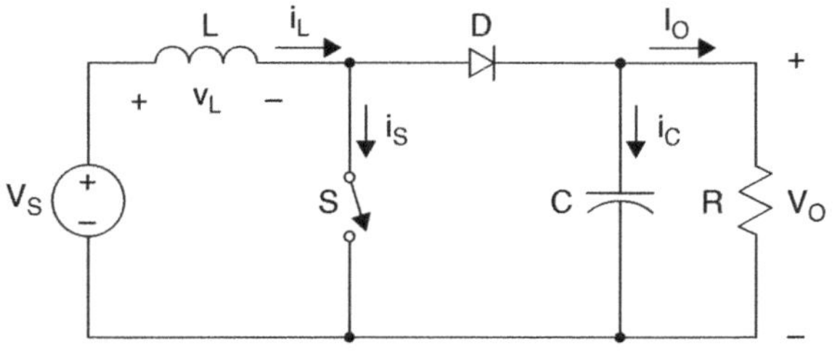

Figure 12.4 Step-up DC–DC converter.

is on, the diode is reverse biased, thus isolating the output stage. The input supplies energy to the inductor. When the switch is off, the output stage receives energy from the inductor as well as from the input. In the steady-state analysis presented here, the output filter capacitor is assumed to be very large to ensure a constant output voltage.

In steady state, the time integral of the inductor voltage over one time period must be zero, so that

$$V_S t_{on} + \left(V_S - V_o\right) t_{off} = 0. \tag{12.10}$$

Dividing both sides by T_S and rearranging terms yields

$$\frac{V_O}{V_S} = \frac{T_S}{t_{off}} = \frac{1}{1-D}. \tag{12.11}$$

Assuming a lossless circuit,

$$P_S = P_O \tag{12.12}$$

$$V_S I_S = V_O I_O \tag{12.13}$$

and

$$\frac{I_O}{I_S} = (1-D) = \frac{V_S}{V_O} \tag{12.14}$$

Buck–Boost Converter

A buck–boost converter can be obtained by the cascade connection of the two basic converters: the step-down converter and step-up converter. The main application of a step-down/step-up or buck–boost converter is in regulated DC power supplies, where a negative polarity output may be desired with respect to the common terminal of the input voltage and the output voltage can either be higher or lower than the input voltage.

In steady state, the output-to-input voltage conversion ratio is the product of the conversion ratios of the two converters in cascade, assuming that switches in both converters have the same duty ratio.

$$\frac{V_O}{V_S} = D\frac{1}{(1-D)} \tag{12.15}$$

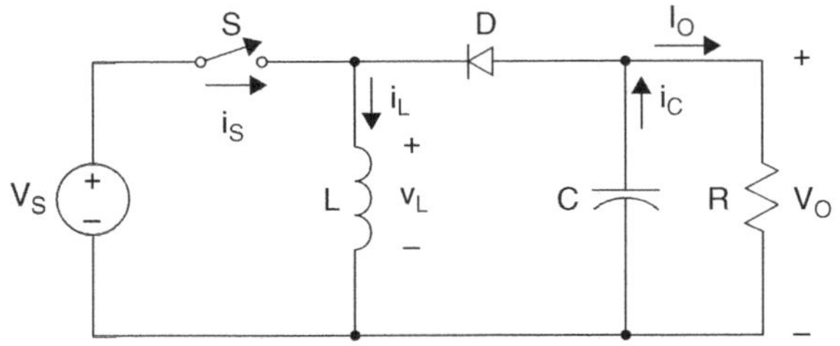

Figure 12.5 Buck–boost converter.

Assuming $P_S = P_O$, then

$$\frac{I_O}{I_S} = \frac{(1-D)}{D}. \tag{12.16}$$

Finally, the output voltage is higher or lower than the input voltage, based on the duty ratio D. The connection of the step-down and step-up converter can be combined into the single buck–boost converter shown in Figure 12.5 [2]. When the switch is open, the energy stored in the inductor is transferred to the output. No energy is supplied by the input during this interval.

Example 1

In a buck–boost converter operating at 20 kHz, L = 0.05 mH. The output capacitor C is sufficiently large and $V_S = 15$ V. The output is to be regulated at 10 V and the converter is supplying a load of 10 W. Calculate the duty ratio.

Solution:

$$I_o = \frac{P_O}{V_o} = \frac{10}{10} = 1\,A$$

Assuming continuous conduction, from Equation 11.15

$$\frac{V_O}{V_S} = D\frac{1}{(1-D)}$$

$$D\frac{1}{(1-D)} = \frac{10}{15}.$$

Therefore, the initial estimate duty ratio, D = 0.4.

Now we know that

$$I_{oB,\,max} = \frac{T_s V_O}{2L} = \frac{0.05 \times 10}{2 \times 0.05} = 5\,A.$$

Using

$$\text{D} = 0.4 \text{ and } I_{oB,\,\max} = 5\,A$$

$$I_{oB} = I_{oB,\,max}(1-D)^2$$

$$I_{oB} = 5(1-0.4)^2 = 1.8\text{ A}.$$

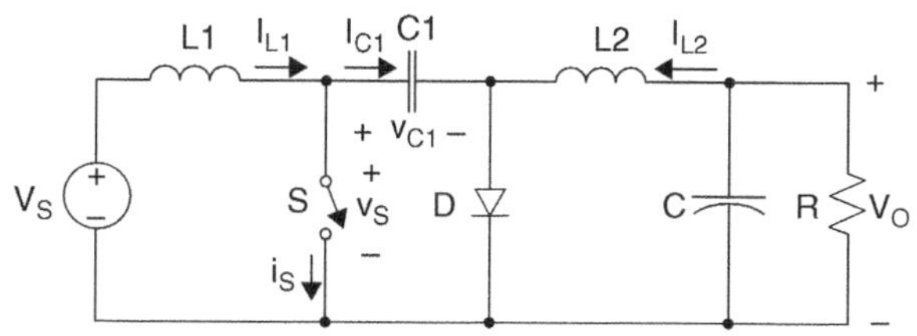

Figure 12.6 Ćuk DC–DC converter.

Since the output current is less than I_{oB}, the current conduction is discontinuous, therefore

$$\text{duty ratio D} = \frac{V_o}{V_S}\sqrt{\frac{I_0}{I_{0B,\,max}}} = \frac{10}{15}\sqrt{\frac{1.0}{5.0}} = 0.3.$$

Ćuk DC–DC Converter

This Ćuk DC–DC converter is obtained by using the duality principle on the circuit of a buck–boost converter. The Ćuk converter (see Figure 12.6 [2]) provides a negative-polarity, regulated-output voltage with respect to the common terminal of the input voltage. Here the capacitor acts as the primary means of storage and transfer of energy from input to output.

In steady state, the average inductor voltages are zero. Therefore, by inspection

$$V_C = V_S + V_0.$$

When the switch is off, the inductor currents flow through the diode. Capacitor C is charged through the diode by energy from both the input and L_2. Current i_S decreases, because V_{C1} is larger than V_S. Energy stored in L_1 feeds the output. Therefore, i_{L_O} decreases.

When the switch is on, V_C reverse biases the diode. The inductor currents i_{L2} and i_{L_1} flow through the switch. Since $V_C > V_O$, C_1 discharges through the switch, transferring energy to the output and L_1. Therefore i_{L2} increases. The input feeds energy to L_2 causing i_{L2} to increase.

The inductor currents are assumed to be continuous. If we assume capacitor voltage V_{C1} to be constant, then equating the integral of the voltages across L_2 and L_1 over one time period to zero yields

$$V_{C1} = V_S\frac{1}{(1-D)} \tag{12.17}$$

and

$$V_{C1} = V_0\frac{1}{(1-D)}. \tag{12.18}$$

From Equations 12.17 and 12.18

$$\frac{V_0}{V_S} = \frac{D}{1-D}. \tag{12.19}$$

Assuming $P_S = P_O$ then

$$\frac{I_O}{I_S} = \frac{(1-D)}{D} \tag{12.20}$$

where

$$I_{L2} = I_S \text{ and } I_{L_1} = I_O.$$

Therefore, the main benefit of the Ćuk converter is that you can control the continuous current at both in the input and output of the converter as it is based on the capacitor energy transfer. It has a low-switching loss making it more highly efficient. The downside of this converter is that it includes a high number of reactive components (L and C) and heavy current stresses on the components. And since the capacitor C provides a transfer of energy, ripples in the capacitor (C_O) current are high.

12.4 INVERTERS

Inverters are broadly classified into two types;

- single-phase inverters
- three-phase inverters

These inverters generally use pulse-width modulation (PWM) control signals for producing AC output voltage. An inverter is called a voltage-fed inverter (VFI) if the input voltage remains constant, a current-fed inverter (CFI) if the input current remains constant, and a variable DC-link inverter if the input voltage is controllable.

12.4.1 Modeling and Analysis of Inverter Technology

The single-phase inverter consists of two choppers. Q_1 and Q_2 are insulated gate bipolar transistors. When only Q_1 is turned on for a time $T_O/2$, the instantaneous voltage across the load V_o is $V_S/2$. If the transistor Q_2 only is turned on for a time $T_O/2$, $-V_S/2$ appears across the load. The logic circuit is designed so that Q_1 and Q_2 are not turned on at the same time. Figure 12.7 [1] shows the waveforms for the output voltage and transistor currents with a resistive load. This inverter requires a three-wire DC source and when a transistor is off, its reverse voltage is V_S instead of $V_S/2$. This is the half-bridge inverter.

The rms output voltage is calculated from

$$V_O = \left(\frac{2}{T_O}\int_0^{T_O/2} \frac{V_S^{\,2}}{4} dt\right)^{1/2} = \frac{V_S}{2}. \tag{12.21}$$

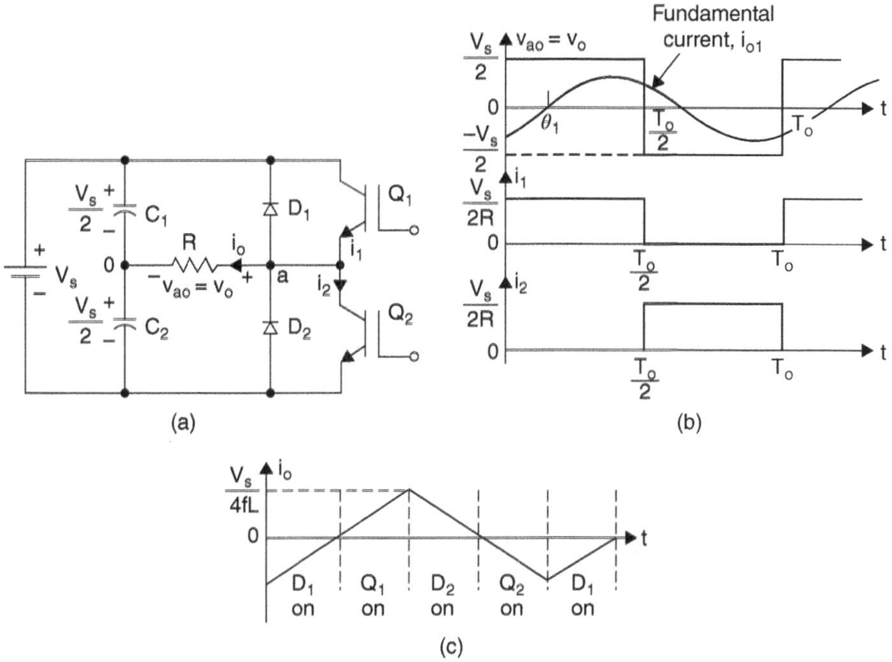

Figure 12.7 Single-phase half-bridge inverter. (a) Circuit. (b) Waveforms with resistive load. (c) Load current with highly inductive load.

The rms value of the fundamental component of the instantaneous output voltage is expressed as

$$V_1 = \frac{2V_S}{\sqrt{2}\,\pi} = 0.45\,V_S. \tag{12.22}$$

If the load is inductive, load current cannot change immediately with the output voltage. If the transistor Q_1 is turned off for a time $T_O/2$, load current continues to flow through D2, load, and lower half of DC source until the current falls to zero. Similarly, when Q2 is turned off at time T_O, load current flows through D_1 load and the upper half of the DC source. When any of the diodes conducts, energy is fed back to the DC source. These diodes are called *feedback diodes.*

If I_{01} is the rms fundamental load current, the fundamental output power (for n = 1) is

$$P_{01} = V_1 I_{01}\cos\theta_1 = I_{01}^2 R = \left[\frac{2V_S}{\sqrt{2}\pi\sqrt{R^2 + (\omega L)^2}}\right]^2 R. \tag{12.23}$$

The single-phase bridge inverter is shown in Figure 12.8 [1]. It consists of four choppers. When transistors Q_1 and Q_2 are turned on simultaneously, the output voltage V_S appears across the load. If transistors Q_3 and Q_4 are turned on at the same time, the voltage across the load is reversed and is $-V_S$.

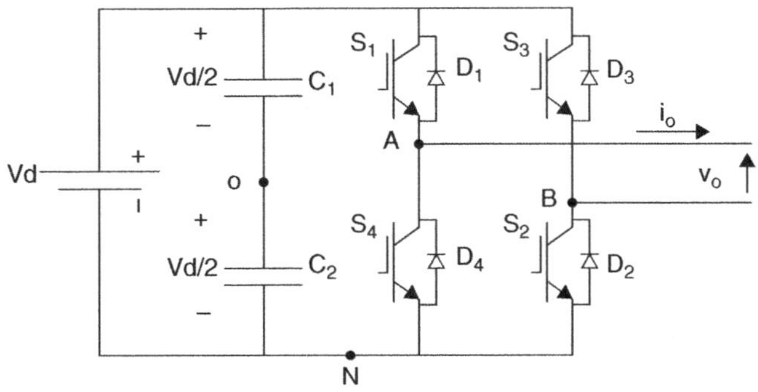

Figure 12.8 Single-phase full-bridge inverter.

The rms output voltage is calculated from

$$V_O = \left(\frac{2}{T_O} \int_0^{T_O/2} V_S^2 \, dt \right)^{1/2} = V_S. \tag{12.24}$$

The rms value of the fundamental component of the instantaneous output voltage is expressed as

$$V_1 = \frac{4V_S}{\sqrt{2}\,\pi} = 0.90\ V_S. \tag{12.25}$$

When diodes D_1 and D_2 conduct, the energy is fed back to the DC source. These are the feedback diodes.

Three-Phase Inverters

Three-phase inverters are normally used for high-power applications. Three single-phase half- or full-bridge inverters can be connected in parallel to form a configuration of a three-phase inverter. The gating signals of single-phase inverters should be advanced or delayed by 120° with respect to each other to obtain three-phase, balanced fundamental voltages.

The transformer primary windings must be isolated from each other, while the secondary windings may be Y connected or delta connected. The Y connection is usually to eliminate triple harmonics (n = 3, 6, 9, …) appearing on the output voltages and the circuit arrangement. This arrangement requires three single-phase transformers, twelve transistors, and twelve diodes. If the output voltages of single-phase inverters are not perfectly balanced in magnitudes and phases, the three-phase output voltages will be unbalanced. A three-phase output can be obtained from a configuration of six transistors and six diodes.

In this setup, two control signals can be applied to the transistors: 180° conduction and 120° conduction.

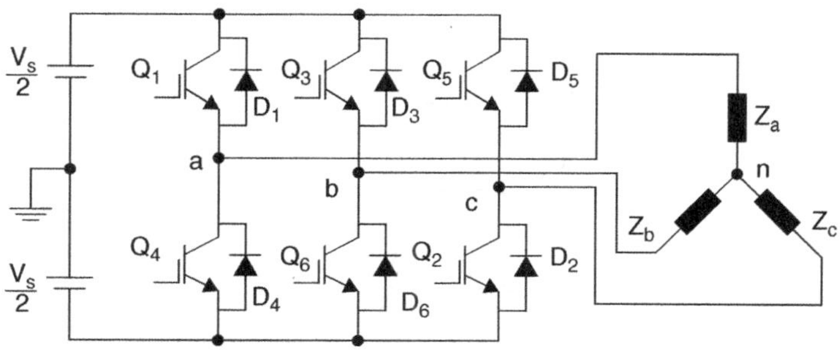

Figure 12.9 Three-phase bridge inverter with Y-connected load.

180° Conduction

Each transistor conducts for 180°. Three transistors remain on at any instant of time. When transistor Q_1 is switched on, terminal a is connected to the negative terminal of the DC input voltage. When transistor Q_4 is switched on, terminal a is brought to the negative terminal of the DC source. There are six modes of operation in a cycle and the duration of each mode is 60°.

The load may be Y connected or delta connected. Figure 12.9 shows a Y-connected load. For a Y connection, the line-to-neutral voltages must be determined to find the line (or phase currents). For the delta-connected load, the phase currents can be obtained directly from the line-to-line voltages. Once phase currents are known, the line currents can be determined.

In the Y-connection configuration, there are three modes of operation in a half cycle.

(a) Mode 1: $0 \leq \omega t < \pi/3$

$$
\begin{aligned}
R_{eq} &= R + \frac{R}{2} = \frac{3R}{2} \\
i_1 &= \frac{V_S}{R_{eq}} = \frac{2V_S}{3R} \\
v_{an} &= \frac{i_1 R}{2} = \frac{V_S}{3} \\
v_{bn} &= -i_1 R = -\frac{2V_S}{3R}
\end{aligned}
\tag{12.26}
$$

(b) Mode 2: $\pi/3 \leq \omega t < 2\pi/3$

$$
\begin{aligned}
R_{eq} &= R + \frac{R}{2} = \frac{3R}{2} \\
i_2 &= \frac{V_S}{R_{eq}} = \frac{2V_S}{3R} \\
v_{an} &= i_1 R = \frac{2V_S}{3} \\
v_{bn} &= v_{cn} = \frac{-i_2 R}{2} = -\frac{V_S}{3}
\end{aligned}
\tag{12.27}
$$

(c) Mode 3: $2\pi/3 \leq \omega t < \pi$

$$R_{eq} = R + \frac{R}{2} = \frac{3R}{2}$$

$$i_3 = \frac{V_S}{R_{eq}} = \frac{2V_S}{3R}$$

$$v_{an} = v_{bn} = \frac{i_3 R}{2} = \frac{V_S}{3}$$

$$v_{cn} = -i_3 R = -\frac{2V_S}{3} \tag{12.28}$$

The line-to-line rms voltage can be found from

$$V_L = \left[\frac{2}{2\pi}\int_0^{2\pi/3} V_s^2 d(\omega t)\right]^{1/2} = \sqrt{\frac{2}{3}}V_s = 0.8165\, V_S \tag{12.29}$$

$$V_{Ln} = \frac{4V_S}{\sqrt{2}\, n\pi}\cos\frac{n\pi}{6}. \tag{12.30}$$

The line-to-neutral voltages can be found from the line voltage

$$V_p = \frac{V_L}{\sqrt{3}} = \frac{\sqrt{2}V_S}{3} = 0.4717\, V_S. \tag{12.31}$$

12.4.2 Performance Analysis in Terms of Quality

The output of practical inverters contains harmonics and the quality of an inverter is normally evaluated in terms of the following performance parameters.

- **Quality 1**: Harmonic factor of nth harmonic, HF_n, a measure of individual harmonic contribution defined as

$$HF_n = \frac{V_n}{V_1}. \tag{12.32}$$

- **Quality 2**: Total harmonic distortion, THD, a measure of closeness in shape between a waveform and its fundamental component:

$$THD = \frac{1}{V_1}\left(\sum_{n=2,3,}^{\infty} V_n^2\right)^{1/2}. \tag{12.33}$$

- **Quality 3**: Distortion factor, DF. THD gives the total harmonic content, but it does not indicate the level of each harmonic component. DF indicates the amount of harmonic distortion that remains in a particular waveform after the

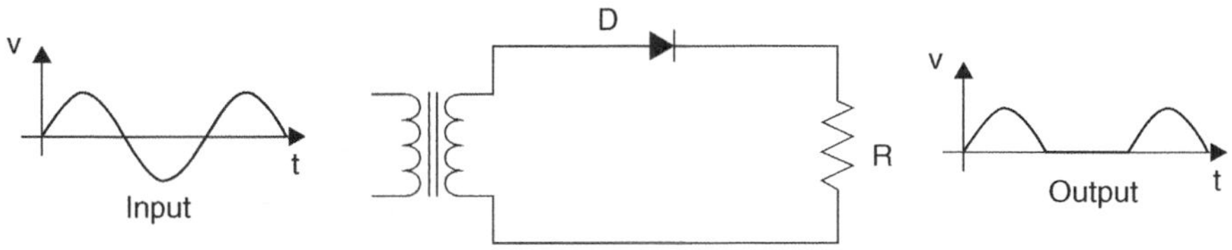

Figure 12.10 Half-wave rectifier.

harmonics of that waveform have been subjected to a second-order attenuation (i.e., divided by n^2)

$$DF = \frac{1}{V_1}\left[\sum_{n=2,3,\ldots}^{\infty}\left(\frac{V_n}{n^2}\right)^2\right]^{1/2}. \tag{12.34}$$

The DF of an individual harmonic component is defined as

$$DF = \frac{V_n}{V_1 n^2}. \tag{12.35}$$

12.5 RECTIFIERS

Physically, rectifiers take a number of forms, including vacuum tube diodes, mercury-arc valves, copper and selenium oxide rectifiers, semiconductor diodes, silicon-controlled rectifiers, and other silicon-based semiconductor switches. Historically, even synchronous electromechanical switches and motors have been used. Early radio receivers, called crystal radios, used a "cat's whisker" of fine wire pressing on a crystal of galena (lead sulfide) to serve as a point-contact rectifier or "crystal detector". Because of the alternating nature of the input AC sine wave, the process of rectification alone produces a DC current that, though unidirectional, consists of pulses of current. Many applications of rectifiers, such as power supplies for radio, television, and computer equipment, require a steady, constant DC current (as would be produced by a battery). In these applications, the output of the rectifier is smoothed by an electronic filter (usually a capacitor) to produce a steady current. There are different configurations of rectifiers:

(a) Single-Phase Rectifiers

- In a **half-wave rectifier**, a single-phase supply, either the positive or negative half of the AC wave is passed, while the other half is blocked. Because only half of the input waveform reaches the output, mean voltage is lower. Half-wave rectification requires a single diode in a single-phase supply, or three in a three-phase supply, Figure 12.10. Rectifiers yield a unidirectional but pulsating DC; half-wave rectifiers produce far more ripple than full-wave

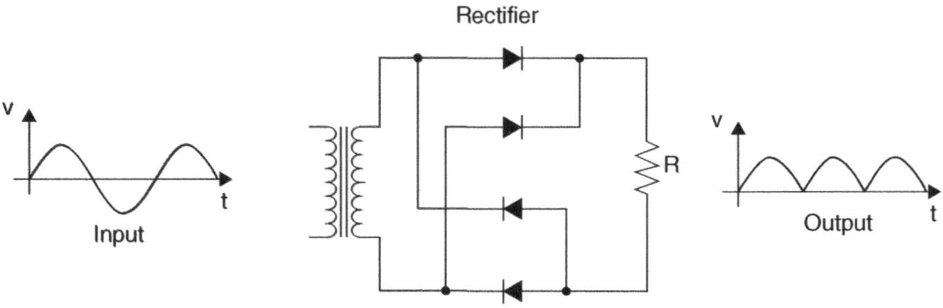

Figure 12.11 Graetz bridge rectifier: a full-wave rectifier using four diodes.

rectifiers, and much more filtering is needed to eliminate harmonics of the AC frequency from the output.

- A **full-wave rectifier** converts the whole of the input waveform to one of constant polarity (positive or negative) at its output. Full-wave rectification converts both polarities of the input waveform to pulsating DC, and yields a higher-average output voltage. Two diodes and a center-tapped transformer, or four diodes in a bridge configuration and any AC source (including a transformer without center tap), are needed [3]. Single semiconductor diodes, double diodes with common cathode or common anode, and four-diode bridges are manufactured as single components. A full-wave rectifier using four diodes is shown in Figure 12.11 [2]

  For single-phase AC, if the transformer is center tapped, then two diodes back to back (cathode to cathode or anode to anode, depending upon output polarity required) can form a full-wave rectifier Figure 12.12. Twice as many turns are required on the transformer secondary to obtain the same output voltage than for a bridge rectifier, but the power rating is unchanged.

(b) Three-Phase Rectifiers

Single-phase rectifiers are commonly used for power supplies for domestic equipment. However, for most industrial and high-power applications, three-phase rectifier circuits are the norm. As with single-phase rectifiers, three-phase rectifiers can take the form of a half-wave circuit, a full-wave circuit using a center-tapped transformer, or a full-wave bridge circuit.

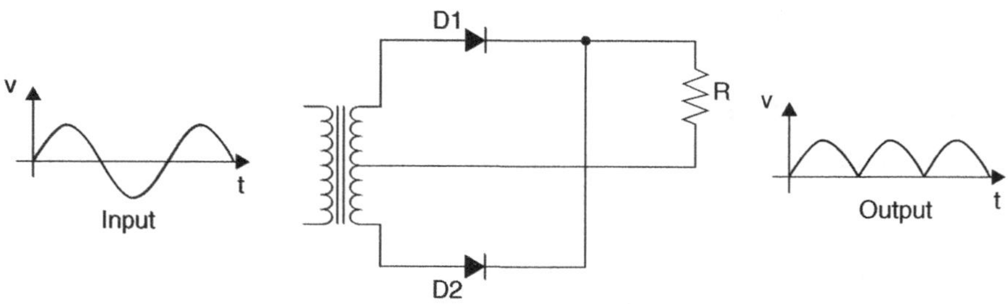

Figure 12.12 Full-wave rectifier using a center tap transformer and 2 diodes.

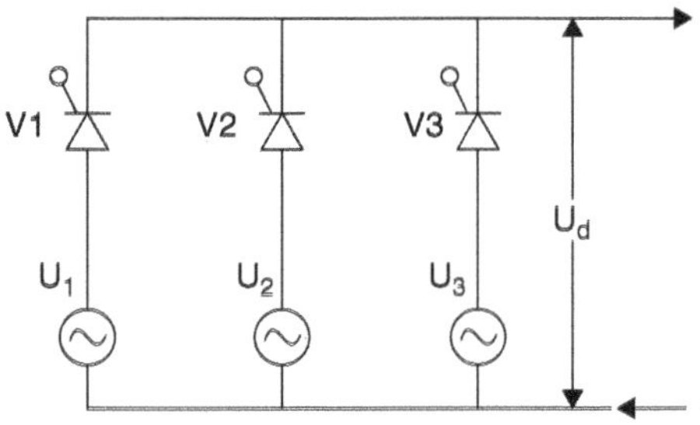

Figure 12.13 Three-phase, half-wave rectifier circuit using thyristors as the switching elements, ignoring supply inductance.

Thyristors are commonly used in place of diodes to create a circuit that can regulate the output voltage. Many devices that provide DC actually *generate* three-phase AC. For example, an automobile alternator contains six diodes, which function as a full-wave rectifier for battery charging.

- An uncontrolled **three-phase, half-wave circuit** requires three diodes, one connected to each phase as shown on Figure 12.13. This is the simplest type of three-phase rectifier but suffers from relatively high harmonic distortion on both the AC and DC connections. This type of rectifier is said to have a pulse number of three, since the output voltage on the DC side contains three distinct pulses per cycle of the grid frequency.
- In a **three-phase, full-wave circuit using center-tapped transformer**, if the AC supply is fed via a transformer with a center tap, a rectifier circuit with improved harmonic performance can be obtained. This rectifier now requires six diodes, one connected to each end of each transformer secondary winding. This circuit has a pulse number of six, and in effect can be thought of as a six-phase, half-wave circuit.

Before solid-state devices became available, the half-wave circuit and the full-wave circuit using a center-tapped transformer were commonly used in industrial rectifiers using mercury-arc valves [4]. This was because the three or six AC-supply inputs could be fed to a corresponding number of anode electrodes on a single tank, sharing a common cathode.

With the advent of diodes and thyristors, these circuits have become less popular and the three-phase bridge circuit has become the most common circuit. Figure 12.14 [2] shows the schematic diagram of a three-phase, full-wave rectifier circuit using a thyristor as a switching circuit.

Voltage-Multiplying Rectifier

The simple half-wave rectifier can be built in two electrical configurations, with the diode pointing in opposite directions. One version connects the negative terminal of the output directly to the AC supply and the other connects the positive terminal of the output directly to the AC supply. By combining both of these with separate output

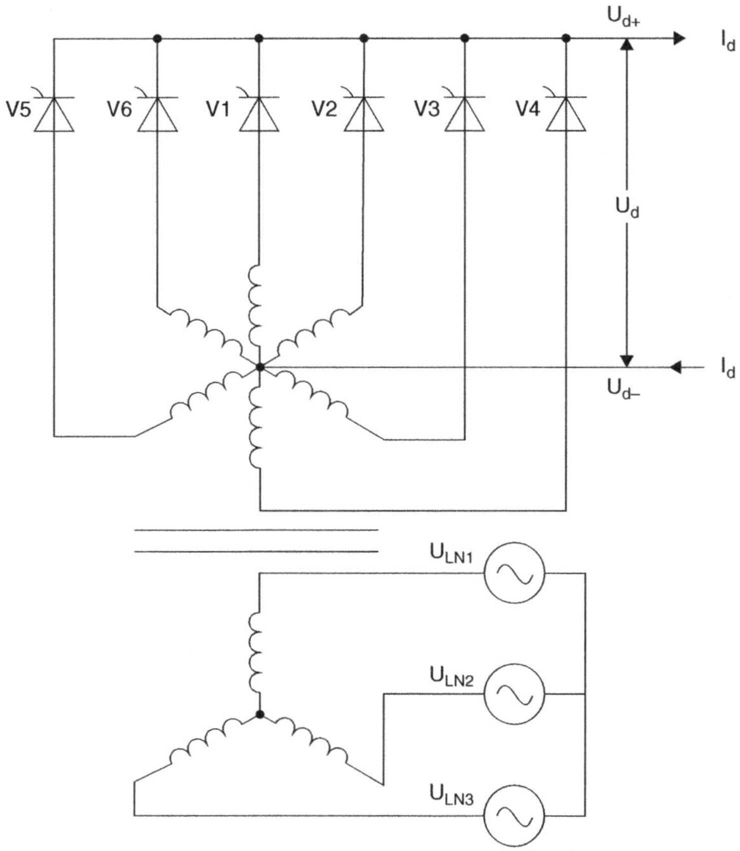

Figure 12.14 Three-phase, full-wave rectifier circuit using thyristors as the switching elements, with a center-tapped transformer, ignoring supply inductance.

smoothing, it is possible to get an output voltage of nearly double the peak AC-input voltage. This also provides a tap in the middle, which allows use of such a circuit as a split-rail power supply.

A variant of this is to use two capacitors in series for the output smoothing on a bridge rectifier then place a switch between the midpoint of those capacitors and one of the AC-input terminals. With the switch open, this circuit acts like a normal bridge rectifier. With the switch closed, it acts like a voltage-doubling rectifier. In other words, this makes it easy to derive a voltage of roughly 320 V (± 15 percent) DC from any 120 V or 230 V mains supply in the world; this can then be fed into a relatively simple switched-mode power supply. However, for a given desired ripple, the value of both capacitors must be twice the value of the single one required for a normal bridge rectifier; when the switch is closed each one must filter the output of a half-wave rectifier, and when the switch is open the two capacitors are connected in series with an equivalent value of half of one of them. A switchable full-bridge circuit is shown in Figure 12.15 [2].

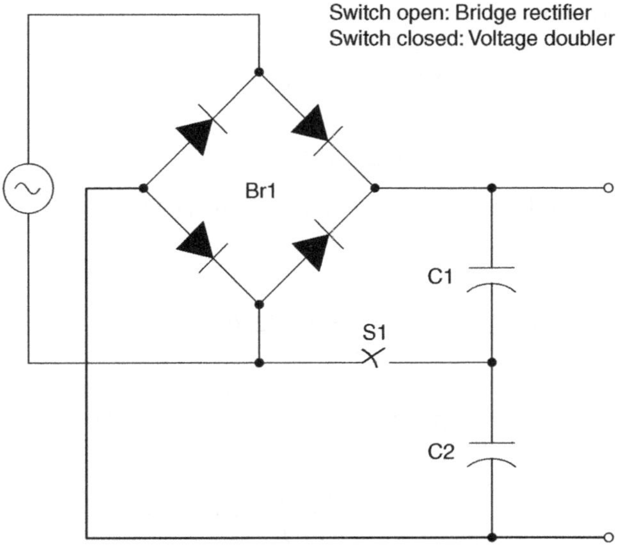

Figure 12.15 Switchable full-bridge/voltage doubler.

Example 1

A three-phase inverter has a Y-connected resistive load of R = 5 Ω and L = 23 mH. The inverter frequency is f = 60 Hz and the DC input voltage is $V_s = 220$ V. Determine:

(a) the rms line voltage V_L

(b) rms line voltage V_{L1} at the fundamental frequency

(c) rms phase voltage V_p

(d) total harmonic distortion

(e) distortion factor

Solution:

(a) $$V_L = \left[\frac{2}{2\pi}\int_0^{2\pi/3} V_s^2\, d(\omega t)\right]^{1/2} = \sqrt{\frac{2}{3}}V_s = 0.8165$$

$$V_S = 0.8165 \times 220 = 179.63\ V$$

(b) The rms line voltage V_{L1} at the fundamental frequency

$$V_{Ln} = \frac{4V_S}{\sqrt{2}\,n\pi}\cos\frac{n\pi}{6}$$

$$V_{L1} = \frac{4V_S}{\sqrt{2}\,\pi}\cos\frac{\pi}{6} = 0.7797\ V_S = 0.7797 \times 220 = 171.53\ V$$

(c) rms phase voltage is given by

$$V_p = \frac{V_L}{\sqrt{3}} = \frac{\sqrt{2}\,V_S}{3} = 0.4717$$

$$V_S = 0.4717 \times 220 = 103.7\,V$$

(d) We know the fundamental value of the rms line voltage is given by

$$V_{L1} = 0.7797\,V_S$$

$$\left(\sum_{n=5,7,11\ldots}^{\infty} {V_{Ln}}^2\right)^{1/2} = \left({V_L}^2 - {V_{L1}}^2\right)^{1/2} = 0.24236V_S$$

$$THD = \frac{\left({V_L}^2 - {V_{L1}}^2\right)^{1/2}}{V_{L1}} = \frac{0.24236V_S}{0.7797\,V_S} = 31.08\%$$

(e)

$$V_{Lh} = \left(\sum_{n=5,7,11\ldots}^{\infty} \frac{{V_{Ln}}^2}{n^2}\right)^{1/2} = 0.00666V_S$$

$$Distortion\,factor\;DF = \frac{V_{Lh}}{V_{L1}} = \frac{0.00666V_S}{0.7797\,V_S} = 0.854\%$$

Example 2

A single-phase, half-bridge inverter has a resistive load of R = 2.4 Ω and the DC input voltage is $V_S = 48$ V. Determine:

(a) rms output voltage at the fundamental frequency V_1

(b) output power P_O

(c) average and peak current of each transistor

(d) total harmonic distortion (THD)

(e) distortion factor (DF)

Solution:

(a) The rms output voltage at the fundamental frequency is given as

$$\mathrm{V_1} = \frac{2\,\mathrm{V_S}}{\sqrt{2}\,\pi} = 0.45\,\mathrm{V_S} = 0.45 \times 48 = 21.6\,\mathrm{V}$$

(b)

$$\mathrm{V_0} = \frac{\mathrm{V_S}}{2} = \frac{48}{2} = 24\,\mathrm{V}$$

$$\text{Therefore, output power} = \mathrm{P_0} = \frac{\mathrm{V_0^2}}{\mathrm{R}} = 240\,\mathrm{W}$$

(c) The peak transistor current

$$I_p = \frac{24}{2.4}\;10\,A$$

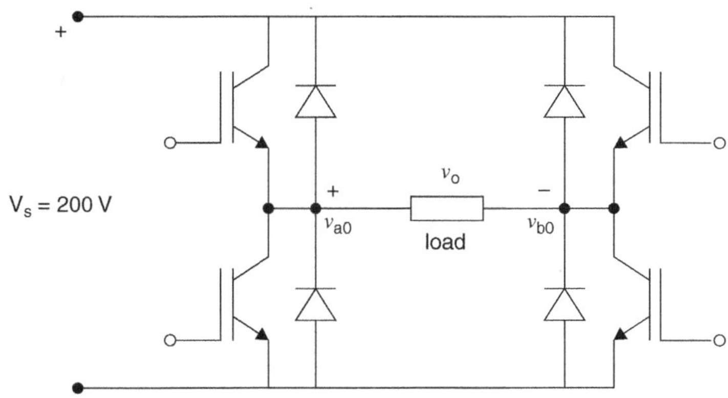

Figure Q12.5.3 Single phase full bridge inverter.

Since each transistor conducts for 50 percent duty cycle, the average current of each transistor is $I_D = 0.5 \times 10 = 5$ A.

(d) The rms harmonic voltage is given as

$$V_h = \left(\sum_{n=3,5,7\ldots}^{\infty} V_n^{\,2} \right)^{1/2} = \left(V_0^{\,2} - V_1^{\,2}\right)^{1/2} = 0.2176\, V_S.$$

Therefore, the THD is

$$\frac{0.2176\, V_S}{0.45 V_S} = 48.34\%.$$

(e) We can find.

$$V_{Lh} = \left(\sum_{n=3,5,\ldots}^{\infty} \frac{V_n^{\,2}}{n^2} \right)^{1/2} = 0.01712\, V_S.$$

Therefore, DF is

$$\frac{0.01712\, V_S}{0.45\, V_S} = 3.804\%.$$

Example 3

Consider the single-phase, full-bridge inverter shown in Figure Q12.5.3 is operated in the quasi-square-wave mode at the frequency f = 100 Hz with a phase-shift of β between the half-bridge outputs v_{ao} and v_{bo}.

(a) With a purely resistive load R = 10 Ω, find β so that the average power supplied to the load is $P_{o,av} = 2$ kW.

(b) With a purely inductive load L = 20 mH and $\beta = 2\pi/3$, find the peak-to-peak value (I_{pp}) of the load current i_o and find the amplitude of the fundamental component (I_{o1}) of i_o.

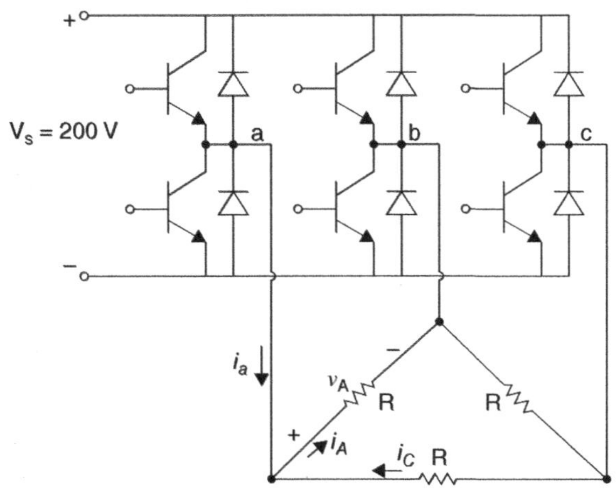

Figure Q12.5.4 Three-phase half-bridge inverter.

Solution:

(a) With a purely resistive load, $I_o = \frac{V_o}{R}$

Instantaneous power $p_o(t) = v_0(t)\, i_0(t) = \frac{V_0{}^2}{R}$

Average Power supplied to the load $P_{o,av} = \frac{1}{T}\int_0^T p_o(t) = \frac{v_1{}^2\beta}{R\pi} = 2,000\ \text{W}$

$$\beta = \frac{R\pi\ 2,000}{v_1{}^2} = \frac{\pi\,(10\times 2,000\,)}{200^2} = \frac{\pi}{2}$$

(b) With a purely inductive load, the peak-to-peak value of the load current is given by

$$I_{pp} = \beta\,\frac{v_i}{\omega L}\,\frac{2\pi}{3}\,\frac{200}{200\pi\times 0.02} = 33.33\ \text{A}.$$

The amplitude of the fundamental component is given by

$$I_{01} = \frac{v_{01}}{\omega L} = \frac{4V_i}{\pi\omega L}\sin{}^{\beta}\!/_{2} = \frac{800}{200\pi^2\times 0.02}\sin{}^{\pi}\!/_{3} = 17.55\ \text{A}.$$

Example 4

The three-phase, half-bridge inverter shown in Figure Q12.5.4 feeds a balanced, Δ-connected resistive load, and is operated in the square-wave mode (pole voltages are square waves) at frequency f_s.

(a) Sketch line current i_a (show all important current and time values) and find its rms value.

(b) Find the total average power supplied to the load (take R = 10 Ω).

Solution:

(a)

$$i_a = i_A - i_C = \frac{v_A - v_c}{R}$$

$$v_A = v_{ao} - v_{bo};\ v_c = v_{co} - v_{ao}$$

$$i_a = \frac{2v_{ao} - v_{b0} - v_{c0}}{R}$$

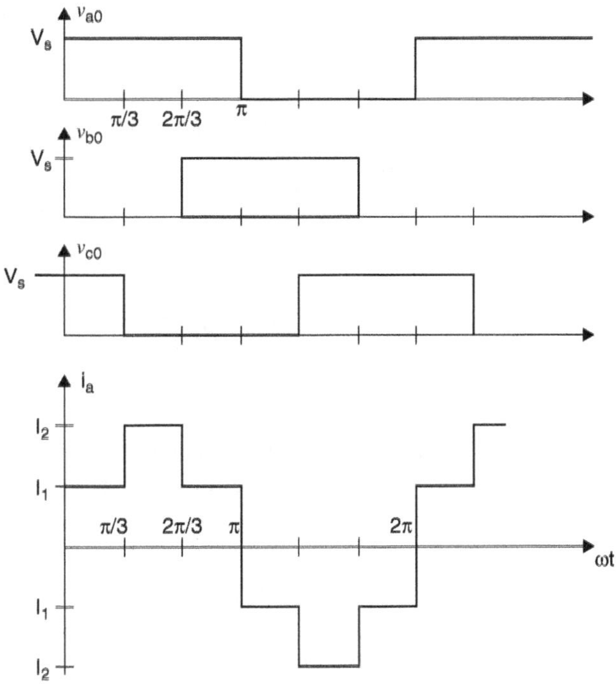

$$I_{a,rms} = \left[\frac{1}{\pi}\ \left(I_1^{\,2}\right)\frac{2\pi}{3} + I_2^{\,2}\frac{\pi}{3}\right]^{1/2} = \sqrt{2}\frac{V_s}{R} = 28.28\,A$$

(b) Instantaneous power in phase A: $p_A\,(t) = \frac{V_A^{\,2}}{R}$

Average power in phase A: $P_A = \frac{1}{R}\ \left[\frac{1}{T}\int_0^{\tau} v_A^{\,2}dt\right] = \frac{V_{A,rms}^{\,2}}{R}$

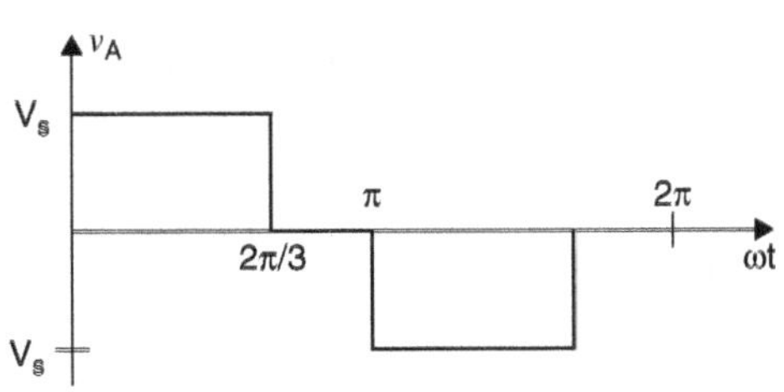

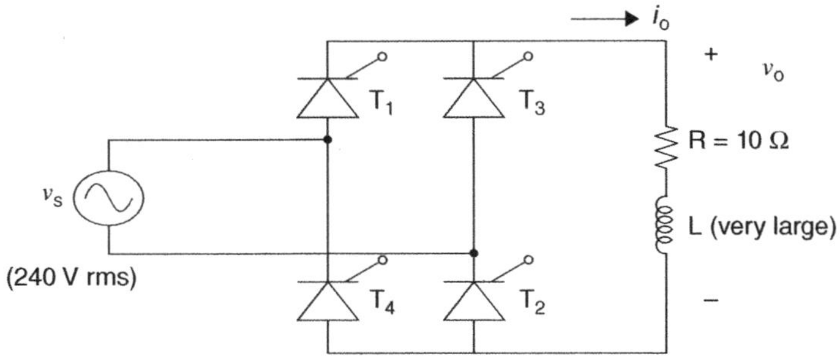

Figure Q12.5.5 Single-phase fully-controlled bridge converter.

$$V_{A,rms} = V_i^2 \left(\frac{2\pi/3}{\pi}\right) = \frac{2}{3} V_i^2$$

$$\textit{Total three} - \textit{phase average power}: P_{av} = 3P_A = \frac{2v_i^2}{R} = 8\,\text{kW}$$

Example 5

A single-phase, fully controlled bridge converter supplies power to a highly inductive load with resistance R = 10 Ω from a 240 V, rms 50 Hz, AC source, as shown on Figure Q12.5.5.

(a) Find the firing angle α so that the average power supplied to the load is $Pav = 1{,}200$ W.

(b) For $\alpha = \pi/6$ find the input power factor of the rectifier.

Solution:

(a) The average power supplied to the load

$$P_{av} = V_{dc}.I_{dc} = \frac{V_{dc}^{\ 2}}{R}$$

$$V_{dc} = \frac{2V_{max}}{\pi} \cos\alpha = \frac{2 \times 240 \times \sqrt{2}}{\pi} \cos\alpha = 216.08 \cos\alpha$$

$$\therefore \quad 1{,}200 = \frac{216.08^2}{10}\ cos^2\alpha$$

$$\cos\alpha = 0.507 \therefore \quad \alpha = 59.54$$

(b) For $\alpha = \pi/6$ the source current is a square wave as shown:

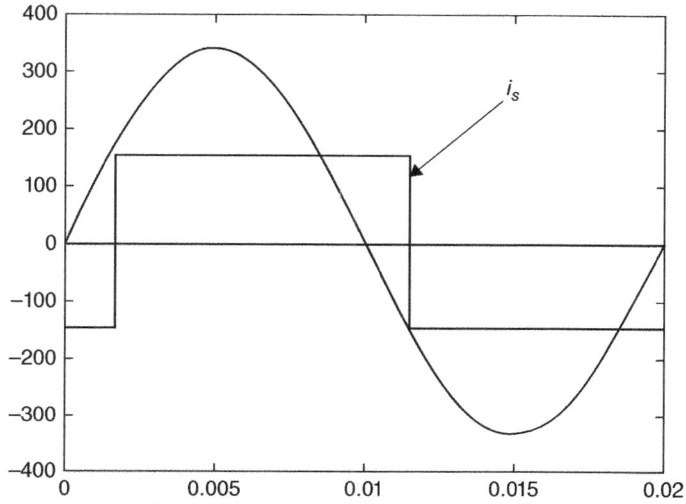

The input power factor is given by

$$PF = Disp.F \times DF$$

$$Disp.F. = \cos \propto \; DF = \frac{I_{sI}}{I_s}$$

I_{s1} : *rms value of fundamental current*

I_s : *rms value of total current*

$$I_{s1} = \frac{2\sqrt{2}}{\pi} I_a$$

$$I_s = I_a$$

$$P.F. = \frac{2\sqrt{2}}{\pi} \cos \alpha = 0.7797$$

Example 6

The three-phase, half-bridge inverter, Q12.5.1 feeds a balanced, Y-connected resistive-inductive load with R = 10 Ω and L = 10 mH, and is operated in the square-wave mode (pole voltages are square waves) at frequency f s = 200 Hz.

(a) Sketch phase **a** voltage v_{an} (show all important voltage and time values) and find its rms value.

(b) Find the amplitude of the fundamental component of the line current i_a.

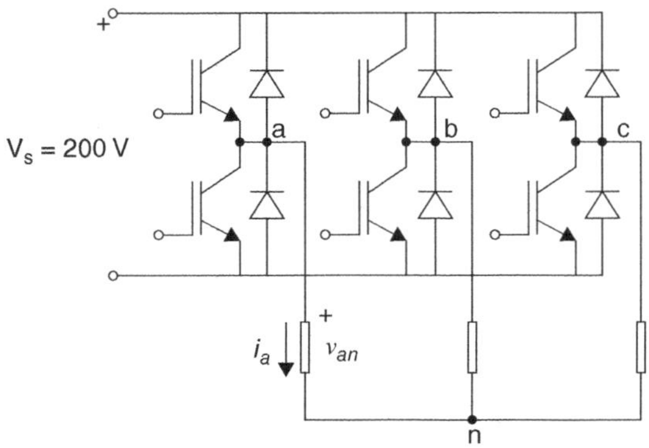

Figure Q12.5.1 Three-phase half-bridge inverter.

Example 7

The three-phase, half-bridge inverter shown in Figure Q12.5.2 feeds a balanced, Y-connected, purely inductive load having inductance L per phase, and is operated in the square-wave mode (pole voltages are square waves) at the frequency f_s.

- **(a)** Sketch phase-to-neutral voltage v_{an} (show all important voltage and time values) and find its rms value $V_{an,rms}$
- **(b)** Find the peak-to-peak value (I_{pp}) of the line current i_a in terms of L and f_s.

12.6 APPLICATIONS

Application of DC–DC Converters

The most typical example of application of DC–DC converters is DC power supplies. However, in most cases, in commercial fields and industries, the power source is AC. In this case, DC–DC converters cannot operate stand-alone. Operating together with

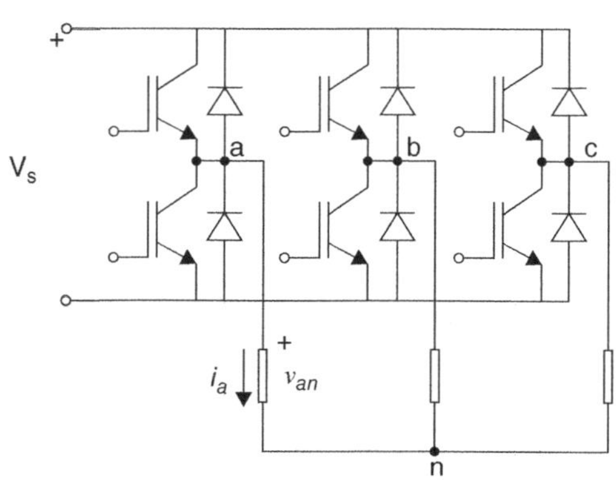

Figure Q12.5.2 Three-phase half-bridge inverter with Purely inductive load.

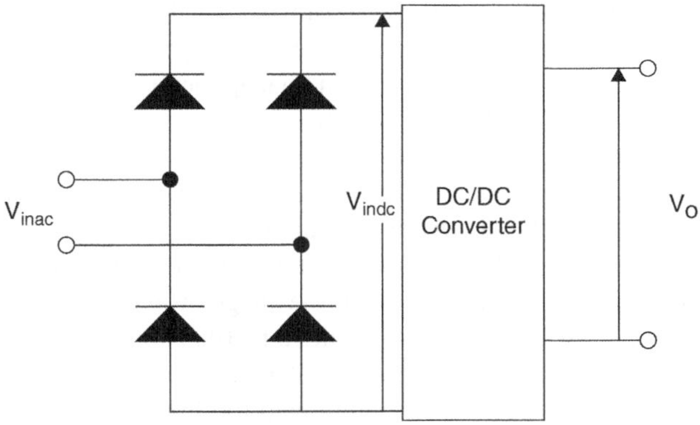

Figure 12.16 Block diagram of DC–DC converter with a rectifier.

other circuits such as rectifiers (Figure 12.16), DC–DC converters are useful in many applications.

Examples of applications of DC–DC converters include:

- DC power supplies for DC or AC input, usually called *switched-mode power supplies* (SMPS)
- battery chargers
- power-factor correction converters (PFCCs)
- motor drives, such as DC drives, brushless DC drives, and switched-reluctance motor (SRM) drives
- electric brakes
- electric vehicles
- most mobile devices (such as cell phones) to maintain voltage at a fixed value whatever the voltage level of the battery
- electronic isolation.
- microgrids, where a power optimizer, is a type of DC–DC converter, maximizes the energy harvest from solar photovoltaic or wind-turbine systems

Application of AC–DC Converters

AC–DC converters (rectifiers) are used every time an electronic device is connected to the mains (computer, television, etc.). These may simply change AC to DC or can change the voltage level as part of their operation.

Rectifiers have many uses, but are often found serving as components of DC power supplies and high-voltage DC power transmission systems. Rectification may serve in roles other than to generate DC for use as a source of power. As noted, detectors of radio signals serve as rectifiers. In gas-heating systems, flame rectification is used to detect presence of a flame.

Rectifier circuits may be single phase or multi-phase (three being the most common number of phases). Most low-power rectifiers for domestic equipment are single phase, but three-phase rectification is important for industrial applications and for transmission of energy as DC (high-voltage DC or HVDC).

The primary application of rectifiers is to derive DC power from an AC supply (AC-to-DC converter). Virtually all electronic devices require DC, so rectifiers are used inside the power supplies of virtually all electronic equipment.

Rectifiers are also used for detection of amplitude-modulated radio signals. The signal may be amplified before detection. If not, a very low-voltage drop diode or a diode biased with a fixed voltage must be used. When using a rectifier for demodulation the capacitor and load resistance must be carefully matched: too low a capacitance makes the high-frequency carrier pass to the output, and too high makes the capacitor just charge and stay charged.

Rectifiers supply polarized voltage for welding. In such circuits control of the output current is required; this is sometimes achieved by replacing some of the diodes in a bridge rectifier with thyristors, effectively diodes whose voltage output can be regulated by switching on and off with phase-fired controllers.

Thyristors are used in various classes of railway rolling stock systems so that fine control of the traction motors can be achieved. Gate turn-off thyristors are used to produce AC from a DC supply, for example on Eurostar trains to power the three-phase traction motors.

Application of DC–AC Converters

DC–AC converters (inverters) are used primarily in uninterruptible power supply (UPS), renewable-energy systems, or emergency lighting systems. Mains power charges the DC battery. If the mains fail, an inverter produces AC electricity at mains voltage from the DC battery. Solar inverters, both smaller string and larger central inverters, as well as solar microinverters, are used in photovoltaics.

(a) DC power source usage: An inverter converts DC electricity from sources such as batteries or fuel cells to AC electricity. The electricity can be at any required voltage; in particular, it can operate AC equipment designed for mains operation, or rectified to produce DC at any desired voltage.

(b) UPS: A UPS uses batteries and an inverter to supply AC power when mains power is not available. When mains power is restored, a rectifier supplies DC power to recharge the batteries. In facilities that require energy at all times, such as hospitals and airports, UPS systems are utilized. In a standby system, an inverter is brought online when the normally supplying grid is interrupted. Power is instantaneously drawn from onsite batteries and converted into usable AC voltage by a voltage-source inverter (VSI) until grid power is restored, or until backup generators are brought online. In an online UPS system, a rectifier DC-link inverter is used to protect the load from transients and harmonic content. A battery in parallel with the DC link is kept fully charged by the output in case the grid power is interrupted, while the output of the inverter is fed through a low-pass filter to the load. High power quality and independence from disturbances is achieved [8].

(c) Harmonic compensation: In power systems, it is often desirable to eliminate harmonic content found in line currents. VSIs can be used as active power filters to provide this compensation. Based on measured line currents and voltages, a control system determines reference current signals for each phase. This is fed back through an outer loop and subtracted from actual current signals to create current signals for an inner loop to the inverter. These signals then cause the inverter to generate output currents that compensate for the harmonic content. This configuration requires no real power consumption, as it is fully fed by the line; the DC link is simply a capacitor that is kept at a constant voltage by the control system [8]. In this configuration, output currents are in phase with line voltages to produce a unity power factor. Conversely, var compensation is possible in a similar configuration where output currents lead line voltages to improve the overall power factor [9].

(d) Electric motor speed control: Inverter circuits designed to produce a variable output voltage range are often used within motor speed controllers. The DC power for the inverter section can be derived from a normal AC wall outlet or some other source. Control and feedback circuitry is used to adjust the final output of the inverter section, which will ultimately determine the speed of the motor operating under its mechanical load. Motor speed-control needs are numerous and include things like industrial motor-driven equipment, electric vehicles, rail transport systems, and power tools.

(e) Power grid: Grid-tied inverters are designed to feed into the electric power distribution system. They transfer synchronously with the line and have as little harmonic content as possible. They also need a means of detecting the presence of utility power for safety reasons, so as not to continue to dangerously feed power to the grid during a power outage.

(f) Solar: A solar inverter is a balance of system (BOS) component of a photovoltaic system and can be used for both grid-connected and off-grid systems. Solar inverters have special functions adapted for use with photovoltaic arrays, including maximum power-point tracking and anti-islanding protection. Solar microinverters differ from conventional converters, as an individual microconverter is attached to each solar panel. This can improve the overall efficiency of the system. The output from several microinverters is then combined and often fed to the electrical grid.

(g) Induction heating: Inverters convert low-frequency main AC power to higher frequency for use in induction heating. To do this, AC power is first rectified to provide DC power. The inverter then changes the DC power to high-frequency AC power. Due to the reduction in the number of DC sources employed, the structure becomes more reliable and the output voltage has higher resolution due to an increase in the number of steps so that the reference sinusoidal voltage can be better achieved. This configuration has recently become very popular in AC power supply and adjustable speed drive applications. This new inverter can avoid extra clamping diodes or voltage-balancing capacitors.

(h) HVDC power transmission: With HVDC power transmission, AC power is rectified and high-voltage DC power is transmitted to another location. At the receiving location, an inverter in a static inverter plant converts the power back to AC. The inverter must be synchronized with grid frequency and phase and minimize harmonic generation. The HVDC transmission method can be useful for things like solar power since solar power is natively DC.

(i) Electroshock weapons: Electroshock weapons and tasers have a DC–AC inverter to generate several tens of thousands of V AC out of a small 9 V DC battery. First the 9 V DC is converted to 400–2,000 V AC with a compact, high-frequency transformer, which is then rectified and temporarily stored in a high-voltage capacitor until a preset threshold voltage is reached. When the threshold (set by way of an air gap or TRIAC) is reached, the capacitor dumps its entire load into a pulse transformer, which then steps it up to its final output voltage of 20–60 kV. A variant of the principle is also used in electronic flashes and insect zappers, though they rely on a capacitor-based voltage multiplier to achieve their high voltage.

(j) Renewable-energy applications: For photovoltaic purposes, the inverter, which is usually a PWM VSI, gets fed by the DC electrical-energy output of a photovoltaic module or array. The inverter then converts this into an AC voltage to be interfaced with either a load or the utility grid. Inverters may also be employed in other renewable systems, such as wind turbines. In these applications, the turbine speed usually varies, causing changes in voltage frequency and sometimes in the magnitude. In this case, the generated voltage can be rectified and then inverted to stabilize frequency and magnitude [8]. Classes of converters are summarized in comparison in Table 12.1.

Illustrative Problems and Examples

1. A three-phase full converter controls the speed of a 150HP, 650 V, 1,750 rpm, and separately excited DC motor. The converter is operating from a three-phase, 460 V, 50 Hz supply. The rated armature current of the motor is 170 A. The motor parameters are Ra = 0.0099 ohms, La = 0.73 mH, and Ka = 0.33 V/rpm. Neglect the losses in the converter system. Determine:

(a) no-load speed of firing angle alpha = 0 degrees, 30 degrees; assume that at no load the armature current is 10 percent of rated current and is continuous

(b) firing angle to obtain rated speed of 1,750 rpm at rated motor current; also compute the supply power factor

(c) speed regulation for firing angle obtained in (b)

TABLE 12.1 Comparison of Classes of Converters

Conversion from/to	Description	Device(s)	Schematic Symbol	Applications
DC to DC	Convert from one DC power level to another DC power level	• buck converter • boost converter • buck–boost converter, • Ćuk converter • flyback converter • forward converter	DC / DC	• DC power supplies • battery chargers • motor drives • electric brakes • microgrids as a power optimizer
DC to AC	Electronic device or circuitry that changes DC to AC	• inverter	DC / AC	• UPS • harmonic compensation, • electric motor speed control • induction heating • electroshock weapons • renewable-energy applications
AC to DC	Converts AC to DC	• rectifier	AC / DC	• detection of amplitude-modulated radio signals • supply of polarized voltage for welding
AC to AC	Converts AC waveforms of specific magnitude and specific frequency into AC waveforms with different magnitude and different frequency	• cycloconverter • matrix converter	AC / AC	• exchange power between power grids of different utility frequencies

Solution:

(a) $\alpha = 30°$ at no load $I_a = 10\% \times 170 = 17A$

$$V_P = 460/3^{0.5} = 265.58\ V$$
$$V_o = 3(6)^{0.5}/\pi \times V_p \times \text{Cos}(\alpha)$$
$$= 3(6)^{0.5}/\pi \times 265.58 \times \text{Cos}(\alpha)$$
$$= 537.97\ V$$
$$\text{Therefore, } V_o = E_a + I_a R_a$$
$$E_a = V_o - I_a R_a$$
$$= 537.97 - 17 \times 0.0099$$
$$E_a = 537.8\ V$$
$$\text{But, } E_a = K\varphi N$$
$$N = E_a/(K\varphi) = 537.8/0.33$$

No-load speed = 1,630 rpm

(b)

$$V = 1,750 \text{ rpm}, \ I_a = 170 \text{ A}$$
$$E_a = K\varphi N$$
$$= 0.33 \times 1,750$$
$$= 577.5 \text{ V}$$
$$V_o = E_a + I_a R_a$$
$$= 577.5 + (170 \times 0.0099)$$
$$= 579.18 \text{ V}$$
$$V_o = 3(6)^{0.5}/\pi \times V_p \times \text{Cos}(\alpha)$$
$$\Rightarrow 579.18 = 3\ (6)^{0.5}/\pi \times 265.58 \times \text{Cos}(\alpha)$$
$$\Rightarrow \alpha = 21.2^\circ$$

For the supply power factor

$$I_a = \left[1/\pi \times 120^2 \times 2\pi/3\right]^{0.5}$$
$$= 138.8 \text{ A}$$
$$S = 3 \times V_p \times I$$
$$= 3 \times 265.58 \times 138.8$$
$$= 110.591 \text{ VA}$$

Assume no loss in converter

$$P_S = V_o I_o$$
$$= 579.18 \times 170$$
$$= 98,460.6 \text{ W}$$

Supply pf = P/S = 98,460.6 / 110.591 = 0.89

(c) Speed regulation

$$\left[N_{NL} - N_{FL}\right]/N_{FL} \times 100$$
$$= [1,750 - 1,630]/1,630 \times 100$$
$$= 7.36\%$$

2. A single-phase full converter is used to control the speed of a 5 hp, 110 V, 1,200 rpm, separately excited DC motor. The converter is connected to a single-phase 120 V, 60 Hz supply. The armature resistance is $R_a = 0.4$ ohm and armature circuit inductance $L_a = 5$ mH. The motor voltage constant is $K_\varnothing = 0.09$ V/rpm.

(a) Assuming the DC machine operates as a motor, runs at 1,000 rpm, and carries an armature current of 30A, and the motor current is ripple free, determine:

(i) firing angle α

(ii) power to the motor

(iii) supply power factor

(b) In inverter operation, the polarity of the motor back emf E_a is reversed, say by reversing the field excitation. Determine:

(i) firing angle to keep the motor current at 30 A when the speed is 1,000 rpm

(ii) power fed back to the supply at 1,000 rpm

$$\begin{aligned}R_a &= 0.4\,\text{ohm}\\ L_a &= 5\,\text{mH}\\ K\varphi &= 0.09\,\text{V/rpm}\\ N &= 1,000\,\text{rpm}\\ I_a &= 30\,\text{A}\\ V_p &= 120\,\text{V}\\ E_a &= K\varphi N\\ &= 0.09\times 1,000\\ &= 90\,\text{V}\end{aligned}$$

(c) **(i)**

$$\begin{aligned}\text{As a vector}: V_o &= E_S + I_aR_a\\ &= 90 + 30\times 0.4\\ &= 102\,\text{V}\end{aligned}$$

But

$$\begin{aligned}V_o &= 2\,(2)^{0.5}/\pi\times V_p\text{Cos}(\alpha)\\ 102 &= 2\,(2)^{0.5}/\pi\times 120\,\text{Cos}\,(\alpha)\\ \alpha &= 19.2^\circ\end{aligned}$$

The firing angle is 19.2°.

(ii)

$$\begin{aligned}\text{Power to the motor} &= I_o^2R_a + E_aI_o\\ &= V_oI_o\\ &= 102\times 30\\ &= 3,060\,\text{W}\end{aligned}$$

(iii)

$$\begin{aligned}S &= V_SI_o\\ &= 120\times 30\\ &= 3,600\,\text{VA}\\ \text{Supply pf} &= P/S\\ &= 3,060/3,600\\ &= 0.85\end{aligned}$$

(d) **(i)** Inverter operation Ie E_a is reversed

$$\begin{aligned}V_o &= -90 + 30\times 0.4\\ V_o &= -78\,\text{V}\\ V_o &= 2(2)^{0.5}/\pi\times V_p\text{Cos}\ (\alpha)\\ -78 &= 2(2)^{0.5}/\pi\times {}_{120}\text{Cos}\ (\alpha)\\ \alpha &= 136.2^\circ\end{aligned}$$

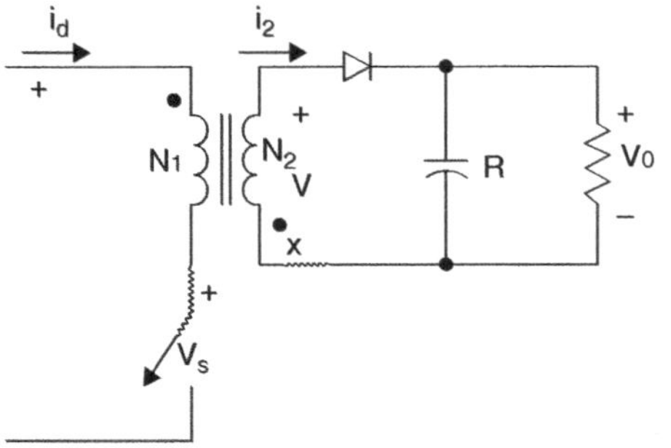

Figure Q12.6.1 Fly back converter.

(ii)

$$
\begin{aligned}
\text{Power from DC machine} &= E_a I_o \\
&= 90 \times 30 \\
&= 2,700\,\text{W} \\
\text{Power loss in } R_a &= I_o^2 R_a \\
&= 30^2 \times 0.4 \\
&= 360\,\text{W} \\
\text{Power fed back to the supply} &= 2,700 - \text{loss} \\
&= 2,700 - 360 \\
&= 2,340\,\text{W}
\end{aligned}
$$

12.7 CHAPTER SUMMARY

This chapter treated the classical and new aspects of inverters and converters. It covered the definition, descriptions, configurations, and performance of the various types of converters: as DC–DC, AC–DC, DC–AC, and AC–AC. Converters function as an interface in the integration of microgrids and traditional grids. A converter interface for DG functions in such a way that it accepts power from the distributed energy source and converts it to power at the required voltage and frequency. The modeling, characteristics, configuration, and application were given. Case studies and exercises using inverters and converters in typical electronic systems were given.

EXERCISES

1. Explain the operation of a fully controlled thyristor bridge converter. Derive the expression for average load voltage and input power factor.

2. A flyback converter is shown in Figure Q12.6.1. The input voltage V_d may vary and the output voltage V_0 should be kept constant by adapting the duty ratio D. Given is:

$$V_d = 12 \ldots 14\,\text{V},\ V_0 = 48\,\text{V},\ N_2/N_1 = 6,\ f_s = 200\,\text{kHz},\ R_0 = 10\,\text{ohm}$$

   (a) Derive the voltage-transfer function V_0/V_d in continuous conduction mode (incomplete demagnetization mode) as a function of the duty ratio D and calculate the duty ratio for $V_d = 12$ V.

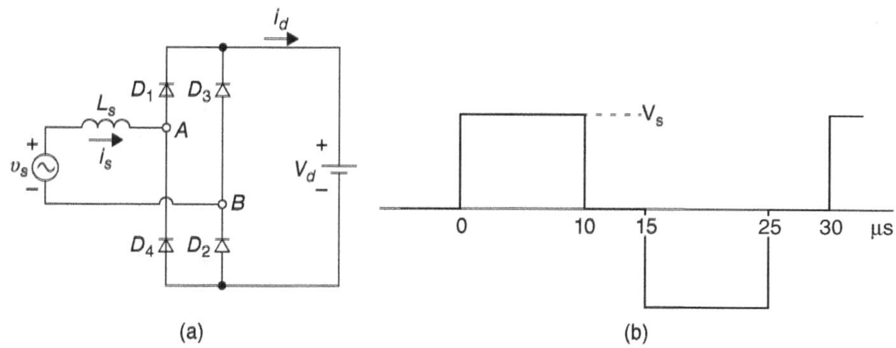

Figure Q12.6.2 Single phase rectifier.

(b) Calculate the minimum value of the transformer inductance L_m required to obtain continuous-conduction mode for all values of V_d in the range 12 to 14 V.

(c) Calculate the maximum voltage across the switch.

(d) Sketch i_d, i_2, v_x, and v_s as a function of time for $V_d = 12$ V.

3. Given is a single-phase rectifier that is connected to a supply voltage v_s and a battery as shown in Figure Q12.6.2. The battery is represented by a DC load voltage V_d.

The voltage V_s has a block-like shape (Figure Q12.6.2(b)) that is produced by a high-frequency inverter (not shown). The rectifier is intended to charge the battery. Depending on the charging state of the battery, the voltage V_d may vary.

Given is V_d, nom = 240 V (nominal voltage of V_d)
$V_s = 300$ V (amplitude of vs as shown in Figure Q12.6.2(b))
$T_s = 30\mu s$ (period of voltage vs)
$f_s = 1/Ts = 33.3$ kHz (frequency of vs)
$L_s = 20\mu H$

(a) Sketch V_L and is as a function of time and indicate relevant values for $V_d = V_d$, nom.

(b) Calculate the peak value of is, the average current I_d, and the average output power P_d for $V_d = V_d$, nom.

(c) What is the range of V_d for discontinuous conduction of id in a steady state?

(d) What will happen if V_d is decreased below the lower limit of this range?

4. At the terminals of a converter, the current and voltage are as shown in Figure Q12.6.3(a) and (b) respectively.

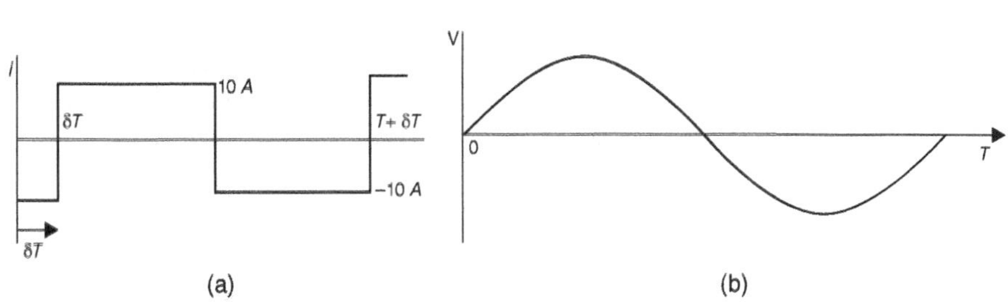

Figure Q12.6.3 Voltage and Current waveform.

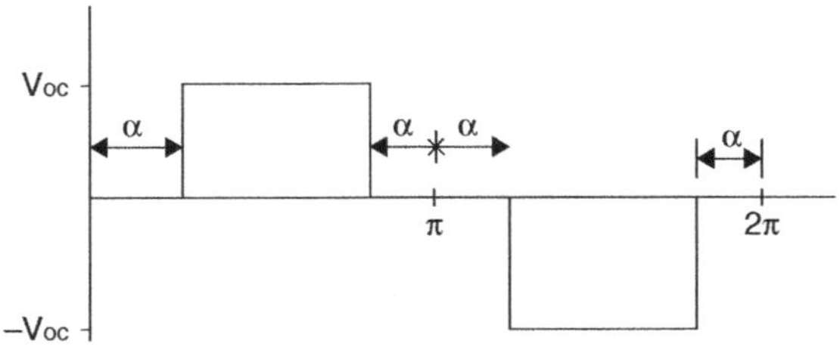

Figure Q12.6.4 Switching scheme for full bridge inverter.

The amplitude of the voltage is $230\sqrt{2}$ V.

(a) Calculate the power at the terminals.

(b) Calculate the rms value of the current i.

(c) Define and calculate the power factor.

(d) Define and calculate the distortion power factor.

5. Given is a switch-mode, DC power supply, where the output voltage V_o should be regulated close to its nominal value. For that purpose, a negative-feedback control system is used to reduce the effect of variations in the input voltage V_d and the load.

(a) Sketch a block diagram of the system, and briefly describe the function of the blocks. Indicate all relevant variables in the diagram.

(b) Briefly describe the steps needed to obtain the small-signal transfer function of the system. Here V_0 is the output voltage and d is the duty ratio. Note that you are not expected to calculate the transfer function.

(c) Obtain an averaged state-space description of an up-converter (the result of part (b)).

6. For a full-bridge inverter, the DC source is 24 V, the load is a series RL connection with $R = 10\,\Omega$ and $L = 35$ mH, and the switching frequency is 60 Hz. Using the switching scheme shown in Figure Q12.6.4 [5], determine the value of the firing angle α to produce an output at 15 V rms.

BIBLIOGRAPHY

[1] N. Mohan, T. Undeland, and W. P. Robbins, *Power Electronics: Converters, Applications, and Design*, John Wiley & Sons, Hoboken, NJ, 2003.

[2] M. H. Rashid, *Power Electronics Handbook: Devices, Circuits, and Applications*, Academic Press, Burlington, MA, 2007.

[3] Wikitronics, "Rectifier," http://electronics.wikia.com/wiki/Rectifier.

[4] J. A. Momoh, Exams for Electronic Power Conversion (ET4119), Howard University, February 2000–January 2007.

[5] D. W. Hart, *Power Electronics*, McGraw-Hill Education, New York, 2010, 343, Figure 8-5a.

[6] N. Mohan, T. Undeland, and W. P. Robbins, *Power Electronics: Converters, Applications, and Design*, John Wiley & Sons, Hoboken, NJ, 2003.

CHAPTER 13

MICROGRID APPLICATION DESIGN AND TECHNOLOGY

13.1 INTRODUCTION TO MICROGRIDS

A microgrid is a group of interconnected loads and distributed resources within clearly defined electrical boundaries that acts as a single controllable entity with respect to the grid, enabling local power generation for local loads. It is comprised of various small power-generating sources that make it highly flexible and efficient. Microgrids can be connected to and disconnected from the grid, which makes it operable in both grid-connected or island mode. Its connection to the utility grid prevents power outages in the system; excess power produced can be sold to the grid. Hence, it provides an answer to the energy crisis and transmission losses.

The advantages of microgrid systems are numerous. First, it is very reliable in case of a blackout of any kind. This is important in places such as hospitals. Also, it combines a variety of energy sources including renewable energy, which makes it environment friendly. Since it is locally operated, it is more in touch with appliances in the community and can perform grid optimization and load priority, and use the heat produced by the plant, which previously went unused, to provide heat to homes. A very important factor is lower costs, which can be achieved because the community microgrid has the power to choose how much power it wants from the megagrid based on what it is producing.

The first power plant created, in 1882, could be called a microgrid since centralized grids had not yet been established. However, as centralized grids were created, the utility industry evolved to a state-regulated monopoly. The idea of microgrid was resurrected in the wake of a July 2012 India blackout that left three hundred million people without electricity. Technology was the key enabler of microgrid development.

The reciprocating engine, fuel cells, solar, and wind-farm development have reached a point where a small network of assorted generators can provide power to neighborhoods, retail areas, and even industrial facilities. In a microgrid-enabled future, you might not have to drive to the hydroelectric dam one state away to see where your electricity comes from. Instead, sources will be found in a refrigerator-sized microturbine behind a house, or in a wind farm on the outskirts of town. An experimental microgrid system in Japan keep a hospital running after an earthquake

Energy Processing and Smart Grid, First Edition. James A. Momoh

that caused weeks-long outages. As a result of successes like this, the US Department of Energy has invested significant research funds into ways to implement microgrids and the smart grid into industry.

Energy markets today are undergoing a transformation. The top-down monopoly system dominated by large-scale fossil and nuclear power plants that send power to passive consumers is shifting. Microgrid systems aim to be dynamic, bidirectional networks increasingly dependent upon distributed resources. Major trends in microgrids include empowering smart meters from back office to the field in real time, marrying renewable power sources, energy storage and smart grid, visualized distribution grid for planning ahead and emerging microgrid controller design elements.

13.2 TYPES OF MICROGRIDS

Microgrids are electricity-distribution systems containing loads and distributed-energy resources—such as distributed generators, storage devices, and controllable loads—that can be operated in a controlled, coordinated way, either while connected to the main power network or while islanded.

13.2.1 Classification by Size

- **Megagrid** occurs at transmission levels serving hundreds of MVA (above 120kV). An example is a large windfarm with storage.
- **Minigrid or milligrid** occurs at a primary- or medium-voltage-distribution level. An example is a small distribution substation.
- **Microgrid** is located at the primary- and secondary- (load) voltage level. An example is a distribution transformer serving industrial or commercial customers.
- **Picogrid** is located at the secondary- or low-voltage level, 100KvA or less. An example is a distribution transformer serving a small group of residential customers.

13.2.2 Classification by Purpose

Most microgrids can be classified into four purposes:

- **Customer microgrids or true microgrids (μgrids)** are self-governed and usually downstream of a single point of common coupling (PCC). Many of the most well-known demonstrations are of this type. They are particularly easy to imagine because they fit neatly into our current technology and regulatory structure. Just as a traditional customer has considerable leeway in the operation of the power system on its side of the meter, so the restrictions on the nature of a μgrid are relatively loose. For this reason, one would expect much of the early deployment of microgrid technology to be of this type.

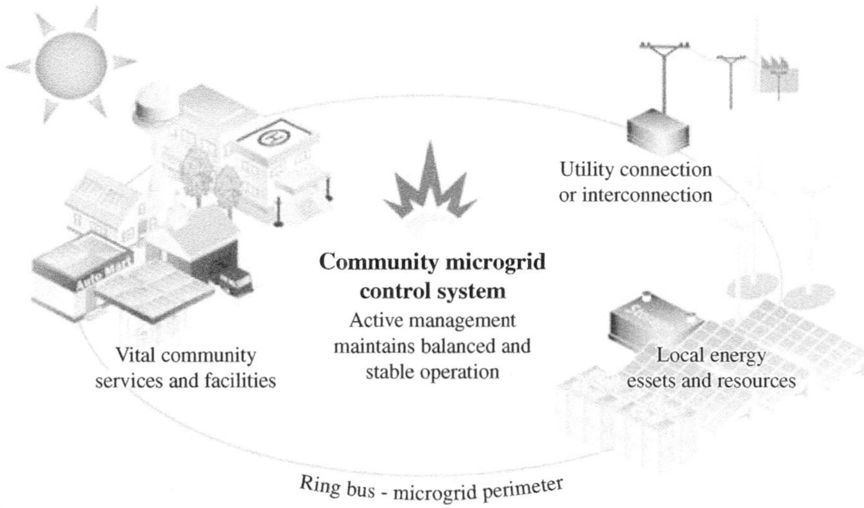

Figure 13.1 Community microgrid.

- **Utility or community microgrids or milligrids (mgrids)** involve a segment of the regulated grid. There are also existing well-known examples such as Sonoma Community Microgrid Initiative (Santa Rosa, CA), Valencia Gardens Energy Storage Project (San Francisco, CA) and Long Island Community Mirogrid Project (East Hampton, NY). While technically not different from μgrids, they are fundamentally different from a regulatory and business model perspective, primarily because they incorporate traditional utility infrastructure. The corollary of this feature is that utility regulation comes much more significantly into play. In other words, any mgrid must comply with existing utility codes or accommodations must be made in the code. A typical community microgrid structure is shown in Figure 13.1 [1].
- **Virtual microgrids (vgrids)** cover distributed energy resources (DERs) at multiple sites but are coordinated such that they can be presented to the grid as a single controlled entity. Very few demonstrations of vgrids exist, but they have been proposed in literature. Note that, to be consistent with the definition above, the system must be able to operate as a controlled island or coordinated multiple islands.
- **Remote power systems (rgrids)** are obviously not able to operate grid-connected, isolated power systems, but involve similar technology and are closely related. From a research point of view, they are so close that they are commonly described as microgrids.

13.3 MICROGRID ARCHITECTURE

Figure 13.2 shows the architecture of a microgrid, which are made up of five major components:

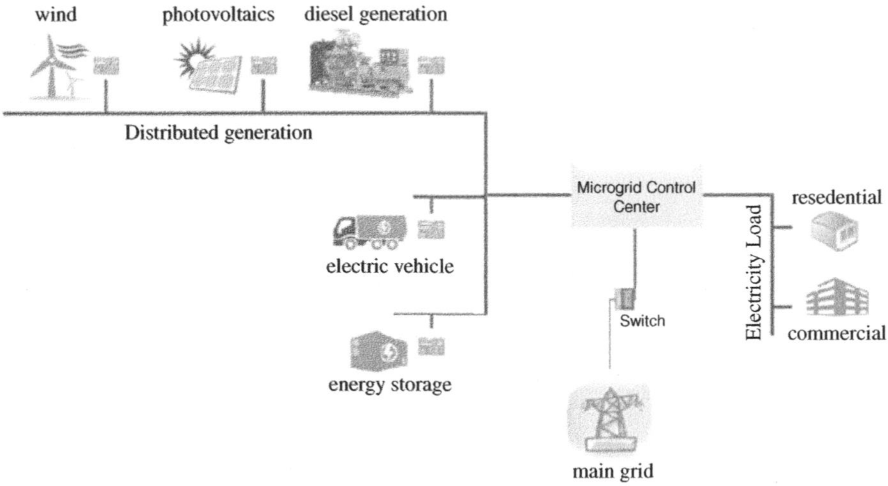

Figure 13.2 Architecture of a microgrid.

(a) Power Source

Sustainable energy is derived from natural sources that replenish themselves over a short period of time—often called *green power*. The future electric grid will invariably feature rapid integration of alternative forms for energy generation as a national priority that requires new optimization for energy resources distributed with interconnected standards. A major component of a modern microgrid is its use of DERs.

There are various types of power sources, usually selected based on cost and efficiency. For example, people in windy areas will favor a wind farm, while people in hot areas of the world will favor solar energy. Several technologies may be considered as probable sources for the generation of electricity. The technology most suitable for each location is unique to that location, depending on the particular application and feasibility as a result of physical location, availability, suitability, source of energy, and cost to deliver or produce energy. Possible energy options for a microgrid are solar photovoltaic cells, biopower, hydroelectric power, wind turbines, fuel cells, and internal combustion engines using natural gas.

Renewable-energy options for the microgrid provide benefits such as remote utilization and storage of renewable-energy resources, enhancement of functionality of grid-connected renewable-energy systems (RES) to facilitate "give and take" of energy from the system, redistribution/relocation of unused power from a grid-connected RES, facilitating storage of grid-generated and renewable-energy resource–generated energy by backup storage technologies at the customer end, and tracking of interactions for billing and study. Enhancement of functionality of electric vehicles and plug-in hybrids and utilization of a vehicle's battery pack as grid-energy storage devices are also among the benefits.

The proportion of electric energy or power supplied from wind turbines or other RES is usually referred to as the *penetration*. It is expressed using annual energy from renewable-energy-powered generators in kWh and total annual energy delivered to loads in kWh as a percentage. When fuel or CO_2 emission savings are considered, the term *average penetration* is used.

$$\textit{Average penetration} = \frac{\text{Annual energy from renewable energy powered generators (kWh)}}{\textit{Total annual energy delivered to loads (kWh)}} \tag{13.1}$$

The term *instantaneous penetration* is used for other purposes, including system control. It is much greater than the average penetration.

$$\textit{Instantaneous penetration} = \frac{\text{Power from renewable energy powered generators (KW)}}{(\textit{Power from renewable energy powered generators } (kW)\,/(\textit{Total power delivered to loads } (kW)} \tag{13.2}$$

(b) Energy Management System
Transmission line provides power management system for microgrid to make it manageable. An inverter is used to change output of the transmission line to the level required by the customer.

(c) Energy Storage System
An energy storage system, such as batteries, is essential to allow the microgrid to balance the demand and supply of electricity, and also to make it accessible.

(d) Electricity Loads
Electricity loads are energy-consuming devices in the microgrid.

(e) Main Grid/Utility Connection
Microgrids are operated as standalone or interconnected mode. When operating in an interconnected mode, it either receives or supplies power to main grid.

13.3.1 Pathway to Design of a Microgrid

Microgrid design configurations are based on numerous sources generating energy that enable the enhancement of energy efficiency and improvement to the total yield. The microgrid has a robust structure depending on the "plug and play" characteristics of energy sources, or the ability of any source to be integrated into existing DG system without requiring any configuration.

A microgrid requires varying technologies depending on whether it is connected to a conventional power grid or stand alone. Among the factors and determinants to consider when planning a microgrid:

- distance from the grid
- dispersion

- type of load
- energy-resource availability
- income level and willingness to pay

Microgrids are designed for robust operation using supervisory and local control and protection schemes in the absence of supervisory control. Supervisory control is used to optimize customer benefits, e.g., performance, operating cost, emissions. Energy-management platforms are economical if they incorporate maximum commonality among various applications. The different controls used are as follows:

(a) Supervisory controls
- optimize electrical and thermal performance and cost
- manage feeder connections to the bulk grid
- manage renewable intermittency

(b) Local controls
- control response based on local measurements
- provide robust response to system disturbances and supervisory-level commands
- provide inherent stability and load sharing for grid-independent and grid-interactive connections

(c) Protections
- provide protection coordination at the PCC for innovative protection functions based on synchronized measurements
- transfer trip issuance of a breaker trip command from one decision locus to another, using standard tools
- provide differential relay protection, specifically of the microgrid, to provide reliable fault detection and location
- used to address issues such as back-feeding from system generation or other sources of power, specifically using voltage-polarized directional overcurrent protection

(d) Design philosophy
- develops control strategies for nonconventional assets constrained by power-ramp rates and current limits (such constraints tend to surface when limited generation capacity exists to service loads)
- manages voltage and power at the point of interconnection (tie-line control)
- achieves multiple control objectives for optimal dispatch control
- determines unit commitment (UC) and economic dispatch (ED) based on benefit objectives, functions, and operation constraints

- integrates combined heat and power (CHP) with thermal loads
- increases controllable loads as an RER
- accounts for renewable generation and stochasticity and forecasting
- utilizes nonlinear control theory to ensure system stability of nonconventional generation in microgrid networks with mixed assets and loads

(e) Control objectives

- allow robust response to system disturbances and supervisory-level commands
- provide inherent stability, load sharing, and fault ride-through for low-inertia grid-independent and grid-interactive connections

(f) Status

- provides a novel control philosophy for nonconventional generation, developed and tested in simulation
- results indicate robust operation for various system contingencies when applied to constrained nonconventional generation equipment

13.3.2 Design of a Microgrid

The steps for planning and designing a microgrid system are as follows:

Step 1: Identify site needs. Clarify the goals of the microgrid and factors impacting microgrid capabilities, availability and feasibility of renewable resources, geographic relationship of facilities, and site electrical distribution system layout/characteristics.

Step 2: Classify loads. Use data such as load profiles, energy, annual and daily peaks, and seasonal behavior to classify loads. Install temporary metering where necessary to record load characteristics, criticality of loads, and sensitivity to power-quality issues. Also consider operational and nameplate equipment data.

Step 3: Classify distributed resources. Consider distributed generation and energy storage requirements, coverage required for the microgrid footprint, and distributed storage considerations such as location, technology, capacity, and duration.

Step 4: Evaluate generation/load match. Decide whether existing/projected generation capacity can meet the proposed peak load and daily operating requirements. Determine whether dispatchable resources can handle transient disturbances on the system while maintaining satisfactory voltage and frequency. Can existing dispatchable generation compensate for the variability of renewable resources?

Step 5: Develop control strategy. Design appropriate control strategies for self-regulation, load, and generation dispatch. Design for both grid-connected and islanded conditions and for control of the utility interface (i.e., static transfer switch) to handle seamless separation and reconnection to utility power. Also consider local/central control schemes to monitor and control DR and loads; integration with existing legacy communication systems/software; cyber and physical security concerns; and integration with existing protection schemes.

Step 6: Determine equipment specifications. Engineering analysis should include modeling and simulation of the microgrid under various scenarios to establish necessary control parameters, devices, infrastructure modifications, and additional generation and storage necessary for microgrid implementation.

13.4 MODELING OF A MICROGRID

A microgrid is basically an active distribution network made up of collection of DG systems and various loads at the distribution-voltage level. Microgrids are generally small, low-voltage, combined heat loads of a small community such as university or school campuses, commercial areas, industrial sites, housing estates, or a suburban locality.

In order to introduce the microgrid to the utility power system as a single controlled unit that meets local energy demands for reliability and security, the micro sources must have power electronic interfaces (PEIs) and controls to provide the necessary flexibility to the semiautonomous entity to maintain the dictated power quality and energy output.

Figure 13.3 shows a typical microgrid connection scheme. Microgrids are connected to a medium voltage (MV) utility "main grid" through the PCC circuit breaker. Micro-source and micro-storage devices are connected to feeders B and C through microgrid controllers (MCs). Some loads on feeder B and C are considered priority loads (i.e., need an uninterruptable power supply), while the rest are non-priority electrical loads. Central controllers (CC) include an energy management module (EMM) and a protection coordination module (PCM). The EMM supplies the set points for active and reactive power output, voltage, and frequency to each MC. This is done by advanced communication and artificial intelligence techniques, while the PCM answers to microgrid and main grid faults and losses of grid situations so that proper protection coordination of the microgrid is achieved.

The model of the microgrid varies for different configurations depending on the type of components used. One important aspect of microgrid modeling is power generation, such as diesel, wind, and solar PV systems.

Diesel engines are a speed-feedback system from a system-control point. After the operator gives a speed command by adjusting the governor setting, the engine governor, also working as a sensor, recognizes the difference between the actual speed and the desired speed and regulates the fuel supply to maintain the engine speed within range. This mathematical relation between fuel flow and mechanical torque is

Renewable Energy

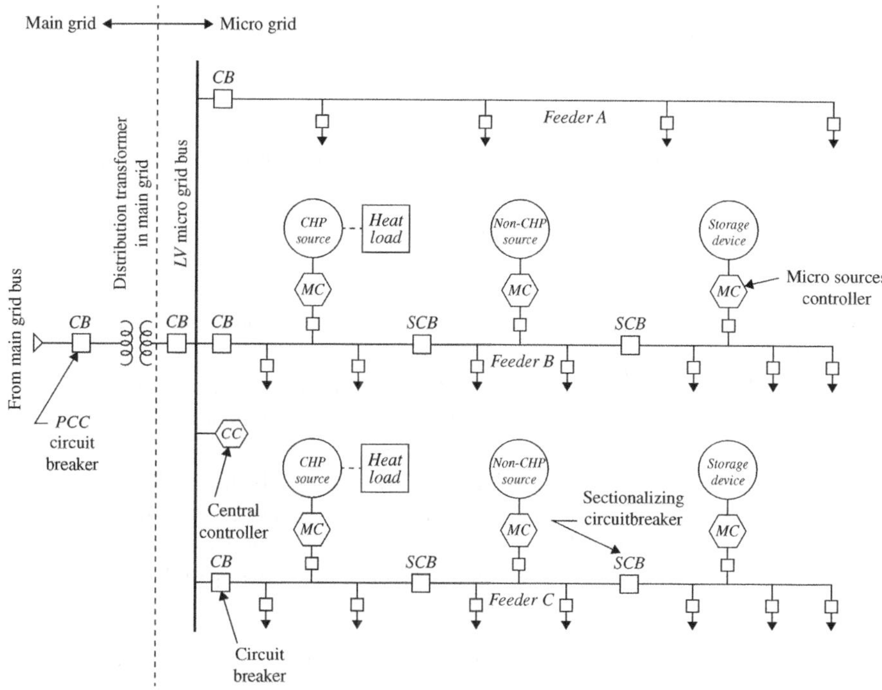

Figure 13.3 Typical microgrid-connection scheme.

given as

$$\phi(\mathrm{s}) = \frac{K_3K_2}{(1+\tau_2 \mathrm{s}}\mathrm{I}(\mathrm{s}) \tag{13.3}$$

$$\mathrm{T}(\mathrm{s}) = \phi(\mathrm{s})K_1\frac{K_3K_2}{\left(1+\tau_2 \mathrm{s}\right)}\mathrm{I}(\mathrm{s}) \tag{13.4}$$

where, $\phi(\mathrm{s})$ is the fuel flow, T(s) is the mechanical torque, I(s) is the input current, τ_2 is the time constant, K_3 is the current driver constant, and K_2 is the gain.

A PV cell can be represented as a diode in parallel with a constant current source and a shunt resistor. Different modeling approaches are followed in a PV system. A simplistic modeling of PV system output using a derating method based on the ratio of the actual insolation to the standard condition is given as

$$\mathrm{P_{PV}} = \mathrm{f_{pv}Y_{pv}}\left(\frac{\mathrm{I_T}}{\mathrm{I_s}}\right) \tag{13.5}$$

where $\mathrm{f_{pv}}$ is the PV derating factor, $\mathrm{Y_{pv}}$ is the Pv array capacity, $\mathrm{I_T}$ is the global solar radiation incident on the PV array, and $\mathrm{I_s}$ is the standard amount of radiation used to rate the capacity.

Wind turbines can be equipped with squirrel-cage-induction generators, doubly fed (wound-rotor) induction generators, and direct-drive synchronous generators. Real power output is given by Equation 13.6:

$$P_m = \frac{1}{2}\rho \Pi R^2 V^3 C_p \tag{13.6}$$

where ρ is the air density (kg/m3), R is the turbine radius (m), Cp is the turbine power coefficient power conversion efficiency of a wind turbine, and V is the wind speed (m/s).

The electrical power output is given by Equation 13.7:

$$Pe = n_o P_m \tag{13.7}$$

where, $n_o = \eta_m$, and η_g η_m and η_g are the efficiency of the turbine and the generator, respectively.

The models developed for these components, and in a similar manner for the remaining components of a microgrid, are used in different applications such as load-flow algorithms and stability studies of microgrids.

For instance, in power-flow modeling of Ac microgrids, the nodes in the network can be generalized to two types: one as the PQ bus and one where the DG units are connected. Buses connected with loads in the system naturally belong to the PQ buses, and the characteristics of voltage and frequency dependency of these buses can be modeled as

$$P_{Di} = P_{oi} |V_i|^{\alpha i} \left(1 + K_{pf,i}\Delta f\right) \tag{13.8}$$

$$Q_{Di} = Q_{oi} |V_i|^{\beta i} \left(1 + K_{qf,i}\Delta f\right) \tag{13.9}$$

where P_{0i} and Q_{0i} are the rated real and reactive power of the load; αi and βi are the exponent coefficients of real and reactive power; Δf is the frequency deviation; and Kpf, i and Kqf, *i* are constants to represent frequency dependency.

The mismatch for the dispatchable DG bus is given as

$$P_i\left(\theta_i, |V_i|\right) + P_{Di}\left(f, |V|_i\right) + P_{Gi}(f) = 0 \tag{13.10}$$

$$Q_i\left(\theta_i, |V_i|\right) + Q_{Di}\left(f, |V_i|\right) + Q_{Gi}(f) = 0 \tag{13.11}$$

Combining the nodal and mismatch expression of PQ bus and dispatchable DG bus, the overall mathematical model of load flow for the overall system can be obtained.

13.5 CHAPTER SUMMARY

In this chapter, the microgrid system was discussed. The basic architectural configuration of a microgrid includes distributed resources, storage devices, utility connection, electrical loads, and power and energy management systems. Identifying site needs, consumer requirements, type and classification of distributed resource availability, and matching with demand and control strategy are among the basic required steps in the planning, design, and operation of a microgrid system. Modeling of a microgrid varies for different configurations depending on the type of components used.

Typically, a microgrid is modeled as a low-voltage power system consisting of distributed resources, commonly solar, wind, and other resources, to supply loads to a small community.

BIBLIOGRAPHY

[1] Microgrid Institute, "Community Energy Development: Unique Value Proposition, Unique Challenge," www.microgridinstitute.org/community-energy-develoment.html.

[2] R. Lamb, HowStuffWorks.com.

[3] Microgrid Institute, "About Microgrids," www.microgridinstitute.org/about-microgrids.html.

[4] Sandia Energy, "Energy Surety Microgrid™," http://energy.sandia.gov/energy/ssrei/gridmod/integrated-research-and-development/esdm.

[5] F. Yanine and E. Sauma, "Review of Grid-Tie Micro-Generation Systems without Energy Storage: Towards a New Approach to Sustainable Hybrid Energy Systems Linked to Energy Efficiency," *Renew. Sust. Energ. Rev.* **26** (October 1, 2013).

[6] A. L. Dimeas and N. D. Hatziargyriou, "Operation of a Multiagent System for Microgrid Control," *IEEE T. Power Syst.* **20**, 3 (2005).

[7] S. Backhaus, G. W. Swift, S. Chatzivasileiadis, et al., *DC Microgrids Scoping Study—Estimate of Technical and Economic Benefits*, Los Alamos National Laboratory, March 2015.

[8] E. Kabalci, E. Hossain, and R. Bayindir, "Microgrid Test-Bed Design with Renewable Energy Sources," *IJRER* **4**, 4 (2014).

[9] M. R. Islam and H. A. Gabbar, "Study of Micro Grid Safety and Protection Strategies with Control System Infrastructures," *Smart Grid and Renewable Energy* **3**, 1 (February 2012).

[10] A. Sinha, R. Lahiri, S. Byabortta, et al. "Formulation of Pricing Mechanism for Microgrid Energy," CIRED Seminar 2008: SmartGrids for Distribution, June 2008.

[11] G. Y. Morris, C. Abbey, S. Wong, et al., "Evaluation of the Costs and Benefits of Microgrids with Consideration of Services beyond Energy Supply," IEEE Power and Energy Society General Meeting, July 2012.

[12] H. Hou, J. Zhou, Y. Zhang, et al., "A Brief Analysis on Differences of Risk Assessment between Smart Grid and Traditional Power Grid," IEEE Fourth International Symposium on Knowledge Acquisition and Modeling, January 2012.

[13] M. Eghbal, T. K. Saha, and N. Mahmoudi-Kohan, "Utilizing Demand Response Programs in Day Ahead Generation Scheduling for Micro-Grids with Renewable Sources," IEEE PES Innovative Smart Grid Technologies Asia, March 2011.

[14] Y. Qiao, Z. Lu, and S. Mei, "Microgrid Reconfiguration in Catastrophic Failure of Large Power Systems," IEEE International Conference on Sustainable Power Generation and Supply, December 2009.

[15] C.-H. Lin, H.-J. Chuang, C.-S. Chen, et al., "Fault Detection, Isolation and Restoration Using a Multiagent-Based Distribution Automation System," IEEE Conference on Industrial Electronics and Applications, May 2009.

[16] J. A. Momoh, S. Meliopoulos, and R. Saint, *Centralized and Distributed Generated Power Systems: A Comparison Approach*, Prepared for the Project "The Future Grid to Enable Sustainable Energy Systems," funded by the US Department of Energy, June 2012.

CHAPTER 14

MICROGRID OPERATIONAL MANAGEMENT

A microgrid is a low-voltage distribution network consisting of different energy components such as wind, solar, storage system, diesel generators known as distributed energy resources (DERs), and controllable loads. With these components, a microgrid can either be connected to the grid or can be used in islanded mode (use DERs to supply the loads). Integrating DERs and controllable loads within the distribution network introduces unique challenges to microgrid management and control, which are implemented by an energy management system (EMS). Figure 14.1 shows the EMS structure for microgrid operation.

14.1 PERFOMANCE TOOLS OF A MICROGRID

(a) Forecasting

Forecasting is a process that uses past and present data for making predictions of future DERs, loads, and the market. Due to intermittency and variability of these factors, forecasting has become a challenging task for the power utilities. Therefore, accuracy of the forecast is crucial for the EMS to balance equality constraints, i.e., supply and demand in the microgrid. Various forecast methods are available in the literature:

- qualitative versus quantitative
- average approach
- drift
- time series
- causal/econometric
- judgmental
- artificial intelligence

Energy Processing and Smart Grid, First Edition. James A. Momoh

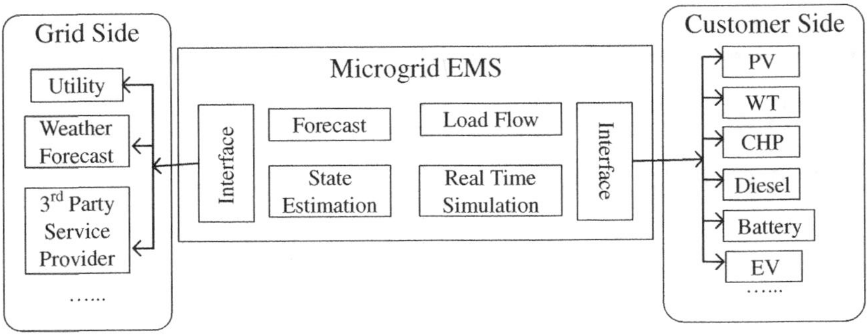

Figure 14.1 Performance structure of a microgrid EMS.

(b) State Estimation

State estimation is a key function in building adequate network models for online monitoring and analyses. State estimation in a distribution system is becoming stringent because of the integration of DERs and the adoption of advanced technology for system modeling and operations for the distribution network. To manage large-scale DERs and load change, the operators in a distribution system need a robust distribution system state estimator (DSSE). The installation of supervisory control and data acquisition (SCADA) system at the feeder level and the widespread adoption of advanced metering infrastructure (AMI) have made the implementation of a reliable DSSE feasible. Various methods are utilized to solve DSSE issues, including:

- nonlinear model: $[z] = [h(x)] + \varepsilon$
- weighted least squares (WLS): $J = \sum_{i=1}^{m} \left(\frac{z_i - h_i(x)}{\sigma_i} \right)^2$
- solution based on normal equations

(c) Load Flow

Load-flow studies play an important role in power system analysis and design. They give the steady-state solution of a power system network for normal operating conditions, helping in continuous monitoring of the current state of the system, and are employed for planning, optimization, and stability studies. All distribution of electrical energy is done by a constant-voltage system. In practice, the following distribution circuits are generally used: radial system, ring main system (mesh system), and interconnected system. Distribution networks are characterized by radial structure with high (Resistance to Reactance Ratio) R/X ratio feeders.

A mesh network (having some loops) can be found in some modern distribution systems. A mesh network may be used to improve the system voltage profile, balance the load, and increase the supply reliability and efficiency. Thus, there is an increasing demand for a generalized load-flow method for distribution systems having more than one feeding source with a mesh-configured network. An efficient and fast load flow method is required. Load flow for a distribution system is divided into two categories:

- forward/backward sweep (FBS), ladder-network-based methods, and loop-impedance methods
- methods that require information on the derivatives of the network equations

(d) Real-Time Simulation

Real-time simulation refers to a simulation or computer model of a physical system that can execute at the same rate as actual "wall clock" time. For example:

- real-time means sixty seconds of simulation time equals sixty seconds
- offline means sixty seconds of simulation time can be five minutes, fifteen minutes, forty minutes, etc., depending on the network size being simulated

Various tools are available to work on real-time distribution systems, such as:

- real-time digital simulator (RTDS)
- real-time simulator (OPAL-RT technologies)
- DIgSILENT PowerFactory

For a deeper understanding, the reader can refer to papers on forecasting [18], state estimation [19], load flow [20], and real-time simulation of the microgrid [22].

14.2 MICROGRID FUNCTIONS

Power quality has become a critical concern for utility customers. Power companies must improve service delivery to retain customers in the deregulated market. For this reason, distribution automation systems (DASs) have been implemented as an intelligent way to improve the reliability and operational efficiency of distribution systems.

DASs are defined by the IEEE as systems that enable an electric utility to coordinate, monitor, and operate distribution network components in real time from remote-control centers. Given that information exchange between the DAS server and field equipment is critical for system operation, a reliable communication network constitutes the core of cutting-edge DAS applications.

An essential feature of a microgrid with advanced smart distribution, unlike a traditional distribution network, is support for a large number of DAS function. Modeling and scheduling control problems in DG sources need to be resolved by DASs in the micro grid. Among the functions achieved by a DAS, fault detection, isolation, and restoration (FDIR) is considered most important.

DAS functions also cover the following areas:

- voltage regulation/var control
- power quality

- demand-side management
- reconfiguration
- restoration
- dispersed generation and storage dispatch
- fault diagnosis/location
- real-time pricing

14.2.1 Control Options for Microgrid Technology

Control options for a microgrid are as follows:

- **Unit power control configuration** tracks requests for power. If extra power is needed demands are made from the grid. Where heat is produced from the grid (p = f (H)) it is used for combined heat and power (CHP) applications.
- **Feeder flow control configuration** tracks requests for frequency, while extra demands for loads are provided by sources. Microgrids become a true dispatchable load as seen from the utility side, allowing for specific demand arrangements.
- **Mixed control configuration** allows some units to track power and others to track frequency. A unit can switch from one mode to another, allowing the system to operate at the most efficient capacity [4].

Voltage/var Control

Voltage control within an identified range of limits and capacitor switching is an effective method of improving voltage profiles, minimizing loss, and delaying construction of new generation within power quality and reliability constraints. In electric power systems, volt-ampere reactive (var) is used to express reactive power in a power system. Reactive power exists in an AC circuit when the voltage and current are not in phase.

Voltage/var control considers a multiphase, unbalanced power distribution system operation, DG, and automated control equipment in large electric systems. Power system distribution automation functions use voltage/var control options to maintain proper parameters throughout the grid based on the requirements of different customers by employing variable capacitors. The improved control and management of voltage/var facilitates increase performance for customer equipment and allow both the consumer and the utility company to realize financial benefits.

To improve efficiency, voltage/var is integrated with the load-management problem under four different objectives, as follows:

(a) Customer outage cost

Min.
$$\mathbf{OC} = \sum_n Y_n Cr_{kn}^{max} CC_k \tag{14.1}$$

where OC is the outage cost, Y_{kn} is the level of curtailable load selection of type k at bus n (pu MW), C_{kn}^{max} is the maximum curtailable MW of type k at the nth bus (pu MW), and CC_{kn} is the curtailment cost of customer type k (\$/pu MW).

(b) Loss minimization

$$\text{Min.} \qquad I^2r = \sum\nolimits_{mn} r_{nm} \frac{P_{nm}^2 + Q_{nm}^2}{|V_i|^2} \tag{14.2}$$

where P_{nm}, Q_{nm} is the transfer power (branch n–m at pu) and V_n is the voltage of bus n (pu).

(c) Load balancing

$$\text{Min.} \qquad \frac{S_n}{S_n^{max}} \tag{14.3}$$

(d) Multiple objective function

$$Z = \text{Min.} \qquad [a + b - c] \tag{14.4}$$

These objectives are subjected to different system constraints, such as:

Branch-flow equation

$$X_{n+1} = F_{n+1}(X_n) \tag{14.5}$$

where

$$X_{n+1} = [P, Q, |V|^2, Y_{k,} \ Q_{s,} \delta_n. \tag{14.6}$$

Branch flow considers the recursive relationships between the successive nodes in the radial-distribution system. The demand in P, Q has an interruptible component, which is the load-management control option Y. In addition, the reactive power equation has the capacity-switching option Q_s.

Voltage limits/current limits:

$$|V|_n^{min} \leq |V|_n \leq |V|_n^{max} \tag{14.7}$$

$$|I|_n^{min} \leq |I|_n \leq |I|_n^{max} \tag{14.8}$$

Capacitor control limits:

$$Q_{sn}^{\ min} \leq Q_{sn} \leq Q_{sn}^{\ max} \tag{14.9}$$

Curtailable load-control limits:

$$P_n \times Y_{kn} \times P_{c_{kn}} \tag{14.10}$$

$$Q_n \times Y_{kn} \times P_{c_{kn}} \tag{14.11}$$

Load Frequency Control

Load frequency control (LFC) is of importance in microgrid system design and operation. The objective of LFC in an interconnected power system is to maintain the frequency of each area within limits and to keep tie-line power flows within some prespecified tolerances by adjusting the megawatt (MW) outputs of the generators to accommodate fluctuating load demands. A well-designed and well-operated power system must cope with changes in the load and with system disturbances, and should provide an acceptably high level of power quality while maintaining both voltage and frequency within tolerance limits. As power load demand varies randomly for an interconnected power system, both area frequency and tie-line power interchange varies also. The primary objective of LFC is to minimize transient deviations in area frequency and tie-line power interchange while ensuring that steady-state errors are as close to zero as possible. Frequency will not change in an interconnected network if there is a balance between customer demand and power supply resources that can also compensate for electrical losses.

The operating point of the power system shifts from its scheduled operating point due to mismatch in generation and demand in the system. This is because without increase in mechanical input, sudden changes in demand in the system will be supplied from the stored kinetic energy of the rotating mass. This results a reduction in kinetic energy of the system, which will further cause the turbine speed reduction of the system. The speed is directly proportional to the frequency, hence frequency will also fail.

The incremental power in the generation and demand is expressed in the system as a change in kinetic energy, increased load consumption, and increased power transferred in the interconnected system. This relation between the incremental power and system dynamics is given as follows.

The stored energy in a generator rotor at the rated frequency and sensitive load dependency in frequency is

$$E_{ke}^{0} = H^{*}\, P_r\, MW - \sec \tag{14.12}$$

$$\frac{d}{dt}(E_{ke}) = \frac{2H\, P_r}{f_0}\frac{d}{dt}\Delta f. \tag{14.13}$$

The incremental power is related to kinetic energy and frequency-dependent loads as shown:

$$\Delta P_m(t) - \Delta P_L(t) = \frac{2H}{f_r}\frac{d}{dt}\Delta f + D\Delta f. \tag{14.14}$$

Taking the Laplace transform of Equation 14.14,

$$\Delta P_m(s) - \Delta P_L(s) = \frac{2Hs}{f_r}\Delta F + D\Delta F \tag{14.15}$$

$$\Delta F = \frac{\Delta\, P_m(s) - \Delta\, P_L(s)}{\frac{2Hs}{f_r} + D} \tag{14.16}$$

$$\Delta F = [\Delta P_m(t) - \Delta P_L(t)]\frac{K_P}{1 + T_P S} \tag{14.17}$$

where, ΔP_m and ΔP_L are the changes in mechanical power and load, Δf is the frequency deviation, H is the inertia constant, D is the load-damping coefficient, K_P is the power system gain given by $K_P = 1/D$, and T_P is the power system time constant $T_P = \frac{2H}{f_r D}$.

The aim is to minimize frequency deviation (ΔF) due to mismatch between generation and load.

Power Quality

Power quality (PQ) is the capability of a power system to operate without causing disturbance or damage to components and loads. PQ is a significant concern because of the sensitivity of modern digital control equipment to PQ distortion and deterioration. A rudimentary requirement for maintaining PQ is balancing demand and supply. Characteristic metrics utilized as indictors of PQ measures include:

- total harmonic distortion
- voltage transient impulse
- voltage sag
- flicker factor

A few accepted PQ standards apply to electrical power distribution systems. The European Norm 50160 standard outlines voltage disturbance standard requirements for acceptable PQ standards. Studies on PQ consider factors such as:

- location of power disturbances
- identification of types and causes of power disturbances
- quantification of power disturbances and their negative impacts on power systems
- real-time measurement of parameters of signal components in power disturbances
- sensitive detection of power disturbances

14.2.2 Demand Response

Practically speaking, electrical power cannot be easily stored on a large scale. Thus, supply and demand must remain in balance in real time. Traditionally, utility companies have leveraged peaking power plants to increase power generation when needed to meet demand. Demand response, however, works from the opposite side of the equation: instead of adding more generation to the system, it pays energy users to reduce their consumption. Utilities pay for demand-response capacity because it is proven to be more flexible and less expensive than bringing additional generation online.

Demand response offers an opportunity for consumers to play a substantial role in the operation of the microgrid by decreasing or shifting their electricity use during peak periods in response to time-based rates or other financial incentives. Demand-response programs are being utilized by electric system planners and operators as resource options for balancing supply and demand. These programs aid in lowering the cost of electricity in wholesale markets, which in turn leads to lower retail rates.

Techniques of engaging customers in demand-response efforts include offering time-based rates such as critical peak pricing, variable peak pricing, time-of-use pricing, critical peak rebates, and real-time pricing. Other techniques include direct load-control programs that provide the ability for power utility companies to cycle loads that are noncritical, such as water heaters and air conditioners, on and off during periods of peak demand in return for a financial incentive and lower electricity bills.

Demand-response programs are becoming an increasingly valuable resource option whose capabilities and potential impacts are expanded by modern microgrids. Microgrids have numerous sensors that can distinguish peak load problems and utilize automatic switching to divert or decrease power in strategic places, which removes the chance of overload and a resulting power failure. Advanced metering infrastructure increases the range of time-based rate programs that may be offered to consumers and smart customer systems such as in-home displays or home-area-networks that make it easier for consumers to change their electricity consumption patterns and reduce peak-period consumption based on information on their power consumption and costs. These programs also possess the potential to assist electricity providers in saving money through reductions in system peak demand, and the ability to defer addition or construction of new DG plants and power-delivery systems—particularly those reserved for use during peak times.

14.2.3 Demand-Side Management

Demand-side management (DSM) is a set of interconnected and flexible programs that allow customers a bigger role in shifting their own demand for electricity during peak periods, reducing their energy consumption overall. DSM program consists of two principal activities: demand response programs or "load shifting," and energy efficiency and conservation.

The evolution of microgrid technology will permit customers to make informed decisions about their energy consumption, regulating both its timing and quantity.

Key levers of successful DSM initiatives include incentives, rates, technology, access to information, controls, public education, and marketing, along with verification and customer insight. Each lever has a distinct influence on customer behavior and depends on the circumstances of the particular utility, such as customer base and geography. Certain combinations of actions within and across levels may produce greater results.

14.2.4 Reconfiguration

In an emergency, a microgrid is frequently required to reconfigure itself as soon as possible in order to support continuous electricity supply to essential loads. Types of operations, including topology switching, generation regulation, and load shedding, are concerned with reconfiguration. Therefore, the problem of reconfiguration turns out to be a typical constrained, nonlinear optimization problem with discrete,

Boolean and continuous variables, which is usually too complex to be solved by regular optimization methods. Reconfiguration is a category of emergency control, which includes topology changes with load-generation control measures that play an essential role in helping microgrids switch from integration mode to island mode.

Dissimilar from large systems, topology reconfiguration is deemed as the final line of defense, only taken into consideration when there is an extreme emergency state to avoid uncontrollable blackout of the entire system—while it is a frequently used control method in microgrids. In an emergency, the microgrid is expected to be operated as an island to protect itself from being infected by faults in the main system. By applying flexible electrical/electronic interfaces, both microgrid and energy storage systems can switch on/off seamlessly and realize "plug and play" functionality.

14.2.5 Fault Analysis

Generally, a microgrid can operate in both the grid-connected mode and the islanded mode, where the microgrid interfaces with the main power system by a fast semiconductor switch called a static switch (SS). It is critical to protect a microgrid in both the grid-connected and the islanded modes of operation against all types of faults. The major problem arises in island operation with inverter-based sources. Inverter fault currents are limited by the ratings of the silicon devices to around two per unit rated current. Fault currents in islanded inverter-based microgrids may not have sufficient magnitudes to use traditional over-current protection systems. This possibility requires an extended protection strategy. The philosophy for protection is to have the same protection strategies for both islanded and grid-connected operation. The SS is designed to open for all faults. With the SS open, faults within the microgrid must be cleared with techniques that do not rely on high-fault currents. The fault analysis of a power system is required in order to provide information for the setting of relays, selection of switchgear, and stability of system operation. Several types of fault analysis are carried out using software packages and data collected from the microgrid system. Faulty-analysis types include symmetrical and asymmetrical, short circuit, stability, and transient.

14.2.6 Energy Pricing

Microgrid is a concept where local energy sources in remote areas, both in renewable (such as PV, small wind generator, etc.) and nonconventional (fuel cells, microturbine, and diesel generator) resources, are tapped and interconnected to form a low-voltage utility system. These small DGs all have different owners, and the supply loads locally with the help of local controllers (μc). A μC controller takes decisions such as scheduling of generation as per load forecast (i.e., unit commitment) and economic dispatch of connection with each DG and microgrid central controller (μcc). In the islanding operation of a microgrid, each source caters to only those loads stipulated for the source. However, when these sources are connected with the microgrid, which is most desirable, the action of the controllers should have a certain degree of intelligence for participation in a common and competitive market. The role of the EMS in the microgrid scenario is for making decisions with regard to the best use of the generator for producing heat and electricity, i.e., CHP operation. Such decisions are based upon the heat requirements of the local establishments, climate, cost of fuel, price of electric power, and many other variables.

The market price is the lowest price obtained at the point of intersection of the aggregated supply and demand curves. At this price both suppliers of generation and customers are satisfied and enough electricity will be supplied from accepted sales bids to satisfy all the accepted purchase bids. The costs incurred in the production of electricity are solely dependent on the type of technologies and fuel used for electricity production. The DG units that use natural gas (microturbines) or diesel have higher marginal costs and these plants are called the *price setting units*, while mini-hydro plants have a low marginal cost. Electricity cannot be stored as long as there are no electricity alternatives for consumers to react to the price signal. These two points lead to an inelastic demand curve.

In the case of renewable sources, the current cost of buying energy is relatively low, with no fuel or operational costs involved compared to the capital cost of buying a wind turbine or PV module. Economic analyses in these cases are performed using a simple payback period. CO_2 emissions are absent and the resulting no-carbon incentives are an added advantage. Since electricity can be produced in many ways using a variety of fuels and applying different technologies, different cost structures apply, with important implications for price formation on short-term electricity markets.

14.3 IEEE STANDARDS FOR MICROGRIDS

14.3.1 IEEE Standards for Synchro Phasors for Power Systems (IEEE C37.118, 2005)

This standard defines synchronized phasor measurements used in power system applications. It provides a method to quantify the measurement, tests to be sure the measurement conforms to the definition, and error limits for the test. It also defines a data communication protocol, including message formats for communicating this data in a real-time system.

14.3.2 IEEE Standard for Interconnecting Distributed Resources with the Electric Power System (IEEE 1547, 2003)

This standard focuses on the technical specifications for, and testing of, the interconnection itself. It provides requirements relevant to the performance, operation, testing, safety considerations, and maintenance of the interconnection. It includes general requirements, response to abnormal conditions, PQ, islanding, and test specifications and requirements for design, production, installation evaluation, commissioning, and periodic tests. The stated requirements are universally needed for interconnection of distributed resources (DR), including synchronous machines, induction machines, and power inverters/converters, and will be sufficient for most installations. The criteria and requirements are applicable to all DR technologies with aggregate capacity of 10MVA or less at the point of common coupling, interconnected to electric power systems at typical primary and/or secondary distribution voltages. Installation of DR on radial primary and secondary distribution systems is the main emphasis of this standard, although installation of DR on primary and secondary network distribution systems is also considered. This standard is written considering that the DR is a 60 Hz source.

14.3.3 Common Information Model for Power Systems (IEC 61968/61970)

The IEC 61970 series of standards deals with the application program interfaces for EMS. The series provides a set of guidelines and standards to facilitate:

- integration of applications developed by different suppliers in the control center environment
- exchange of information to systems external to the control center environment, including transmission, distribution, and generation systems external to the control center that need to exchange real-time data with the control center
- provision of suitable interfaces for data exchange across legacy and new systems

14.3.4 Communication Networks and Systems in Substations (IEC 61850, Ed. 1, 2009)

High-speed peer-to-peer IEC 61850-8-1 (International Electrotechnical Commission (IEC) 61850-8-1) generic-object-oriented substation event (GOOSE) and IEC 61850-9-2 sampled values (SVs)–based information exchange among IEDs in modern IEC 61850 substations have opened the opportunity for designing and developing innovative all-digital protection applications. The transmission reliability and real-time performance of SVs and GOOSE messages, over the process-bus network, are critical to realize all-digital IEC 61850 substation automation systems (SASs) protection applications. To address the reliability, availability, and deterministic delay performance needs of SASs, a novel IEC 61850-9-2 process-bus based substation communication network (SCN) architecture is proposed in this standard.

14.3.5 Advanced Metering Infrastructure System Security Requirements (AMI–SEC, June 2009)

This standard provides the utility industry and vendors with a set of security requirements for advanced metering infrastructure (AMI) to be used in the procurement process, and represents a superset of requirements gathered from current cross-industry-accepted security standards and best practice guidance documents.

14.3.6 American National Standards Institute (ANSI) End-Device Data Tables (ANSI C12.19, 2008)

This standard defines a table structure for utility application data to be passed between an end device and a computer. The end device is typically an electricity meter, and the computer is typically a handheld device carried by a meter reader, or a meter communication module that is part of an automatic meter-reading system. C12.19 does not define end-device design criteria or specify the language or protocol used to transport that data. There are however related ANSI standards that do specify the transportation of these tables. ANSI C12.18 describes the communication of C12.19 tables over an optical port. ANSI C12.21 describes the communication of C12.19 tables over a modem. ANSI C12.22 describes the communication of C12.19 tables over a network.

14.4 MICROGRID BENEFITS

Many microgrid projects have been developed with government or research funding, and many are in various stages of planning and development. Reasons for this interest in microgrids primarily revolve around a number of benefits that the microgrid can provide to stakeholders that go beyond energy provision alone. These may include a variety of technical and social benefits. Due to the comparatively high cost of renewable and small thermal generation, energy storage systems, and other equipment necessary to develop a microgrid, microgrids may not be cost-competitive on the basis of energy price alone, but a host of benefits may result from the various operational advantages of the DR units on which the microgrid is based (e.g., efficiency, low emissions, and close proximity to loads) and microgrid functionality (e.g., ability to island and various "smart grid" functionalities such as demand response and SCADA). Some of these benefits are valued as services by the market or contract in a few jurisdictions.

- Microgrid customers could benefit from reductions in energy costs and improvements in reliability and PQ. Energy-cost reductions can come from reduced energy use, for example from increased efficiency and demand response, or from reductions in peak charges and cost per energy consumed.
- Grid customers may in some cases benefit from improved reliability and PQ.
- Independent power producers (IPPs) or microgrid developers could benefit from energy-sales profits, and may benefit from the proceeds of contractual agreements for provision of other services or participation in other markets.
- Distribution network operators (DNO) could benefit from reduced operation and maintenance costs, deferred investment and upgrade costs, reductions in contractual compensation for poor reliability and PQ, if applicable, and possibly reduced or avoided energy purchases.
- Society accrues all external benefits, which may include reductions in pollutants, reductions in resource use, reductions in infrastructure footprint, downward pressure on electricity prices, and increases in employment.

The division of benefits depends on the ownership model and the regulatory and market environment, and so before undertaking an analysis of benefits of the microgrid, it is critical to understand the ownership model and the goals of the microgrid developer. In general, the microgrid can be owned by the DNO or utility, the customer, or an IPP. In each case, benefits will be valued differently. DNOs would tend to value technical benefits such as reduced peak loading, whereas the customer would likely value energy-cost reduction. A free-market IPP-owned model, on the other hand, could be operated with the interests of all stakeholders in mind, insofar as they are represented by financial incentives. It has been suggested that with the proper incentives in place, profit maximization for an IPP could automatically optimize the benefits for other stakeholders.

14.4.1 Cost–Benefit Analysis in Microgrids

- Distribution automation has been a promising area of development in support of distribution systems.
- The potential benefits and costs associated with these functions are quantifiable.
- Research has been carried out to determine the costs and benefits to justify the feasibility of undertaking the automation of distribution system networks.
- Benefits to customers may not be obvious or justifiable based on the existing technology and the present costs associated with it.

14.4.2 Cost–Benefit Analysis Methodology

The revenue-requirements model is the most rigorous methodology for electric utility cost–benefit analysis. This technique represents the complete financial environment of a utility, accounting for taxes, depreciation, and time/value costs of money, actual capital investment, and operation and maintenance expenses. Two or more alternatives, each of which may involve differing life cycles and cash flows, can be compared on a common basis. With the revenue-requirement methodology, several alternatives are defined, and the cash flows for capital investment and operation and maintenance expenses are determined.

Illustrative Problems and Examples

Problem 1

(a) Distinguish between the smart grid, microgrid, and super grid.

(b) Briefly discuss the performance matrix of a microgrid compared with CG in terms of reliability, economy of scale, and sustainability.

Problem 2

At approximately 8:50 a.m. Saturday morning, PEPCO reports a service interruption affecting approximately 17,000 customers in northwest Washington, DC, including the main campus of Howard University. Service is restored at approximately 11:30 a.m. to all buildings on the main campus. Such outages can cause Howard University a loss of $400,000 per annum at an initial rate of 5 percent. To prevent further outages, the university is looking into a microgrid alternative on a five-year project plan. The various costs incurred and estimates and the derived benefits are summarized in Table Q2

TABLE Q2 Summary of Estimated Cost and Derived Benefits

Quantity	Average Value (in millions of dollars)
*Capital investment (one-time cost)	3.000
*Installation (one-time cost)	0.800
Maintenance and operations	0.007
Reduction in losses	0.035

(a) Calculate the total losses in dollars for the project time if there is no microgrid alternative plan.

(b) Calculate for the proposed microgrid alternative:

  (i) net-present value (NPV)

  (ii) discounted cost–benefit ratio

  (iii) payback period

(c) Write a short paragraph discussing the cost benefit of implementing the microgrid project plan.

Problem 3

Discuss the control options for the proposed Howard University microgrid in Problem 2 to combat voltage/var, frequency variation and PQ problems.

Problem 4

The following are utilized for real-time data measurement in a microgrid system: power management units, smart meters, DSSE. Provide a:

(a) detailed description of each device

(b) advantages and limitations of each device

Problem 5

Use a simple optimization process to construct the algorithm for load-flow implementation. Write the power-flow equations in terms of rectangular coordinates. For example, express complex voltage at bus n as

$$V_n = E_n + jF_n.$$

Problem 6

Discuss the importance of the demand response and demand-side management in distribution automation.

Problem 7

Discuss different techniques for forecasting. Selecting any one of the techniques, perform the forecasting for DERs or on-load demand.

Problem 8

What is PQ disturbance and what are its causes?

14.5 CHAPTER SUMMARY

It is envisioned that the microgrid will integrate renewable energy sources, particularly solar and wind, with conventional power plants, in an intelligent and coordinated way by utilizing energy-conversion technologies. It is expected that energy conversion devices will allow for improved reliability, dependability, and service continuity while also effectively reducing energy consumption and considerably reducing carbon emissions. All these requirements dictate the modernization of the power grid by installing intelligent electronic and conversion devices such as sensors, switches, power electronic converters and inverters, and advanced communication for data acquisition and interactive software for real-time control that optimizes the operation of the entire electrical system and makes more efficient use of the grid assets. This chapter detailed the technology and tools needed for design of microgrid systems in support of future network designs.

BIBLIOGRAPHY

[1] R. Lamb, "How Smart Grid Will Work," https://science.howstuffworks.com/environmental/green-science/smart-grid.htm.

[2] Microgrid Institute, "About Microgrids," www.microgridinstitute.org/about-microgrids.html.

[3] Sandia Energy, "Energy Surety Microgrid^{TM}," http://energy.sandia.gov/energy/ssrei/gridmod/integrated-research-and-development/esdm.

[4] F. Yanine and E. Sauma, "Review of Grid-Tie Micro-Generation Systems without Energy Storage: Towards a New Approach to Sustainable Hybrid Energy Systems Linked to Energy Efficiency," *Renew. Sust. Energ. Rev.* **26** (October 1, 2013).

[5] A. L. Dimeas and N. D. Hatziargyriou, "Operation of a Multiagent System for Microgrid Control," *IEEE T. Power Syst.* **20**, 3 (2005).

[6] S. Backhaus, G. W. Swift, S. Chatzivasileiadis, et al., *DC Microgrids Scoping Study—Estimate of Technical and Economic Benefits*, Los Alamos National Laboratory, March 2015.

[7] E. Kabalci, E. Hossain, and R. Bayindir, "Microgrid Test-Bed Design with Renewable Energy Sources," *IJRER* **4**, 4 (2014).

[8] M. R. Islam and H. A. Gabbar, "Study of Micro Grid Safety and Protection Strategies with Control System Infrastructures," *Smart Grid and Renewable Energy* 3, 1 (February 2012).

[9] A. Sinha, R. Lahiri, S. Byabortta, et al. "Formulation of Pricing Mechanism for Microgrid Energy," CIRED Seminar 2008: SmartGrids for Distribution, June 2008.

[10] G. Y. Morris, C. Abbey, S. Wong, et al., "Evaluation of the Costs and Benefits of Microgrids with Consideration of Services beyond Energy Supply," IEEE Power and Energy Society General Meeting, July 2012.

[11] H. Hou, J. Zhou, Y. Zhang, et al., "A Brief Analysis on Differences of Risk Assessment between Smart Grid and Traditional Power Grid," IEEE Fourth International Symposium on Knowledge Acquisition and Modeling, January 2012.

[12] M. Eghbal, T. K. Saha, and N. Mahmoudi-Kohan, "Utilizing Demand Response Programs in Day Ahead Generation Scheduling for Micro-Grids with Renewable Sources," IEEE PES Innovative Smart Grid Technologies Asia, March 2011.

[13] Y. Qiao, Z. Lu, and S. Mei, "Microgrid Reconfiguration in Catastrophic Failure of Large Power Systems," IEEE International Conference on Sustainable Power Generation and Supply, December 2009.

[14] C.-H. Lin, H.-J. Chuang, C.-S. Chen, et al., "Fault Detection, Isolation and Restoration Using a Multiagent-Based Distribution Automation System," IEEE Conference on Industrial Electronics and Applications, May 2009.
[15] J. A. Momoh, S. Meliopoulos, and R. Saint, *Centralized and Distributed Generated Power Systems: A Comparison Approach*, Prepared for the Project "The Future Grid to Enable Sustainable Energy Systems," funded by the US Department of Energy, June 2012.
[16] Association for Computing Machinery, "2017 ACM Master Article Template," https://www.acm.org/publications/proceedings-template.
[17] I. Moghram and S. Rahman, "Analysis and Evaluation of Five Short-Term Load Forecasting Techniques," *IEEE T. Power Syst.* **4**, 4 (Nov. 1989).
[18] A. Primadianto and C.-N. Lu, "A Review on Distribution System State Estimation," *IEEE T. Power Syst.* **32**, 5 (Sep. 2017).
[19] M. H. Haque, "A General Load Flow Method for Distribution Systems," *Electr. Pow. Sys. Res.* **54** (2000).
[20] Y. Li, D. M. Vilathgamuwa, and P. C. Loh, "Design, Analysis, and Real-Time Testing of a Controller for Multibus Microgrid System," *IEEE T. Power Electr.* **19**, 5 (Sep. 2004).

CHAPTER 15

THE SMART GRID: AN INTRODUCTION

15.1 EVOLUTION, DRIVERS, AND THE NEED FOR SMART GRID

The purpose of this chapter is to study the application of digital processing and communications to the power grid, as data flow and information management are central to the smart grid. Various capabilities result from the deeply integrated use of digital technology with power grids, and integration of the new grid information flows into utility processes and systems is one of the key issues in the design of smart grids. Electric utilities now find themselves making three classes of transformations: improvement of infrastructure (called the *strong grid* in China), addition of the digital layer (the essence of the smart grid), and business-process transformation (necessary to capitalize on the investments in smart technology). Much of the work that has been ongoing in electric-grid modernization, especially substation and distribution automation, is now included in the general concept of the smart grid, but additional capabilities are evolving as well.

Smart-grid technologies emerged from earlier attempts at using electronic control, metering, and monitoring. In the 1980s, automatic meter reading was used for monitoring loads from large customers, and evolved into the advanced metering infrastructure of the 1990 s, using meters that could record how electricity was used at different times of the day. Smart meters add continuous communications so that monitoring can be done in real time, and can be used as a gateway to demand response-aware devices and "smart sockets" in the home. Early forms of such demand-side management (DSM) technologies were dynamic demand-aware devices that passively sensed the load on the grid by monitoring changes in the power supply frequency.

It is expected that the new smart grid will provide:

- greater penetration of renewable resources
- effective and extensive communication overlay from power generation customers

Energy Processing and Smart Grid, First Edition. James A. Momoh

- utilization of high-speed control and cutting-edge sensors to make the grid more robust
- superior operating efficiency
- superior resiliency against natural disasters and attacks
- rapid service restoration after storms
- effective automated metering
- real-time as well as time-of-use pricing of electrical power by users
- increased customer participation in selling and generation of electrical energy generated by through renewable resources

A smart grid utilizes innovative services and products together with intelligent control, monitoring, communications, and self-healing technologies.

15.2 COMPARISON OF SMART GRID WITH THE CURRENT GRID SYSTEM

The comparative analysis between todays Power Grid and Smart Grid is summarized on Table 15.1

A working definition of a smart grid should include the following attributes:

- assess grid health in real time
- predict behavior, anticipate requirements
- adapt to new environments like distributed resources and renewable-energy resources
- handle stochastic demand and respond to smart appliances
- provide self-correction, reconfiguration, and restoration
- handle randomness of loads and market participants in real time
- create more complex interactive behavior with intelligent devices, communication protocols, and standard and smart algorithms to improve smart communication and transportation systems

In this environment, smart control strategies will handle congestion, instability, or reliability problems. The smart grid will be a cyber system that is secure, resilient, and able to manage shock to ensure durability and reliability. Additional features include facilities for the integration of renewable and distribution resources, and obtaining information to and from renewable resources and plug-in hybrid vehicles. New interface technologies will make data-flow patterns and information available to investors and entrepreneurs interested in creating goods and services.

Thus, the working definition of a smart grid becomes:

> The smart grid is an advanced, digital, two-way, power flow, power system capable of self-healing, is adaptive, resilient, and sustainable, with foresight for

TABLE 15.1 Comparison of Current Grid to Smart Grid

Preferred Characteristics	Current Grid	Smart Grid
Active consumer participation	Consumers are uninformed and do not participate	Informed, involved consumers; demand response and distributed energy resources
Accommodation of all generation and storage options	Dominated by central generation; many obstacles exist for distribution energy resources interconnection	Many distributed energy resources with plug-and-play convenience; focus on renewables
New products, services and markets	Limited, poorly integrated wholesale market; limited opportunities for consumers	Mature, well-integrated wholesale market; growth of new electricity markets for consumers
Provision of power quality for the digital economy	Focus on outages; slow response to power-quality issues	Power quality is a priority with a variety of quality/price options; rapid resolution of issues
Optimization of assets and efficient operation	Little integration of operational data with assets management and business processes	Greatly expanded data acquisition of grid parameters; focus on prevention and minimizing impact to consumers
Anticipation of responses to system disturbances (self-healing)	Responds to prevent further damage; focus is on protecting assets; flowing fault	Automatically detects and responds to problems; focus on prevention and minimizing impact to consumers
Resiliency against attacks and natural disasters	Vulnerable to acts of terror and natural disasters	Resilient to attacks and natural disasters with rapid restoration capabilities

prediction under different uncertainties. It is equipped for interoperability with present and future standards of components, devices, and systems that are cyber-secured against malicious attack.

15.3 ARCHITECTURE OF A SMART GRID

The reference architecture describes the structure of a system with its elements and their structure, as well as their interaction types, among each other and with the environment. Figure 15.1 shows the architecture of the smart grid.

15.4 DESIGN FOR SMART-GRID FUNCTION FOR BULK POWER SYSTEMS

Electric power grids are highly complex dynamical systems vulnerable to several disturbances in day-to-day operations, which make the system challenging. Hence, to handle the complexity and challenges in the system, an electric grid should be

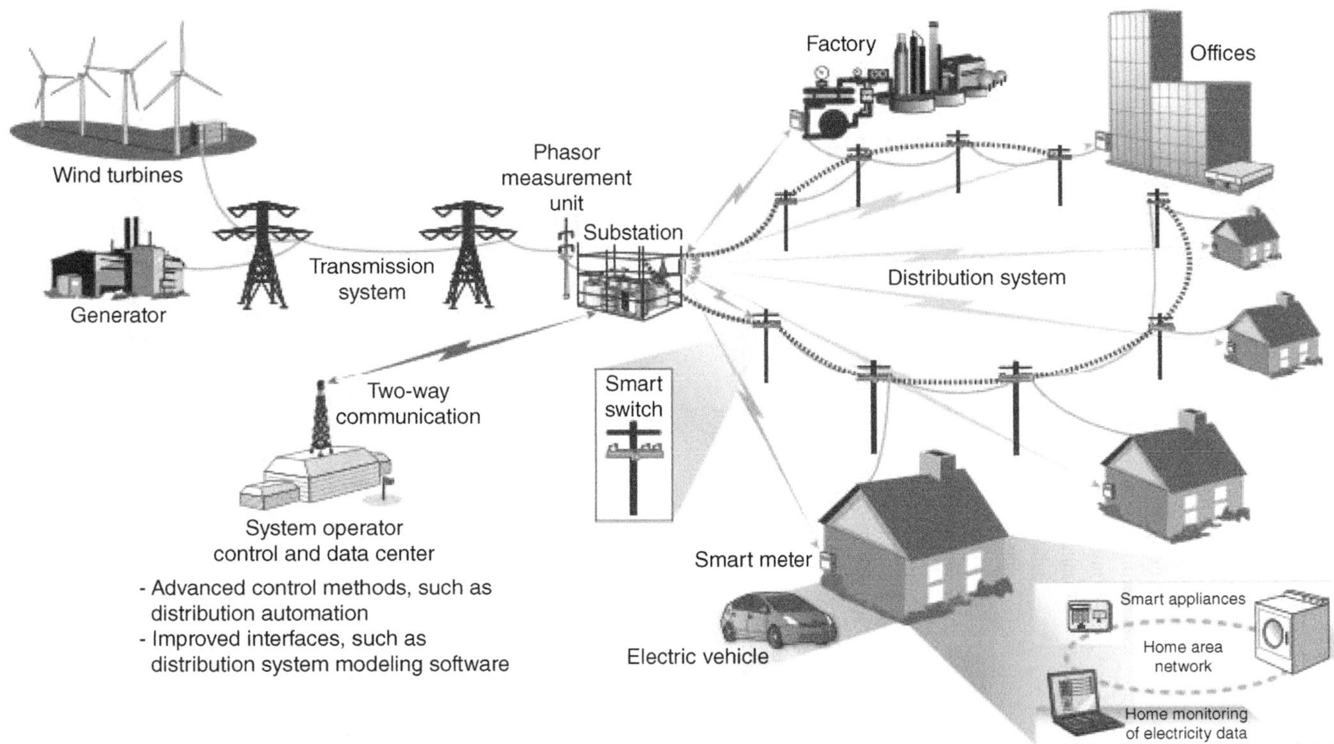

Figure 15.1 Architecture of the smart-grid design.

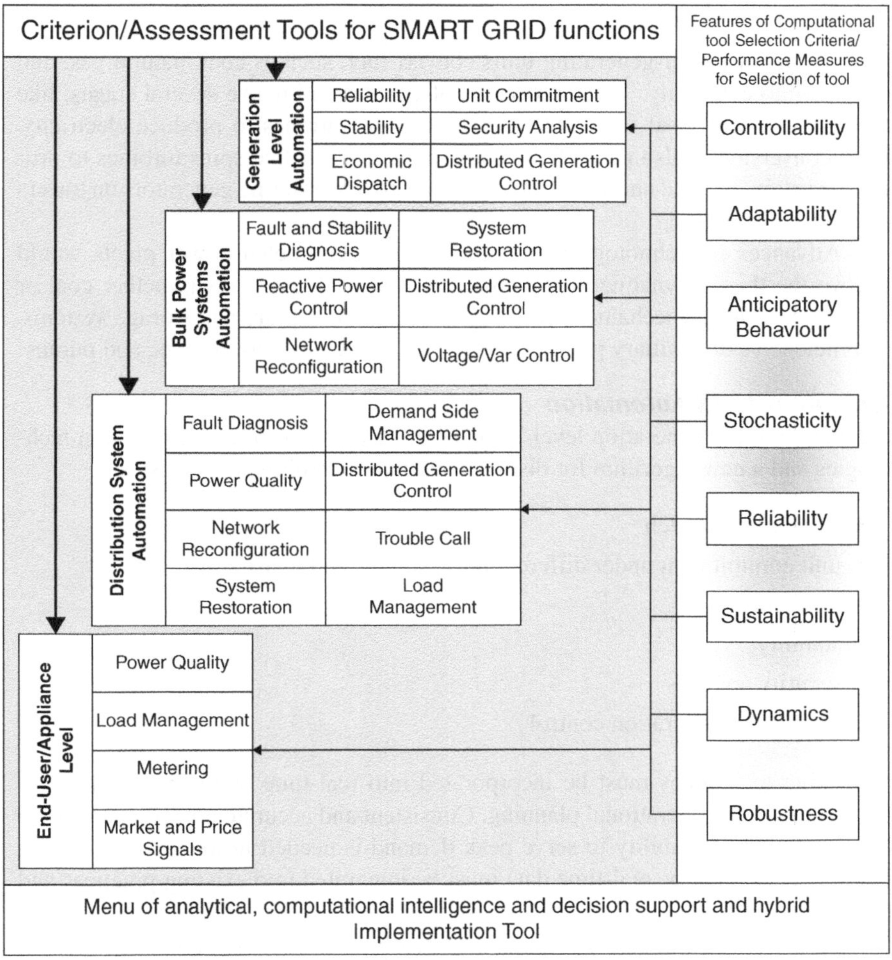

Figure 15.2 Smart grid using advanced optimization and control techniques.

smart. The design of the smart grid involves the coupling of tools, technologies, and techniques for three different subsystems:

- generation
- transmission
- distribution

Figure 15.2 illustrates the four advanced optimization and control techniques required to meet the criteria for smart-grid performance.

15.4.1 Generation

Power plants or power-generating units convert fuel, such as coal, natural gas, and uranium, into electricity. These processes of generation involve several stages, like heating water with coal to create steam that spins turbines to produce electricity. Direct conversion is also possible by using flow of water that spins turbines to produce electricity or wind and solar, such as through wind-rotating generators on towers and solar PV panels.

Advances in technology or more efficient fossil-fueled power plants would improve the thermodynamic efficiency of converting fossil fuels (whether coal or gas) into electricity, mechanical efficiencies of carbon capture and storage systems, and efficiencies of auxiliary power loads at plants, such as fans, motors, and pumps.

Generation-Level Automation

Automation at the generation level involves the use of advanced computation technologies and a new algorithm for dispatch and unit commitments to ensure:

- economic dispatch
- unit commitment under different uncertainties and constraints
- reliability
- stability
- security analysis
- distributed generation control

Forecasting techniques must be incorporated into real-time operating practices as well as day-to-day operational planning. Consistent and accurate assessment of variable generation availability to serve peak demand is needed in longer-term system planning. High-quality, real-time data must be integrated into existing practices and software.

Economic dispatch is a computational process where the total required generation real and reactive power, including renewable-energy resources, is allowed to vary within certain limits, so as to meet a particular load demands with minimum fuel cost. Mathematically, objective function of a load dispatch problem can be formulated as

$$Min\ F_c = \sum_{k=1}^{N_G} F_k(P_k) \tag{15.1}$$

where, F_c is the total operating cost of the system, N_G is the number of generating units, and $F_k(P_k)$ is the fuel cost of the generating unit k for real power P_k.

Traditionally, the economic-dispatch problem is formulated as an optimization with cost as the quadratic function of generating units. Mathematically, this formulation can be expressed as:

$$F_k(P_k) = \alpha_k(P_k)^2 + \beta_k(P_k) + \gamma_k \tag{15.2}$$

where, α_k, β_k, and γ_k, denote the fuel cost.

This is subjected to different constraints as follows:

- Power balance constraints: Total power output from generating units should exactly satisfy load demand for that hour and corresponding network losses.

$$\sum_{k=1}^{N_G} P_k = P_D + P_{loss} \tag{15.3}$$

where, P_D is total load demand and P_{loss} represents losses in transmission network. Kron's formula is used to calculate total losses P_{loss}, calculated using B-coefficients, given by

$$P_{loss} = \sum_{k=1}^{N_G} \sum_{i=1}^{N_G} P_k B_{ki} P_i + \sum_{k=1}^{N_G} B_{0k} P_k + B_{00} \tag{15.4}$$

where, P_i and P_k are the real power injection at i^{th} and k^{th} buses, respectively, and B_{ki} is the loss coefficient, which can be assumed to be constant under normal operating conditions.

- Thermal constraints

$$P_k^{min} \leq P_k \leq P_k^{max} \tag{15.5}$$

$$V_k^{min} \leq V_k \leq V_k^{max} \tag{15.6}$$

$$P_i^{\max}(t) = \min(P_i^{\max}, P_i(t-1) + \tau.RU_i) \quad \forall i \in N_G \tag{15.7}$$

$$P_i^{\min}(t) = \max(P_i^{\min}, P_i(t-1) - \tau.RD_i) \quad \forall i \in N_G \tag{15.8}$$

where, RU_k/ RD_k are ramp-up/down rate of the kth generating unit and τ is a (Unit Commitment) UC time step.

15.4.2 Transmission

Advanced transmission operations apply advanced digital technologies and power electronics devices to increase the performance of the system, enable interconnection of inaccessible power systems, and increase the size and capacity of existing transmission assets to increase the ability of system operators to control the system.

Automation of the Smart Grid at Transmission Level

The automation of different functions of the transmission system is important for achieving resilience and sustainability of the system. The following functions are evaluated, and the appropriate intelligent technology proposed:

- congestion minimization of the transmission line
- voltage stability, collapse detection, and prevention using intelligent database
- monitoring and controlling reactive power using intelligent controls
- based on intelligent-switching operations performing fault analysis and reconfigurations

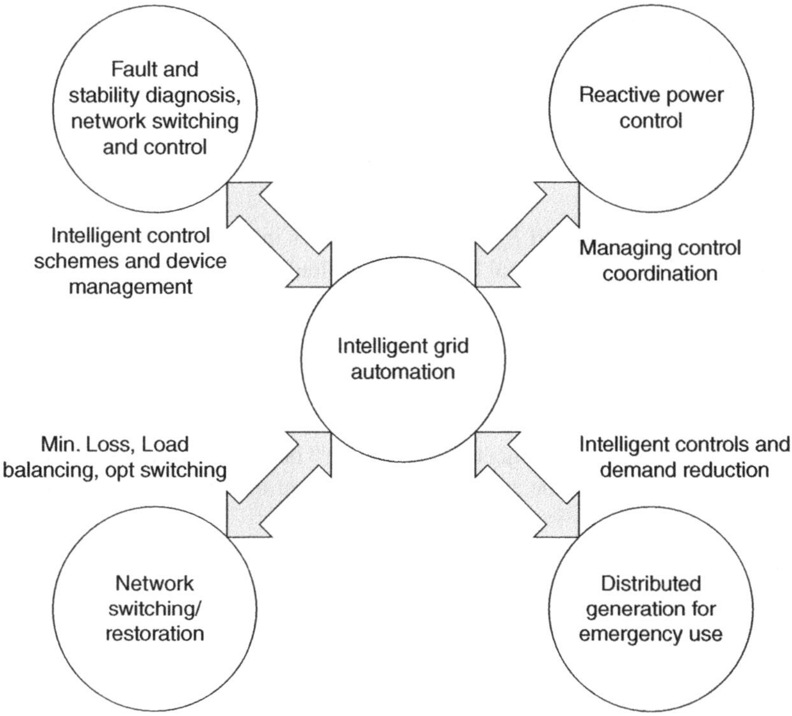

Figure 15.3 Automation functions at transmission level.

- load balance and power generation via intelligent switching operations and minimizing demand interruption
- DG and DSM via DR strategy

The integration of these functions is shown in Figure 15.3.

Congestion minimization of the transmission line can be done utilizing two different approaches. The first approach uses rescheduling of generating units and prioritization and curtailment of loads/transactions. The second approach utilizes operation of transformer taps, phase shifters, or FACTS devices. FACTS devices assume importance in the context of power system restructuring since they can expand the usage potential of transmission systems by controlling power flows in the network. FACTS devices are operated in a manner so as to ensure that the contractual requirements are fulfilled as far as possible by minimizing line congestion.

The mathematical formulation can be stated as:

$$min\left\{ \sum_{n\in S_g^P} A_n^P\left(P_{gn}^P\right) - \sum_{n\in S_d^P} B_n^P\left(P_{dn}^P\right)\right\} \tag{15.9}$$

subject to constraints such as:

$$P_n(\delta, V) - P^P_{gn} + P^P_{dn} - \sum_{i \in I} P^M_{gi,n} + \sum_{i \in I} P^M_{di,n} = 0 \tag{15.10}$$

$$Q_n(\delta, V) - Q^P_{gn} + Q^P_{dn} = 0 \tag{15.11}$$

The illustrated equations are well known as power-balance equations. After Thyristor-Controlled Series Capacitor (TCSC) location, the achieved equations are:

$$P_n(\delta, V) - P^P_{gn} + P^P_{dn} - \sum_{i \in I} P^M_{gi,n} + \sum_{i \in I} P^M_{di,n} + P^F_n = 0 \tag{15.12}$$

$$Q_n(\delta, V) - Q^P_{gn} + Q^P_{dn} + Q^F_n = 0 \tag{15.13}$$

where P^F_n and Q^F_n are the injected active and reactive powers of TCSC to the bus n. The other constraints are:

$$V^{min}_n \leq V_n \leq V^{max}_n \tag{15.14}$$

$$X^{min}_c \leq X_c \leq X^{max}_c \tag{15.15}$$

where superscript i is the bilateral transaction index; n and m are the bus indices, P and M signify pool transaction and bilateral (or multilateral) transaction, P^P_{gn} and P^P_{dn} show the pool real power generation and demand at bus n, $P^M_{gi,n}$ and $P^M_{di,n}$ are the bilateral injection and extraction of agent i at bus n, Q^P_{gn} and Q^P_{dn} are the pool-reactive power generation and load at bus n, V^{min}_n and V^{max}_n are the lower and upper limits of voltage at bus n, and X^{min}_c and X^{max}_c are the lower and upper limits of capacitive reactance of TCSC, respectively.

15.4.3 Distribution

Advanced distribution operations allow a grid that can self-heal through auto-detection of faults and auto-reconfiguration of circuits, with greater efficiency due to innovative system modeling and analysis and voltage/var control, which is more flexible and situationally aware due to cutting-edge sensing, data acquisition, and automated controls. The distribution system of the future smart grid will permit the incorporation of distributed energy resources and will also operate with novel circuit configurations required for microgrids.

Distribution Automation Function and Relationship to the Smart Grid

The IEEE defines distribution automation (DA) as a system that enables an electric utility to remotely monitor, coordinate and operate distribution companies in real-time mode from remote locations [1]. Strategies such as various DMS applications are utilized to automate and manage electric grid operations. Moreover, DA optimizes a utility's operations and directly improves the reliability of its distribution power system.

TABLE 15.2 Equipment, Technology, and Control Packages for DA

Equipment	Technology	Control Packages
On Load Tap Changer (OLTC)	Substation/distribution voltage and current sensors	Transformer load tap changers
Distribution capacitor banks	Supervisory control and data acquisition (SCADA) and advanced metering infrastructure (AMI) communication	Capacitors/voltage regulators
Distribution voltage regulators	Smart meters	Distribution management systems and voltage optimization software

The goal of DA is real-time adjustment to changing loads, generation, and failure conditions of the distribution system, usually without operator intervention. However, accurate modeling of distribution operations supports optimal decision-making at the control center and in the field. The following functions are evaluated, and the appropriate intelligent technology is proposed:

- **Voltage/var control technologies** monitor power, determine control settings, and physically adjust voltage and reactive power. Equipment, technology, and control packages for change, measurement, and control in voltage and power factors are given in Table 15.2. Problem formulation for voltage/var control is discussed in Section 14.4.1.
- **Power quality** refers to the ability of electrical equipment to consume the energy being supplied to it. Several power quality issues impact the efficiency of electrical equipment, including:
  - electrical harmonics
  - poor power factors
  - voltage transient impulses
  - voltage sag
  - total harmonic distortion
  - flicker factors
- **Network reconfiguration** of a distribution system has long been identified as a useful method for the improved performance of the system. A better reconfiguration scheme would take care of these issues to maximize the benefit of network reconfiguration in the distribution system. The objectives of network reconfiguration may be formulated as:
  - minimize power loss
  - maximize sag voltage during fault or switching
  - minimize harmonic distortion
  - minimize system unbalances

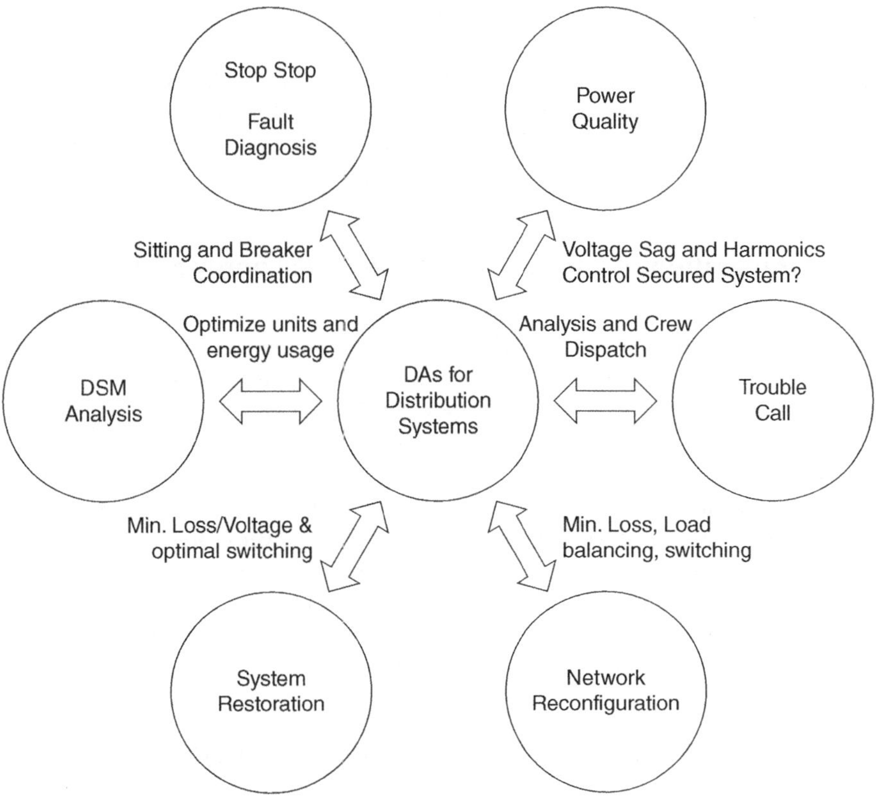

Figure 15.4 Automation functions at distribution level.

- **DSM or demand-side response (DSR)** provides an opportunity for consumers to play a significant role in the operation of the power network by reducing or shifting their demand during peak periods in response to time-based rates or other forms of financial incentives. DSM may be aimed at addressing the following issues:
  - cost reduction
  - reliability and network issues
  - improved markets
  - environmental and social improvement

  For distribution automation functions, DSM is classified into three categories:
  - energy-reduction programs
  - load-management programs
  - load growth and conservation programs

Figure 15.4 shows the distribution automation schemes for distribution systems.

15.5 SMART-GRID CHALLENGES

Significant progress has been made toward the growth and employment of a smart grid, nevertheless, there are challenges still to be addressed. Various road maps and reports have defined the technical issues and potential methodologies for overcoming them from the industry, state, federal, and even worldwide perspectives.

15.5.1 Integration of Utility-Scale Renewable Energy Sources onto the Grid

Huge, utility-scale renewable-energy (RE) systems (approximately 50MW or larger) can potentially make a significant influence on future domestic power supplies. A wide assortment of RE sources may be considered for large-scale grid integration, such as wind, geothermal, solar, and hydropower. Even though progress has been made in the deployment of renewable energy technologies, there are still unique challenges associated with the integration of these sources in large-scale systems. Challenges have developed when attempting to integrate power sources that are variable and nontraditional, and sources that are non-base load power onto the grid. The variable nature of the output has to be considered when attempting to optimize the operation of the entire grid. These new sources will require that conventional power plants, along with other system assets, operate differently.

15.5.2 Integration of Energy Storage and Distributed Generation on the Grid

While advancement has been made in the accommodation and deployment of DER technologies, there are distinctive challenges related to integrating these resources. Issues related to economics, regulation, and technology are inhibiting larger integration of DERs. For instance, communication between DER, dispatchers, loads, and utilities is generally inadequate for an optimized smart grid. By and large, new devices and technologies are outstripping the ability to integrate them on the grid at full functionality.

15.5.3 Technical Challenges

Among the technical challenges for the smart grid are:

- Decision tools for operators are required to increase visibility and situational awareness, aid planning and forecasting, and offer logic for decision-making.
- Communications infrastructure currently is insufficient and must be enhanced to allow interconnection between various components and systems, public networks, and devices, along with operations and planning functionalities.
- Performance metrics are lacking to fully comprehend, control, and manage performance and flexibility of the system.

- Analytics and data management and are not sufficient for effectively collecting, storing, and interpreting the enormous volumes of data potentially to be collected.
- Robust operational and business models are desirable to facilitate effective planning and operations to integrate diverse sources of generation, storage options, and models for flexibility.

15.5.4 Nontechnical Challenges

Nontechnical challenges to the smart grid also remain, including:

- Privacy of information is still unreliable, and it is necessary to assure consumers that personal data is protected and its release is tightly controlled.
- Coordination of regulations and policies on the smart grid is lacking at the federal level and between states, along with utilities. The result is uncertainty and a business environment not fully supportive of risk taking and innovation.
- Market fragmentation for smart-grid technologies exists due to a lack of a common vision among utility standards organizations for the smart grid and the residential sector. This leads to manufacturers having to address numerous different solutions and types of technologies when entering the market.
- The business proposition for smart-grid technologies is still ambiguous and lacks clarity. Investments are further convoluted by uncertainties over who should bear the cost for upgrades (e.g., consumers, utilities) and how and when these costs should be recouped.

15.6 DESIGN STRUCTURE AND PROCEDURE FOR SMART-GRID BEST PRACTICES

The complexity of smart grid projects adds to the challenges of electric power system modernization as utilities will be required to make significant investments in information communication technology, an area largely outside their core competencies. The smart grid will be as integral to the core systems as enterprise resource planning is to the manufacturing industry, and it will be as geographically diverse as the new telecommunications network. Successful deployment will necessitate robust coordination across customary organizational boundaries, substantial process change, and with rigorous governance. The attributes of a good smart grid as follows:

- unconditional reliability of power supply
- optimal utilization of bulk electric power storage and generation in tandem with distributed energy resources and dispatchable/controllable consumer loads to guarantee lowest cost
- minimal environmental impact of electricity production and delivery
- reduction in the generation of electricity and an increase in the efficiency of the power delivery system and effectiveness of end use

- resiliency of supply and delivery from physical and cyberattacks and major natural disasters
- assuring optimal power quality for all consumers who require it
- monitoring of critical power system components to enable automated maintenance and prevention of outages

Illustrative Problems and Examples

Problem 1

(a) Distinguish between the smart grid, intelligent grid, and microgrid, using all their attributes.

(b) Describe the functions or criteria for a smart grid and suggest the technology needed to achieve these functions.

Problem 2

(a) Compare the smart grid to the prevalent traditional grid.

(b) What are the attributes of the smart grid?

Problem 3

Discuss the existing challenges and benefits to the application of the concept of smart grids. How do we overcome them?

Problem 4

Consider a generating station that contains three generating units. The fuel costs of these units are given by

$$f_1 = \frac{0.8}{2}P_1^2 + 10P_1 + 25\ Rs/h$$

$$f_2 = \frac{0.7}{2}P_2^2 + 5P_2 + 20\ Rs/h$$

$$f_3 = \frac{0.95}{2}P_3^2 + 15P_3 + 35\ Rs/h$$

The generation limits of the units are

$$30\text{MW} \le P_1 \le 500\text{MW}$$

$$30\text{MW} \le P_2 \le 500\text{MW}$$

$$30\text{MW} \le P_3 \le 250\text{MW}$$

The total load that these units supply varies between 90MW and 1,250MW. Assuming that all three units are operational all the time, compute the economic operating settings as the load changes.

Problem 5

What about cybersecurity? Does the smart grid make us less or more secure?

Problem 6

Discuss the following terms that relate to the smart grid:

- self-healing

- interoperability
- adaptiveness
- cybersecurity

Problem 7

What is the simplest thing about the smart grid? What is the most complex thing about it? Why is it important for states and localities to build a smart-grid infrastructure?

Problem 8

Review different test beds for a smart grid design and validate the experimental set-up at CESaC. Develop processes and procedures for achieving an economically viable test bed for studies of different functions of smart-grid concepts outline in the syllabus.

15.7 CHAPTER SUMMARY

In this chapter we discussed the concept of smart grid, its evolution and drivers, the need as well as energy sources and architecture of smart grid. A comparative analysis has been made between smart grid and current power grid. The components forming smart grid are addressed with automation function applications on transmission and distribution of the grid. Design procedures with challenges of integrating DGs are briefly discussed. The materials presented in this chapter are designed to equip the student with fundamental knowledge needed about smart grid functions, controls, and interfaces, which are necessary in smart-grid design.

BIBLIOGRAPHY

[1] D. Gruenmeyer, "Distribution Automation: How should it be evaluated?," *Rural Electric Power Conference,* 1991.

[2] W. Wang, Y. Xu, and M. Khanna, "A Survey on the Communication Architectures in Smart Grid," *Comput. Netw.* **55**, 15 (2011).

[3] J. H. Chow, F. F. Wu, and J. A. Momoh, eds. *Applied Mathematics for Restructured Electric Power Systems: Optimization, Control, and Computational Intelligence*, Springer Science and Business Media, New York, 2005, pp. 11–24.

[4] Workshop on Technology, Measurement, and Standards Challenges for the Smart Grid, August 13–14, 2012, Boulder, Colorado.

[5] Fujitsu and GTM Research, *Next-Generation Communications Protocols for Advanced Smart Grid Applications*, GTM Research Whitepaper, April 2012, www.fujitsu.com/global/Images/wp.pdf.

[6] V. C. Güngör, D. Sahin, T. Kocak, et al., "Smart Grid Technologies: Communication Technologies and Standards," *IEEE T. Ind. Inform.* **7**, 4 (November 2011).

[7] X. Lu, W. Wang, and J. Ma, "An Empirical Study of Communication Infrastructures Towards the Smart Grid: Design, Implementation, and Evaluation," *IEEE T. Smart Grid* **4**, 1 (Mar. 2013).

[8] A. Hahn, A. Ashok, S. Sridhar, et al., "Cyber-Physical Security Testbeds: Architecture, Application, and Evaluation for Smart Grid," *IEEE T. Smart Grid* **4**, 2 (June 2013).

[9] SCE-Cisco-IBM SGRA Team, *Smart Grid Reference Architecture, Volume 1: Using Information and Communication Services to Support a Smarter Grid,* July 14, 2011, www.pointview.com/data/files/1/636/2181.pdf.

[10] Kentucky Smart Grid Roadmap Initiative, *Kentucky's Smart Grid Roadmap: Recommendations on a Vision and Direction for the Future of the Electric Power Grid in the Commonwealth*, nd, http://energy.ky.gov/generation/Documents/KYSGRM_Final.pdf.

[11] A. Asthana, A. Booth, and J. Green, *Best Practices in the Deployment of Smart Grid Technologies*, McKinsey and Company, June 1, 2010.

[12] T. Gönen, *Electrical Machines with MATLAB*, 2nd ed., Taylor & Francis Group, Boca Raton, FL, 2011.

[13] J. Momoh, *Smart Grid: Fundamentals of Design and Analysis*, IEEE Press, John Wiley & Sons, Hoboken, NJ, 2012.

CHAPTER 16

SMART-GRID LAYERS AND CONTROL

16.1 INTRODUCTION

This chapter aims to evaluate various control systems involved in a smart grid. The layers of a smart grid, including the physical, real-time measurement, communication, application, and cost–benefit analysis will be comprehensively reviewed. State-of-the-art tools for control and communications are evaluated, as are related issues and challenges. Smart grid as a case study of a cyber-physical system will also be described.

16.2 CONTROLS FOR THE SMART GRID

Smart-grid control is important so that the following can be achieved:

- New micro sources can be integrated into the system without modification of existing equipment.
- The microgrid can choose operation point independently.
- The microgrid can connect to or island itself from the traditional grid quickly and seamlessly.

A key element of smart-grid operation is the grid energy management system (EMS). It includes control functions that define the grid as a system that can manage itself, operate autonomously or grid connected, and seamlessly connect to and disconnect from the main distribution grid for the exchange of power and the supply of ancillary services.

Hierarchical levels of control for micro grids may be categorized as primary, secondary, and tertiary, Figure 16.1. Primary control is the level in the control hierarchy that is based exclusively on local measurements, which includes islanding detection, output control, and power sharing (and balance control). Secondary control, the energy management system (EMS), is responsible for microgrid operation in either the grid-connected or islanded mode. Tertiary control is the highest level

Energy Processing and Smart Grid, First Edition. James A. Momoh

Device Level Control Functions	Local Area Control Functions	Supervisory Control Functions
Voltage/Frequency Control, Reactive Power Control, Electric Vehicle Control, Energy Storage Control, Islanding Detection, Fault Protection	Sequence Logic/Status control, Load Management, Building Energy Management, Plant Controller, AGC, Fast Load Shedding, Resynchronization, Disturbance Recording.	Forecasting, Data Management and Visualization, Optimization (e.g. Volt/VAR, Economic Dispatch), Dispatch, State Estimation, Emergency Handling, Generation Smoothing, Spinning Reserve, Topology Change Management, Black Start, Protection Coordination

Figure 16.1 Hierarchical levels of control in microgrid.

of control and sets long-term and "optimal" set points depending on the host grid's requirements.

16.2.1 Device-Level Control

Voltage/frequency control enables prevention of reductions in the power factor or efficiency of a motor in a wide range of variable speed operations for changes in the frequency for speed control by outputting a voltage (V/f characteristic) corresponding to the frequency set by a parameter in an inverter. The revolution speed of an induction motor is proportional to the frequency, which can reduce the power factor and efficiency of a motor even with a variable frequency because changes to the frequency cause the internal impedance of a motor to change. Therefore, you must change the voltage corresponding to the frequency. V/f control reduces the torque in low-speed operation with the primary resistance voltage drop even though it attempts to keep the torque stable regardless of the frequency.

Fault protection involves monitoring the microgrid system, identifying when a fault has occurred, pinpointing the type of fault and its location and setting into motion-control schemes for recovery from the fault. The major protection problem in microgrid is a result of a large difference between fault currents in the grid-connected and islanded modes. An optimal smart-grid protection system must be very sensitive to faults and be able to selectively sectionalize the microgrid, especially in the case of renewable-energy resources with power electronic interfaces (low-fault-current levels). A decision must be made by the integrated device control whether to sectionalize/isolate the microgrid or to shut it down entirely. This decision, in the event of a fault, is dependent on the needs of the customers and on cost benefits. Cost benefits in this case imply whether the cost involved in protection and communication could be justified for benefits gained by sectionalizing the microgrid instead of shutting it down entirely. The benefit in sectionalizing the microgrid involves reducing the end-consumer service interruption time.

Droop control enhances the power-quality parameters, like active and reactive power controls both in grid and autonomous operation. Droop-control techniques have been employed in voltage (V) and current (I) to provide power-quality control in the microgrid. Droop control is designed for batteries and generators to coordinate the sources and regulate the system frequency.

16.2.2 Local-Area Control

Peer-to-peer strategy implies that every distributed source maintains the same status in the microgrid. The benefit is that extra sources can be integrated into the microgrid without modifications to the control and protection of units that are already part of the system. This implies that every controller must be able to respond effectually to system changes without requiring data from other sources or locations, resulting in an improvement in the reliability of the smart grid and reducing system cost.

Peer-to-peer strategies can be realized by droop control. There are two control methods based on droop control:

- P-f/Q-V droop control produces reference active power and reactive power by measured system frequency and DG output voltage amplitude.
- P-f/V-Q droop control produces output frequency and voltage amplitude by measured output active power and reactive power. P-f/V-Q control is the more wide applicable peer-to-peer strategy.

Master–slave strategy refers to the presence of a master controller in the smart grid while others are slave controllers that obey the rules of the master controller through a communication link. When in a grid-connected mode, the microgrid requires no frequency regulation because the traditional grid can stabilize the system frequency. However, when islanded from the main grid, the master controller takes up the task of maintaining system frequency, voltage, and power balance separately by V/f control and regulating its own output.

Resynchronization control: While connected to the grid, micro sources can maintain a high level of output and high efficiency. In islanded mode, the master micro source switches from power-quality control to V/f control to provide frequency and voltage support to the microgrid, while other micro sources are still power-quality controlled. The regulation of this control strategy can only involve micro sources' output power, but not the transfer power between the microgrid and the traditional grid.

Seamless transfer control is an important control technology to affect a smooth transition of the microgrid between islanded and grid-connected modes. This smooth transition is done by controlled triggering of antiparallel thyristor (SCR) considering bulk grid state, microgrid operation mode and direction of power flow.

16.2.3 Supervisory Control Functions

The supervisory control system is the master control of a microgrid. It organizes and schedules the midterm and long-term energy flow in the power system, and aims to achieve an optimal operation of the system as a whole. It manages the dispatching and control of all the active components of a microgrid. Supervisory control may be regarded as a piece of intelligence-embedded software, a computer algorithm, or similar that aims to achieve optimal system operations. These operations depend on a set of goals, rules, and constraints, which are to be defined before the system is developed. According to the International Energy Agency, following supervisory control goals is of primary importance.

Effective supervisory control should consider economics, reliability and service delivery factors discussed below.

- **Most economical operation**: This strategy aims to minimize operation and maintenance (O&M) costs. The generating sets will generally operate at best efficiency levels, and energy dumping will be avoided. This strategy may not achieve full use of the input from the solar generator nor minimize the system's greenhouse emissions.
- **Highest reliability**: Prioritizing system reliability generally involves maintenance of backup contingency generating sets in a multi-generating sets system, and frequent maintenance. This may impact service delivery.
- **Lowest carbon footprint (or highest renewable-energy fraction)**: This strategy optimizes the use of renewable generator input to minimize diesel-generating operations. In some cases, this can lead to increased start-stop cycles of the generating sets.
- **Service-delivery optimization**: This strategy prioritizes service delivery, and is likely to imply a greater amount of generator run time. In a multi-generator system, a larger amount of spinning reserve may be maintained. This schedule implies greater O&M costs and reduced system lifetimes.
- **Component lifecycle optimization**: This strategy will configure the system to maximize component lifetime, particularly batteries. Component sizing and generator control will optimize battery cycling for efficient life-cycle operations. This strategy may include service-delivery restrictions.
- **Load optimization through demand-side management**: This strategy aims to optimize system operation through controlling the load via demand-side management. By deferring loads to periods of solar input where possible, energy-storage demands can be minimized and generators can operate at better efficiencies.
- **Best supply quality**: In this case, electricity-quality variables such as voltage range and harmonic distortion are prioritized. This may require the use of generators as spinning reserve and/or compensation nodes, resulting in lower system efficiencies, and increased costs.

Operation control strategies generally incorporate multiple objectives in their design.

16.2.4 Evaluation of Control Types Taking Area into Consideration

Development of Concepts, Models, and Performance Index

The smart grid is a virtual concept and there are just a few working smart grids in the world. But before it becomes a more widespread reality, a model needs to be developed that can mimic its behavior, issues, challenges, and operation in real time. This model should take into consideration all characteristics of the smart grid and must be inclusive of all the mathematical relations that exist—from simple line parameters to state estimation and power management units. One such model under review is Smart Grid Maturity Model (SGMM), a management tool that helps utilities plan smart-grid implementation, prioritize options, and measure progress.

Conceptual models for the smart grid map out generation, transmission, and distribution, plus markets, operations, service providers, and consumption. Each of these domains has its own terms, some reflecting a century of use, with others invented daily. For instance, the acronym HEMS (home energy management systems) did not exist until a few years ago, but now is commonly used in the dynamic business sector focused on software and hardware for consumption solutions and services. Other emerging industries, like renewables and energy storage, are also producing new words and acronyms. So there is a need for a unique model that is inclusive of both longtime and emerging standards and norms for the sake of simplicity and easy understanding.

A performance index for the smart grid also needs to be calculated and distributed. Different researchers and industry experts calculate performance measures differently; no unique terms exist that can universally state smart-grid performance.

Controls for Operation of a Smart-Grid System

The following functions in the operation of a smart-grid system can be controlled:

Frequency control uses load frequency control methods to control for regulation of load, i.e., transfer of load, load shedding, and utilization of available storage options, including the connection of distributed generation (DG) if available. Hence, by regulating the load at the point, the frequency can be controlled.

Voltage control uses var compensators, switching of load, capacitors, new transformers, new substations, and new lines. Real-time voltage stability indices (VSI) may also be calculated using optimal power flow (OPF) or advanced measuring devices like PMUs. VSI provide stability limits for the system, which enables the determination of system instability.

Protection control can be achieved by implementing microprocessor relays, circuit breakers, sectionalizers, and other devices. The operation of these protection devices is coordinated by a SCADA and EMS for improved dynamic protection.

Stability control is achieved by monitoring and maintaining system performance indices, which provide measures of system stability such as a reliability index, angle stability index, VSI, or vulnerability indices.

Supervisory control and data acquisition (SCADA) is a computerized system used for collecting data and controlling one or more remote stations from a master control station. It is comprised of sensing and digital control equipment, telemetry equipment, and two-way communications channels between the centralized master and remotely controlled stations. SCADA digital control equipment includes computers and terminals for data entry and display. Telemetry and sensing equipment includes sensors, analog-to-digital (ADC) and digital-to-analog (DAC) relays, actuators, and converters used at remote stations to sense normal operating and alarm conditions as well as to activate equipment remotely. Communications equipment includes modems for transmitting data and a communications link (power line, phone line, radio-microwave link). Typical functions performed by the SCADA include:

- control and position indication of a device, such as a circuit breaker or motor driven switch
- state indication without control features such as transformer fans off or on
- control without indication, for instance, capacitors switched out or in
- manipulation of a set point of a remote-control station, such as the nominal voltage for an automatic tap changer
- alarm sensing for a fire or the performance of non-command and control functions
- control operators to initiate an operation at a remote station from a centralized control station
- recognize and initiate sequences of events, for example, routing of electrical power around a transformer that may be malfunctioning by closing and opening circuit breakers, or sectionalizing a bus that is faulted
- metering and data acquisition, typically via analog/digital converters and communication links

With the modern automation of substation functions, a SCADA system can perform the following functions remotely at a substation:

- automatic sectionalizing of buses
- automatic reclosing after the concurrence of a fault
- monitoring of system synchronization
- protection of substation equipment
- reporting of faults
- load balancing of transformers

- reactive power and voltage control
- monitoring of statuses
- acquisition of data
- monitoring of equipment condition
- logging of system data

Load control and demand response: Both the customer side and the supply side of the meter are mutually significant when considering energy efficiency (EE), demand response (DR), and load control (LC) implications of the smart grid. Variations in the types of loads are also a critical consideration. The vital load types that impact EE, DR, and LC of the smart grid comprise industrial, commercial and residential loads. The smart grid has commenced exploration into real-time pricing, demand bidding, DR, and other consumer-response programs. The market for these programs is currently fragmented, and there is a lack of a clear vision and developed policy to guide the implementation of technology, which blurs the direction that the smart grid will take. While regulators are supportive of the promotion of energy efficiency, the impact of technologies and applications for future development are not fully understood.

Substation controllers: In the future, remote terminal units will provide system protection, local operation, graphical user interfaces, and data gathering/concentration from other systems. This will be in addition to customary station telemetry functions and control to the master station. The substation controllers will provide functionalities such as:

- applications for data collection
- applications data processing
- control database

16.3 LAYERS OF SMART GRID WITHIN THE GRID

Distributed energy resources (DER) comprise DG and energy storage devices that can help increase grid reliability and help manage peak loads. Sufficient DER allows a utility company to defer capital investments in the generation, distribution, and generation layers in addition to lowering emissions and improving system security. However, some are not yet cost competitive when compared with traditional sources of generation, and need more sophisticated integration strategies as penetration rates rise and a new pricing model for electricity.

The smart grid is the incorporation of communication infrastructures into the electrical grid to enhance the information exchange and achieve fully automated management in power systems. From this point of view, smart-grid architecture is divided into five dominant layers: Physical layer, Real-time measurement layer, Communications layer, Application layer and Cost–benefit layer.

This system architecture for the smart-grid paradigm is detailed in Figure 16.2.

- The physical power layer includes power generation units, transmission systems (delivering power from the plants to the substations), and distribution systems (delivering power from the substations to the consumers).
- The real-time measurement layer involves advanced sensing technologies, measurement devices, controls, and monitoring equipment, such as smart sensors, phasor measurement units, SCADA, and actuators.
- The communications layer provides reliable, secure, and effective information exchange between the layers.
- The application layer supports all services provided to end customers and utilities, such as automated metering and broadband access.
- The cost–benefit analysis layer provides justification of the smart grid over the classical grid.

16.3.1 Physical Power Layer

There are various energy options in powering a smart grid that are consistent with powering any grid, that are part of the Physical Layer on Figure 16.3.

- gas or diesel cogeneration/CHP
- fuel cells and microturbines
- photovoltaic (PV) modules
- wind, biomass, small hydro
- storage capacity

16.3.2 Real-Time Measurement Layer

Real-time measurements, monitoring, and control are the specialty of the smart grid, and involve the latest communication and real-time devices to make control operations faster and facilitate the grid with self-healing and self-restoration qualities.

Core duties are evaluating congestion and grid stability, monitoring equipment health, energy theft prevention, and control strategies support. Technologies include: advanced microprocessor meters (smart meters) and meter-reading equipment, wide-area monitoring systems, dynamic line ratings (typically based on online readings by distributed temperature sensing combined with real-time thermal rating (RTTR) systems), electromagnetic signature measurement/analysis, time-of-use and real-time pricing tools, advanced switches and cables, backscatter radio technology, and digital protective relays. The following are details of some of these real-time measurement layer tools:

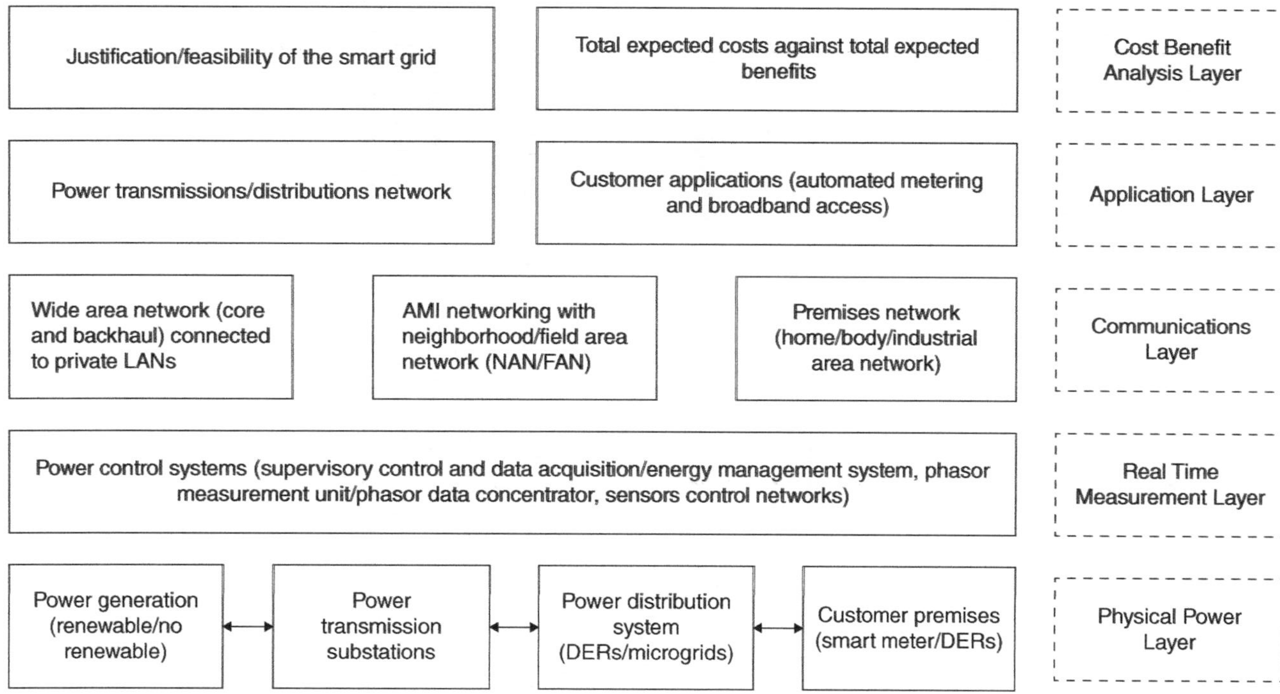

Figure 16.2 Smart-grid layers.

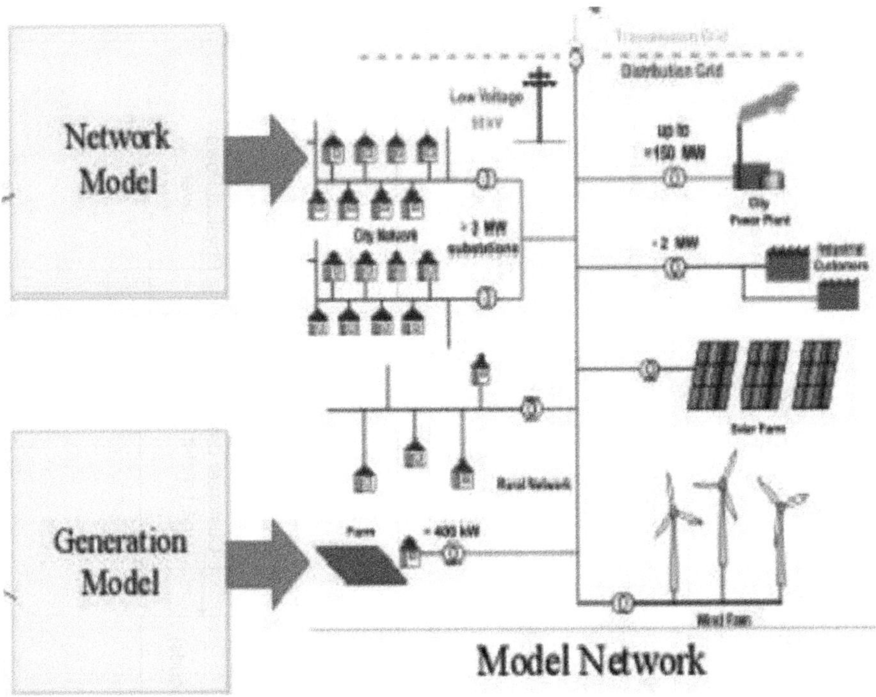

Figure 16.3 Physical layers.

(a) Smart Meters

A smart meter, shown in Figure 16.4 [1], is an electronic device that records consumption of power in intermissions of an hour or less and communicates that information at least daily back to the utility for monitoring and billing.

Automated meter reading (AMR)—the replacement of conventional, mechanical meters read once a month with smart meters that can be remotely read every ten minutes or so—is often the starting point for smart grid implementation. AMR, and its

Figure 16.4 Smart meter.

two-way enhancement advanced metering infrastructure (AMI), is often the costliest component of a smart grid because of the hundreds of thousands of meter end-points. In some parts of the world, there has been a tendency to consider an AMR/AMI system and its communications requirements as an independent solution rather than as part of the entire smart-grid infrastructure.

(b) Phasor Measurement Units

High-speed sensors called phasor measurement units (PMUs) distributed throughout a network can be used to monitor power quality and in some cases respond automatically. Synchronized PMUs (synchro-phasors) provide real-time measurement of electrical quantities from across a power system. Applications of synchro-phasor measurements include validating system models, determining stability margins, maximizing stable system loading, islanding detection, systemwide disturbance recording, and visualization of dynamic system response. The basic system building blocks are GPS satellite-synchronized clocks, a phasor data concentrator (PDC), and visualization software. Figure 16.5 [2] shows a sample phasor measurement unit.

The accuracy, speed, and reliability of PMUs are critical to fulfilling the promise of the smart grid. However, the current standard that governs PMU performance is quite broad, and testing by the National Institute of Standards and Technology has shown that PMUs from various manufacturers can report dynamic conditions very differently. For example, the primary transducer that converts high line voltages to smaller voltages used by the measuring system can introduce inaccuracies. From the primary transducer, signals pass through an anti-aliasing filter that also generates some level of error. Next, the signal goes through an analog/digital (A/D) converter that can introduce magnitude or gain errors and channel phase shifts. Signal processing is carried out to evaluate the phasor magnitude and phase angle, which may create additional error; the phase is reported with respect to the global time reference so errors in the global positioning system (GPS) source and cable delays may introduce synchronization errors.

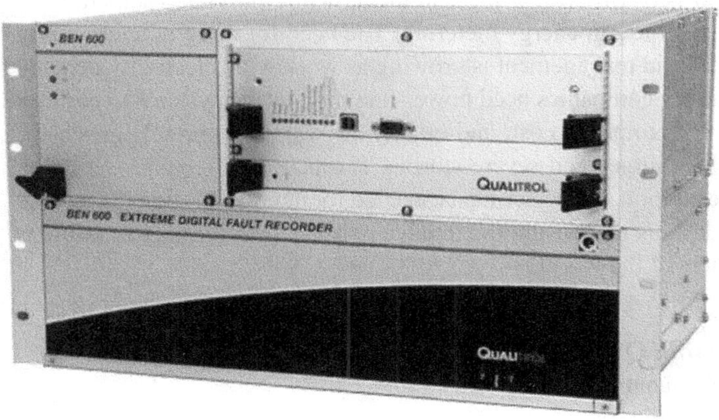

Figure 16.5 Phasor measurement unit.

The goal of PMU calibration is to ensure that all PMUs have consistent performance across the system, are interoperable regardless of their makes and models, and comply with system application requirements. A general rule of thumb is that a calibration system should have less than one-fourth of the total uncertainties of the system under calibration.

(c) Geographical Information Systems

Spatial data systems, including geographical information systems (GISs), have been used successfully by utility companies, government agencies, and many industries for more than three decades. Systems based on spatial technology have had a positive impact on a wide variety of business processes, ranging from marketing to environmental management. Spatial systems bring geography into play within the business processes of a company or organization while enabling data to be visualized more effectively. Smart-grid analytical software such as distribution management systems (DMSs) and outage management systems require high-quality data to provide the right answers. By using the functionality inherent in GIS, smart-grid technology will start with better-quality data and should, in turn, produce better-quality analytics and operational decisions.

(d) Remote Terminal Units

A remote terminal unit (RTU) is a microprocessor-controlled electronic device that interfaces objects in the physical world to a distributed control system or SCADA by transmitting telemetry data to the system, and by using messages from the supervisory system to control connected objects. The primary source of real-time data in a smart grid is the remote terminal unit (RTU). While most utilities have long had RTU technology in almost every transmission substation, these devices have historically been far less common in distribution substations.

(e) Analog to Digital/Digital to Analog

Superior in quality, reliability, and performance, analog devices' energy-metering ICs combine analog-to-digital converters (ADCs) with fixed-function digital signal processors (DSPs) to perform critical measurements while providing unparalleled functionality and ease of use in energy metering. The need for more efficient utility substations and smart-grid management is growing as worldwide electricity demand increases. Electric utility companies need power-line monitoring systems to monitor and control energy consumption, cost, and quality, as well as to protect expensive equipment from power surges and severe storms. The new simultaneous-sampling ADCs provide the resolution and performance needed for next-generation power-line monitoring system designs that ensure the reliable delivery of electricity to millions of people in all corners of the world.

Real-Time Measurable Quantities

Real-time measurable quantities include

- power
- voltage

- current
- frequency

16.3.3 Communications Layer

The communications layer has premises networks, neighborhood/field-area networks (NAN/FAN), and wide-area networks (WAN). The premises network, such as the home-area network (HAN), building-area network (BAN), and industrial-area network (IAN), provides access to applications in customer premises. The NAN/FAN provides a relatively long-distance communication link between smart meters, field devices, DERs, customer premises' networks, and WAN. The WAN provides very long-distance communication links between the grid and the utility via a core network and FAN/NAN. However, the coverage area and data-rate requirements for the customer premises' network, FAN/NAN, and WAN vary for different communication standards and protocols.

Communication Networks

Network security is an important consideration for any smart-grid deployment. Integrating communications technologies and control functionalities to traditionally isolated grid equipment raises apprehensions among utilities, regulators, and consumers.

Distributed automation necessitates additional security procedures, as these protection and control functions alter how distribution networks are configured, and eventually the behavior of the electricity that is delivered to the customer. Guaranteeing the integrity of messages to and from distributed control equipment is paramount to detecting and reacting to incomplete and tampered communication signals. Just as with AMI, encryption, authentication, authorization, and accounting are important in order to limit access and the ability to compromise the network.

The grid of the future will rely extensively on communication technologies for critical functionalities such as protection, monitoring, and control of electrical equipment. Hence the security of such a connected structure from any cyberattack is of paramount importance. Security against any physical attack is also a concern. In the event of a security breach, it is expected that the grid will detect and isolate such a breach to minimize its effect while raising an alarm to prompt service restoration.

(a) Wide-Area Networks

Wide-area networks (WANS) form the communication backbone for connecting the highly distributed smaller area networks that serve power systems at various locations. If control centers are located far from substations or end consumers, real-time measurements taken at the electric devices are transported to the control centers via WANs and, in the reverse direction, WANs perform the instructions communicated from control centers to the electric devices.

(b) Field-Area Networks

Field-area networks (FAN) create the communication facilities for electricity distribution systems. Electrical sensors on transformers and distribution feeders, IED devices capable of carrying out control commands from DMS, DERs in the distribution systems, plug-in electric vehicle (PEV) charging stations, and smart meters at customer premises form the key sources of information to be monitored and controlled by the DMS at the control centers. Power system applications operating in the distribution domain exploit FANs in order to exchange and share information.

(c) Home-Area Networks

Home-area networks (HANs) are essential in the customer domain to perform control and monitoring of smart devices inside customer premises, and to employ new functionalities such as DR and AMI. Within customer premises, a secure two-way communication energy services interface (ESI), acts between the customer and utility. The ESI supports several types of interfaces, including the utility-secured interactive interface used to secure two-way communications and the utility public-broadcast interface that allows one-way receipt of price and event signals located at customer devices.

(d) Power-Line Communication

Power lines are primarily used for electrical power transmissions; however, they can also be used to transmit data. Power-line communication (PLC) systems function by sending modulated carrier signals over power transmission cables. Normally, data signals cannot propagate through transformers, so the PLC signal is restricted to line segments between transformers.

(e) Wireline Networks

Dedicated wireline cables can be used to create data-communication networks that are independent of electrical power lines. Those these dedicated communication networks require additional investment for cable deployment and installation, they offer greater communication capacity and reduced communication delays. Subject to the transmission medium used, wireline networks may include Ethernet, SONET/SDH, coaxial cable access networks, and DSL. SONET/SDH networks transmit high-speed data packets using optical fiber cables with supported data rates between 155 Mbps and 160 Gbps.

(f) Wireless Networks

Innovation in wireless networking technology has enabled the connection of devices wirelessly, eliminating the need to install physical wires or cables. Generally, wireless signals are subject to transmission attenuation as well as environmental interference. The result is that wireless networks usually provide short-distance connections with relatively low data rates. The 802.11 networks are the most popularly used for local-area wireless networks, communicating at a maximum data rate of 150 Mbps and a maximum distance of 250 m.

Communication Protocols and Standards

The Open Systems Interconnection (OSI) model conceptually characterizes and standardizes the communication functions of a telecommunications or computing system without regard to their underlying internal structure and technology. Its objective is the interoperability of diverse communication systems through standard protocols. The model partitions a communication system into several abstraction layers. The original model defined seven layers:

1. Physical layer
2. Data link layer
3. Network layer
4. Transport layer
5. Session layer
6. Presentation layer
7. Application layer

The smart grid features widespread interconnections of power equipment to enable two-way flows of electricity and information for various intelligent power-management applications, such as accurate relay protection and timely DR. To fulfill such pervasive equipment interconnects, a full-fledged communications infrastructure is of great importance in the smart grid. There has been extensive work on different layouts of communication infrastructures in the smart grid by surveying feasible wired or wireless communication technologies, such as power-line communications and cellular networks. A variety of smart-grid applications demand robust communications capabilities that span across a wide range of often-conflicting requirements.

Proper networking protocols can provide strong support to help mitigate many of the challenges associated with AMI and DA applications. These protocols can incorporate standards, help perform preliminary processing, help monitor and optimize network reliability, and incorporate varying security features for different applications.

By using standards-based communications protocols, the flexibility and longevity of the network can be enhanced. Utility communications should be compliant with TCP/IP and UDP/IP to allow the utility to make use of the widest variety of standards, including IEEE 61850 and distributed network protocol (DNP), whether IPv4 or IPv6. Compliance with IP will also ease developers' concerns regarding the creation of new applications and process modeling. Providing an internet protocol (IP)-based solution also eases concerns about multi-vendor network interoperability for electric utilities.

Various standards have been recommended to guide the development of smart-grid power systems. These standards cover several technical features of the power systems, including power equipment, electricity services, management automations, and system control and protection. Standards that can be used to guide the development of communication systems are as follows:

(a) Distributed Network Protocol

The distributed network protocol (DNP) went through many revisions to before the current version (DNP3). The DNP3 standard sets the protocols used for substation automation through LAN and WAN. It is utilized when implementing substation SCADA control systems. Devices compatible with the DNP3 protocols exchange control and status information to automate substation management. DNP3 uses the IP suite to transport packets. Hence, it adopts a layered architecture with all the DNP3 packets encapsulated into TCP or UDP packets throughout the transmission. Security controls are provided by virtual private network (VPN) and IPsec, which are derivatives of the IP suite. DNP3 is used in substations for equipment monitoring and control. It uses the basic functions of an automated system to communicate equipment states to the control station and convey configuration commands to electric equipment.

(b) IEEE Standards

IEEE has recommended several standards associated with communications in power systems, such as C37.1, 1379, 1547, and 1646.

- **IEEE C37.1** designates the functional requirements of IEEE for automation and SCADA. This standard provides specification, definition, performance analysis, and application of automation and SCADA systems in substations. It describes the system function and architecture in a substation as well as protocols, human machine interfaces, and implementation. It also stipulates network-performance requirements for availability, reliability, security, maintainability, changeability, and expandability.
- **IEEE 1379** outlines implementation practices and guidelines for interoperation and communication of IEDs and RTUs in substations. It gives examples of communication support in substations by use of existing protocols. Primarily, it describes the communication protocol stack mapping of the substation network to IEC 60870-5 and DNP3. Processes are also discussed to expand the objects and data elements used in substation communications to improve the functionality of the network.
- **IEEE 1547** specifies and defines the microgrid power system that interconnects with distributed resources. It comprises three parts: information exchange, electric power system, and compliance test. In the power system aspect, the standard stipulates the requirements for different power-conversion technologies and the requirements for their interconnection to provide the best quality electricity services. General guidelines are defined to ensure response to abnormal electric power system conditions, power quality, and subsystem islands. The information-exchange section specifies the requirements for power-system control and monitoring through data networks. Critical network aspects are defined, including interoperability, extensibility, and performance. Protocol and security issues are considered in the standard as well.
- **IEEE 1646** stipulates the requirements for communication delivery times within and external to a substation. Given the variety of communication types, this standard classifies substation communication into different categories, and outlines the communication delay requirements for each category. For example, system protection messages should be transmitted within 4 ms while operation maintenance messages are transmitted within 1 s. It outlines the communication delay as the time spent in the network between the applications running at the two end systems. Hence, the packet-processing time should be considered in the delay, so that the combination of transmission and processing times do

not exceed the required delay bound. Since delays are introduced into both the end-system processing phase and the network-transmission phase, the standard details system and communication capabilities required to deliver information on time, including message priority, real-time support, system interfaces, and data criticality.

(c) IEC Standards

The International Electro-Technical Commission (IEC) has recommended a number of standards for the communication and control of microgrid electric power systems. Standard 60870 outlines the communication system used for power-system control. Through the standard, electric equipment is able to interoperate to achieve automated management. This standard comprises six parts, which specify the general requirements for power-system performance and interoperability. Standard 61850 focuses on substation-automated control. It states comprehensive communication functions and system-management requirements to facilitate substation management. The management perspectives include system reliability, availability, security, maintainability, integrity, and general environmental conditions. Standards 61968 and 61970 provide common information models for data exchange between networks and devices in the power-distribution and power-transmission domains, respectively. Cybersecurity of the IEC protocols is addressed by standard 62351, which stipulates the requirements for achieving different security objectives, including data confidentiality, data authentication, intrusion detection, and access control.

(d) NIST Standards

The National Institute of Standards and Technology (NIST) has published standards that provide guidance to the construction of smart grids. The NIST Special Publication 1108 designates a road map for the standards on smart-grid interoperability. It states the vision and importance of the smart grid, explains the conceptual reference model, classifies implementation standards, recommends priority action plans, and stipulates security assessment procedures. Specifically, it presents the expected services and functions in the smart grid and the application and requirements of communication networks in the operation of a smart grid. NIST report 7628 focuses on information-security issues of the smart grid. It gives details on the critical security challenges of the smart grid, presents security architectures, and stipulates security requirements. Security objectives and strategies are mentioned, including cryptography, privacy, key management, and vulnerability analysis. Report 7628 seeks to ensure reliable and trustable communications for automated energy management in the smart grid.

Integrated Communications

Some communications are up-to-date but are nonuniform because they have been developed in an incremental fashion and not fully integrated. In most cases, data is being collected via modem rather than direct network connection. Areas for improvement include: substation automation, DR, distribution automation, SCADA, energy-management systems, wireless mesh networks, power-line carrier communications, and fiber optics. Integrated communications will allow for real-time control, information, and data exchange to optimize system reliability, asset utilization, and security.

16.3.4 Application Layer

Smart-grid concepts encompass a wide range of technologies and applications. We describe a few below that are currently in practice with the caveat that, at this early stage in the development of smart grids, the role of control, especially advanced control, is limited.

(a) **AMI** is a vision for two-way meter/utility communication. Two fundamental elements of AMI have been implemented. First, AMR systems provide an initial step toward lowering the costs of data gathering through use of real-time metering information. They also facilitate remote disconnection/reconnection of consumers, load control, detection of and response to outages, energy-theft responsiveness, and monitoring of power quality and consumption. Second, meter data management (MDM) provides a single point of integration for the full range of meter data. It enables leveraging of that data to automate business processes in real time and sharing of the data with key business and operational applications to improve efficiency and support decision making across the enterprise.

(b) **DMS** software mathematically models the electric distribution network and predicts the impact of outages, transmission, generation, voltage/frequency variation, and more. It helps reduce capital investment by showing how to better utilize existing assets, enabling peak shaving via DR, and improving network reliability. It also facilitates consumer choice by helping identify rate options best suited to each consumer and supports the business case for renewable generation solutions (DG) and for electric vehicles and charging-station management.

(c) **GIS** technology is specifically designed for the utility industry to model, design, and manage critical infrastructure. By integrating utility data and geographical maps, GIS provides a graphical view of the infrastructure that supports cost reduction through simplified planning and analysis and reduced operational response times.

(d) **Outage management systems (OMSs)** speed outage resolution so power is restored more rapidly and outage costs are contained. OMSs eliminate the cost of manual reporting, analyze historical outage data to identify improvements and avoid future outages, and address regulatory and consumer demand for better responsiveness.

(e) **Intelligent electronics devices (IEDs)** are advanced, application-enabled devices installed in the field that process, compute, and transmit pertinent information to a higher level. IEDs can collect data from both network and consumer facilities (behind the meter) and allow network reconfiguration either locally or on command from the control center.

(f) **Wide-area measurement systems (WAMS)** provide accurate, synchronized measurements from across large-scale power grids. They have been implemented in numerous power systems around the world, following initial developments within the Western Electricity Coordinating Council through the early 1990s. WAMS consist of PMUs that provide precise, time-stamped data, together with phasor data concentrators that aggregate the data and perform event recording. WAMS data plays a vital role in post-disturbance analysis, validation of system dynamic models, FACTS control verification, and wide-area protection schemes. Future implementation of wide-area control schemes is expected to build on WAMS.

(g) **EMSs** at customer premises can control consumption, onsite generation and storage, and potentially electric-vehicle charging. EMSs are in use today in large industrial and commercial facilities and will likely be broadly adopted with the rollout of smart grids. Facility energy management can be seen as a large-scale optimization problem: given current and (possibly uncertain) future information on pricing, consumption preferences, DG prospects, and other factors, how should devices and systems be used optimally? Smart-grid implementations are occurring rapidly, with numerous projects under way around the world.

The following are application functions of a smart-grid:

- **Power quality (PQ) control** is the capability of a power system to operate without causing disturbance or damage to components and loads. PQ is a significant concern because of the sensitivity of modern digital control equipment to PQ distortion and deterioration. A rudimentary requirement for maintaining PQ is balancing demand and supply. Characteristic metrics utilized as indictors of PQ measures include total harmonic distortion, voltage transient impulse, voltage sag, and flicker factor. A number of accepted power-quality standards apply to electrical power distribution systems. The European Norm 50160 standard outlines voltage disturbance standard requirements for acceptable power-quality standards. Studies on PQ considers factors such as:
  - location of power disturbances
  - identification of types and causes of power disturbances
  - quantification of power disturbances and their negative impacts on power systems
  - real-time measurement of parameters of signal components in power disturbances
  - sensitive detection of power disturbances
- **Voltage/var control** within an identified range of limits and capacitor switching is an effective method of improving voltage profiles, minimizing loss, and delaying construction of new generation within the power-quality and reliability constraints. In electric power systems, var is used to express reactive power in a power system. Reactive power exists in an AC circuit when the voltage

and current are not in phase. Voltage/var control considers multiphase, unbalanced power-distribution system operation, DG, and automated control equipment in large electric systems. Power-system distribution automation functions use voltage/var control options to maintain proper parameters throughout the grid, based on the requirements of different customers by employing variable capacitors. The improved control and management of voltage/var facilitates increase performance for customer equipment and allow both the consumer and utility company to realize financial benefit.

- **Load frequency control (LFC)** is of importance in microgrid system design and operation. The objective of LFC in an interconnected power system is to maintain the frequency of each area within limits and to keep tie-line power flows within some prespecified tolerances by adjusting the megawatt (MW) outputs of the generators so as to accommodate fluctuating load demands. A well-designed and well-operated power system must cope with changes in the load and with system disturbances, and should provide an acceptably high level of power quality while maintaining both voltage and frequency within tolerance limits. As power-load demands vary randomly for an interconnected power system, both area frequency and tie-line power interchange varies also. The primary objective of LFC is to minimize transient deviations in area frequency and tie-line power interchange while ensuring that their steady-state errors are as close to zero as possible. Frequency will not change in an interconnected network if there is a balance between customer demand and power supply resources that can also compensate for electrical losses.

16.3.5 Cost–Benefit Layer

Cost–benefit analysis (CBA) is a systematic approach to estimating the strengths and weaknesses of alternatives and is used to determine options that provide the best approach to achieve benefits while preserving savings. If all the equipment that needs to be installed for communication and control operation are considered, the CBA for the smart grid can be calculated by considering costs for such items as PMUs, smart meters, relative communication infrastructure, relative protection systems, and RTUs, and comparing them to the benefits obtained through installation of these devices. Worst-case analysis should also be carried out to have a better idea of CBA, e.g., consider that after installing everything if grid performance does not change much then benefits obtained will be minimal. Physical benefits can also be converted into societal benefits or utility economic benefits. CBA should be carried out with different contingencies applied. Considering the economic potential of the smart grid and the requisite considerable investments, it is important to estimate the costs and benefits of control and communication of smart grids. There are four major guidelines for performing a comprehensive CBA:

(a) Definition of Assumption and Setting of Critical Parameters

This involves definition of boundary conditions (demand growth forecast, discount rates, local grid characteristics) and of implementation choices (roll-out time, chosen controls, communication functionalities). Furthermore, for CBA studies input information such as critical data related to the project, social discount rate and project value in terms of weight of importance is required.

(b) Steps of Classical CBA

The assessment framework is expressed by the following seven steps based on the approach developed by the Electric Power Research Institute (EPRI). The benefit of charting assets to functionalities and functionalities to benefits is that they assist in thinking of sources of benefits, making a complete set of estimated benefits more likely.

Evaluation of the impact of a project is made easy. Progress measurement, toward attaining characteristics of the comprehensive smart grid control is facilitated. Progress is measured through some previously established standard key performance indicators such as enhanced efficiency and better service in electricity supply and grid operation, and voltage quality performance of electricity grids (e.g., voltage dips, voltage and frequency deviations). The relationship between assets and benefits through functionalities is more than likely not to be easily deduced and requires a good deal of brainstorming. However, execution of these steps adds clarity to a CBA process.

Step 1: Review and define project goals.

This step involves summarizing and describing the elements and goals of the project. The goals must be concise and clearly defined. This may involve collation of the following information:

- scale and dimension of the project (e.g., consumers served, energy consumption per year)
- engineering features (e.g., the control technologies adopted and the functionalities of the main components, the local characteristics of the grid control elements)
- relevant stakeholders (e.g., whose costs and benefits count?)
- strong statement of the project's objectives and its expected socioeconomic impacts
- regulatory context and its impacts on the project

Step 2: Define and map assets to functionalities.

Determining which smart-grid functionalities are activated by the control assets proposed by the project is an important early step in a CBA for smart-grid projects. Smart-grid assets provide different types of functionalities that enable benefits. If the assets deployed and/or functionalities enabled by the project are unclear, the analysis is likely to be incomplete.

Step 3: Map functionalities to benefits.

Each functionality should be considered individually and assessed on how it could contribute to any of the benefits. This analysis should be thorough and applied to all applicable functionalities.

Step 4: Define the baseline.

This involves the formal definition of the "control state" that reflects the smart-grid condition that would have occurred had the project not taken place. This is the baseline condition against which all other setups are compared. The CBA is based on the variance between the costs and benefits associated with the "do-nothing" scenario and those associated with the implementation of the project. These two scenarios are:

- Scenario 1, the baseline conditions that depict what the grid condition would have been without the control and communication installations
- Scenario 2, the realized and measured conditions with the smart-grid controls and communication system installed

Step 5: Monetize the benefits.

In this step, data required to quantify the benefits in equivalent monetary forms are gathered. The data may be already analyzed or may be in its raw form. The benefits to be calculated and the baseline scenarios needed to calculate those benefits determine the type of data collected. Each set baseline and established benefit requires compilation of data for computation. Data should be collected both before and after implementation of the control and communication project. Conceptually, the monetary value of a benefit can be calculated as:

$$\text{Value (\$)} = [\text{Condition}]_{\text{Baseline}} - [\text{Condition}]_{\text{Project}}$$

For example, the estimate of the benefit reduction in meter-reading costs could be done by comparing the total cost of local meter readings for an established baseline that do not have smart meters installed (i.e., the "do-nothing" scenario) to the cost of local meter readings under project conditions for the fraction of customers who use smart meters (communication scenario).

Next, the benefits quantified are expressed in equivalent economic terms and benefits are allocated to different beneficiaries, like consumers, retailers/aggregators, and society at large. A concluding step would be to carefully assess any uncertainty. This is because it may not be possible to provide a good estimation of some control and communication benefits—for instance those based on environmental or socioeconomic factors—with the same level of precision and clarity as other benefits.

Step 6: Quantify cost.

The costs of a project are incurred in implementing the project, relative to the predefined baseline scenario. This step involves evaluation of the cost-benefits of the project and the advantage is that it facilitates calculation of a project's return on investment (ROI).

Step 7: Compare costs and benefits.

Once costs and benefits have been estimated, there are several methods to compare them to evaluate the cost-effectiveness of the project.

- The **net present value (NPV) method** consists of estimating the sum of NPVs of individual cash flows of the smart-grid project for the entire study period. The project needs to subtract estimated costs from benefits for each year, discount these annual net benefit amounts, and sum the discounted values. The NPV can be defined as the net benefit "brought back" to the baseline year by applying a discount rate, thereby accounting for the time value of money. The NPV is calculated as follows:

$$\sum_{t=0}^{n} \frac{R_t}{(1+i)^t} \tag{16.1}$$

where t is the time of the cash flow, I is the discount rate, $R_t$is the net cash flow (cash inflow minus cash outflow at time t), and n is the total number of periods (typically years) under consideration. A small discount rate sets future values close to present ones, while a larger discount rate incrementally devalues present values of future cash flows, usually indicating the presence of an alternative option of investment in financial markets.

- The **benefit-cost ratio** is the net present value of project benefits divided by the net present value of project costs. A project is accepted if the benefit-cost ratio is equal to or greater than one. It is used to accept independent projects, but it may give incorrect rankings and often cannot be used for choosing among mutually exclusive alternatives. A project's value can also be represented as a ratio of benefits to costs (either on a present value basis or on an annual basis). This method is a simple way of representing the size of the benefits relative to the costs. If the ratio is greater than one, the project is cost-effective.

(c) Sensitivity Analysis
Sensitivity analysis is performed to determine the range of variables that would result in a positive outcome of a CBA. It requires identifying the value that would have to occur for the NPV of the project to become zero, or more generally for the outcome of the project to fall below the minimum level of acceptability.

(d) Evaluation of Merit and Social Impact
A qualitative-impact analysis is performed to evaluate the social impacts, typically representing a significant portion of the project externalities. These include:

- job impact
- safety
- environmental impact
- social acceptance
- time lost/saved by consumers
- enabling new services and applications and
- market entry to third parties

- reduction of the gap in skills and personnel
- privacy and security

Value Proposition (Return on Investment) for Smart Grid

The costs and benefits of smart-grid systems include both tangible and intangible benefits. The companies have estimated costs and benefits for individual smart-grid programs and DR through their respective project applications. With DR, the projected benefits by far exceed the estimated program costs, primarily due to capacity deferral.

With AMI, the tangible quantifiable benefits are comparable to the costs; however, not all tangible benefits can be quantified. Across the nation, there are many examples of business case analyses that were developed for utility smart-grid projects, as well as by organizations such as the EPRI, the Edison Electric Institute (EEI), the Utilities Telecom Council (UTC), and numerous consultants.

In many of these cases, utility smart-grid projects show net benefits. Analysis of the costs and benefits of these business cases will help increase understanding of all the benefits and costs of smart-grid projects and better inform our business case development. That said, it is essential to develop business cases for each smart-grid project or integrated set of projects specific to each system and customer base. Developing accurate assumptions before having specific data is difficult. Data from the companies' pilots, demonstrations, and targeted deployment projects will help validate some of the benefits derived from implementing a smart grid.

Smart-grid goals include implementing targeted deployments focused on increasing the ability to accommodate increased levels of renewable energy, and on improving the efficient utilization of electricity (energy efficiency) all the way to customer end uses, while ensuring grid reliability and service quality are maintained or improved. The pace of smart-grid implementation will depend on the business case for each smart-grid application, availability of commercially viable solutions, and availability of resources across utility companies.

16.4 COMMAND, CONTROL, AND COMMUNICATION APPLICATIONS IN REAL TIME

16.4.1 Need for Command, Control, and Communication

Smart grid is the term generally used to describe integration of the elements connected to the electrical grid with an information infrastructure to offer numerous benefits for both the providers and consumers of electricity. It is an intelligent future electricity system that connects all supply, grid, and demand elements through an intelligent communication system. The backbone of a successful smart-grid operation is a reliable, resilient, secure, and manageable standards-based, open communication infrastructure that provides for intelligent linkages between the elements of the grid while participating in the decision-making that delivers value to the utility and supply-and-demand entities connected to it. The general purpose of a smart grid is to monitor electricity consumption and to manage its delivery to avoid blackouts or brownouts. RER integration and other advanced real-time communication devices have added complexity into the system and posed many questions, such as:

- How can we retrofit and engineer a stable, secure, robust, and resilient grid with large numbers of such unpredictable power sources?
- What roles will asset optimization, novel control algorithms, increased efficiency, control/coordination of energy storage, advanced power electronics, power quality, electrification of transportation, cybersecurity, policies, and technologies play in the smart grid of the future?
- How can robust controls and observers be developed that use secure sensing to identify and build realistic models and appropriate responses?
- Will renewable energy sources be able to adapt, control, and mitigate disturbances to achieve the goals for reliability and stability of overall system?

The above questions must be answered by researchers in related fields, industry experts, and standards organizations before advancing the grid. In addition, certain technical areas concentrating in controls should be evaluated, such as:

- optimization of DR under variable pricing
- integration of DG and storage in automation and control strategies
- distributed state estimation of power systems and elements, with limited sensing and communication
- cybersecurity in smart-grid control systems
- new power market designs that exploit feedback and information integration
- stability issues, especially given latency and uncertainty in communication networks
- self-monitoring and self-healing in intelligent transmission systems
- modeling and analysis of multiscale, complex grids

Real-Time Measurable Quantities

A grid that is considered "smart" uses information technology in the operation of the transmission grid as much as it does in the distribution network. The lifeblood of any smart-grid network is inherently the data and information that drive the applications that, in turn, enable new and improved operational strategies to be deployed. Data collected at any level of the system, from customer metering to distribution, transmission, generation, and market operations, may be pertinent to improving operations at any other level. Thus, sharing data in a timely fashion, approaching real-time, with all those with a need or right to know, is an essential ingredient of a smart grid. The smart grid involves many real-time devices: PMUs, smart meters, SCADA, DMS, EMS, and many more including real-time sensors and power electronics devices. These all work together to achieve smart-grid objectives.

Overvoltage Limits

A smart grid should have self-healing and self-restoration qualities and should monitor at each single corner of the system for abnormalities in order to self-heal or self-restore. So, it needs to check voltage, current, phase angle, and frequency at each node of the grid. It also should compare the obtained quantities with the overvoltage and overcurrent quantities at that node. If abnormalities are found, the control action should be initiated immediately and automatically. For global control, PMU can help to measure voltage and phase angle for wide areas. For local control, CTs and PTs, various meters, and transducers connected at each level can help to initiate relays, circuit breaker, or any other protective device operation.

Moreover, smart grids will be inclusive of stochastic behavior due to RER and randomly changing loads. So, to carry out control operation in this environment requires continual dynamic optimal power flow calculations. This can be achieved with the help of next-generation optimization tools such as adaptive dynamic programming (ADP), stochastic programming (SP), mixed-integer programming (MIP), evolutionary programming (EP), or heuristic programming.

Security Measures

Security measures involve addition of the latest devices into the system such as PMUs, which can measure quantities of voltage and phase angle for a large area and compare quantities of one area with another to help catch abnormalities immediately. They also allow monitoring through graphs, signals, and charts, making monitoring easy and convenient. Other optimization tools to can calculate optimal power flow to enhance the system with the ideal conditions and easily locate abnormal values at each end so they can be corrected before becoming an issue of major concern. While managing a system with RER, we should include advanced storage options such as supercapacitors, hydrogen storage, hydro storage, and flywheels so that, in case of fault, these sources can help compensate for shortcomings in power needs.

Contingency Analysis

Contingency analysis is an important part of control system design and deserves extra effort. Contingency analysis gives the extreme values (max, min) for line parameters that a system can withstand. Also, through contingency analysis one can rank such values to see whether one or more abnormalities are going on in parallel and which should be attended to first.

16.4.2 Application of Control and Communication to the Smart Grid in Real Time

Power systems are fundamentally dependent on control, communications, and computation for ensuring stable, reliable, efficient operations. Generators rely on governors and automatic voltage regulators (AVRs) to counter the effects of disturbances that continually buffet power systems, and many would quickly lose synchronism without the damping provided by power system stabilizers (PSSs). FACTS devices, such as static var compensators (SVCs) and HVDC schemes rely on feedback control to enhance system stability. At a higher level, EMSs use SCADA to collect data from expansive power systems and sophisticated analysis tools to establish secure, economic operating conditions. Automatic generation control (AGC) is a distributed, closed-loop control scheme of continental proportions that optimally reschedules generator power set points to maintain frequency and tie-line flows at their specified values.

The use of time-synchronized, high-data-rate sensor technology is widely viewed as a critical enabler for increasing the reliability of the power grid while allowing the integration of many more stochastically variable renewable-energy sources such as solar radiation and wind.

PMUs are capable of sampling frequency, voltage, and current thousands of times per second and outputting accurate, time-stamped measurements 30–120 or more times per second. It is difficult, however, for utilities to take full advantage of these devices due to a lack of tools for designing and evaluating the control systems that exploit them. Furthermore, the behavior of such control systems also depends on the performance of the wide-area communications systems that connect the sensors, control logic, and actuators, and whose design and specifications are themselves still evolving.

The smart grid will consist of intelligent devices that use advanced communication methods, which in turn will improve situational awareness and enable networked control. Networked control will increase efficiency, reliability, and safety; enable plug-and-play asset integration, such as in the case of DG and alternative energy sources; support market dynamics; and reduce peak prices and stabilize costs when supply is limited.

Networked smart devices currently include smart meters at the customer site and PMUs (including synchro-phasors) in the transmission and distribution grids. In the future, smart appliances will adjust their functions depending on instantaneous electricity costs, and a host of communicating devices will enable the feedback-loop networked control of the smart grid.

Smart-grid communication should protect users' privacy and security. Networked control requires adequate, real-time performance and reliability. Communication should use existing infrastructure whenever possible, due to:

- high cost of deploying a new communications infrastructure
- substantial investments made in the past (e.g., dark fiber) or currently under way in hardware (e.g., universal broadband) and software (e.g., secure real-time middleware)
- potential for widespread technology adoption, which will result in increased grid responsiveness and larger economic benefits

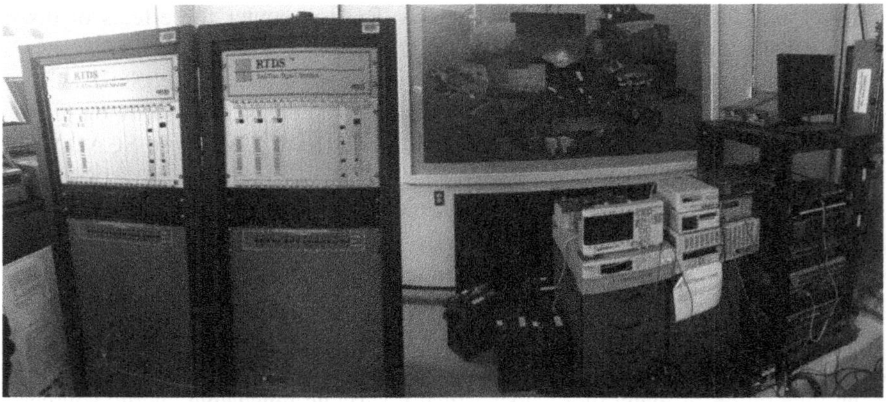

Figure 16.6 Howard University microgrid test bed with OPAL-RT.

Smart-grid devices can communicate over a variety of substrates with different properties. Furthermore, different media may be appropriate in different circumstances. For example, smart appliances in the home can exploit existing Ethernet home LANs, 802.1I home wireless LANs, or power line home networks. At the other end, if a synchro-phasor is deployed in the transmission grid, it may be better served by wide-area wireless, such as a mesh network or a power-line wide-area network. Since the smart grid will exploit different types of data from a variety of devices communicating over heterogeneous networks, communication should be possible across various media, which in turn requires a convergence layer, such as the IP.

16.5 HARDWARE-IN-THE-LOOP FOR ENERGY PROCESSING AND THE SMART GRID

Figure 16.6 shows the Howard University microgrid test bed with OPAL-RT, which uses the very latest technology—an advantage to the power-system engineer as speed, flexibility, and creativity is encouraged and enhanced. The introduction of the OPAL-RT real-time digital simulator to the array of research and educational equipment at the Center for Energy System and Control (CESaC) at Howard University through the US Department of Energy has added capacity for improvement in research and educational output.

16.6 CYBER-PHYSICAL SYSTEMS FOR SMART GRIDS

Cyber-physical systems (CPS) are the next generation engineered system with tools such as communication, control and computation working together. By definition,

CPS is a mechanism that is controlled by computer-based algorithms, closely integrated with the internet and its users. the terms in CPS are defined as follows:

- Cyber: a computation, communication and control that is discrete, logical and switched. It involves computation and communication that performs some function which requires a physical platform to run into.
- Physical: a natural and human made system governed by laws of physics and operating in continuous time. It is a functional and interconnected entity.

Cyber-Physical is then deeply embedded computing and communication platform.

Practical applications of CPS include manufacturing environment, healthcare environment, transportation environment, renewable energy environment, smart grids and others. Where it provides a system with new capabilities, reduced cost of building and operating system and provide safer and efficient system.

In smart grids and microgrids, the CPS provides secured system against attacks by providing attack and counter attack schemes. These capabilities CPS will be realized by deeply embedding computational intelligence, communication, control, and new mechanisms for sensing, actuation, and adaptation into the physical system with active and reconfigurable components.

The capabilities are added to the physical system as shown in Figure 16.7. For the design of CPs, a theoretical foundation will be created consisting of

- system model
- communication model
- cyber model
- optimization model

The system model and communication model have a real-time data from a test bed, consisting of renewable-energy sources, storage, meters, and PMUs that capture real voltages of the system. Research computes the system model from the signals, which are the voltage stability index, frequency control index, and power quality index.

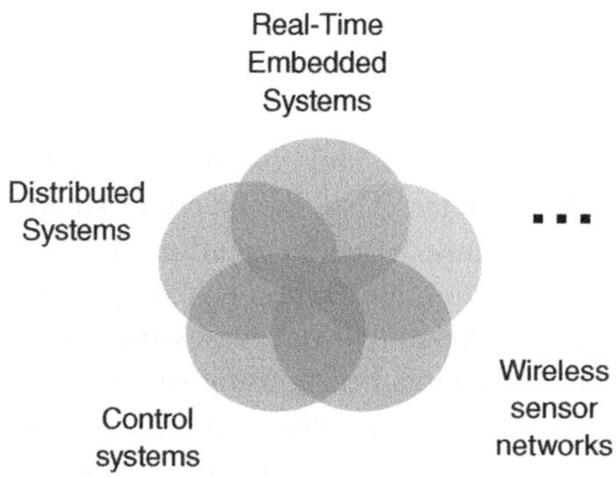

Figure 16.7 Convergence of computation, communication, information, and control.

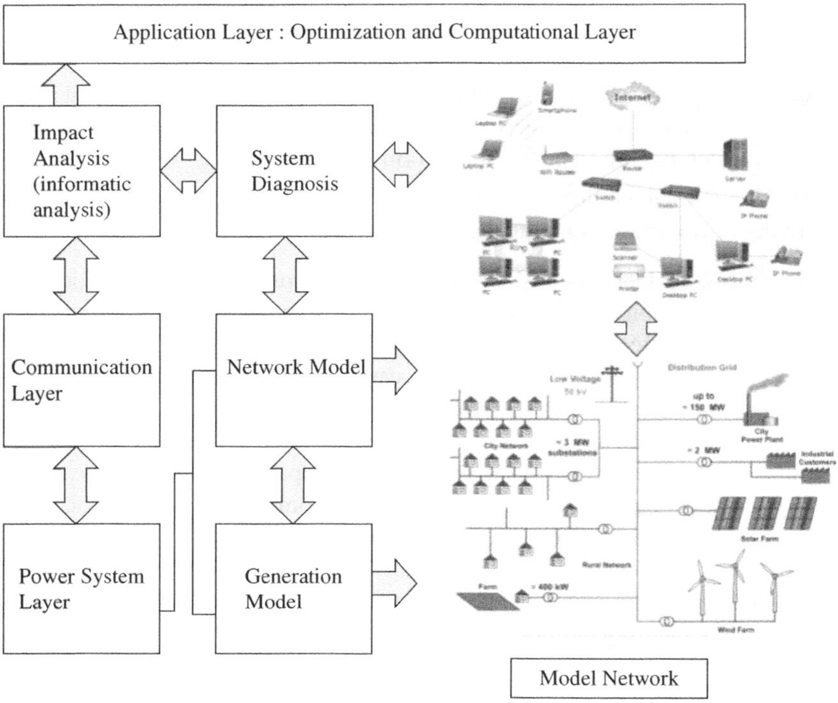

Figure 16.8 Smart grid with CPS attributes.

Communication network failures can be caused by a noisy environment, congestion, bandwidth problems, or link failures and have a two-way communication with an effective communication interface for effective control of power-quality problems, load-flow problems, and voltage/var problems. Figure 16.8 shows summary of Smart grid Attributes with cyber physical system.

16.7 CHAPTER SUMMARY

The role of real-time metering, communication, command, and control in efficient running of smart grids was one objective of this chapter. Also, the layers of smart grids were comprehensively treated. Smart grids operate based on coordinated control of numerous functions, devices, and sensors to achieve the central goal of quality, reliable, available power. This is may be achieved through real-time assessment and utilization of results of real-time measurement and data gathering using smart communication devices and metering systems. An effective smart grid requires various functions—such as demand-side management, power-quality control, load-frequency control, voltage/var control, load reconfiguration, energy resources management, and monitoring and control of protective devices—to be carried out simultaneously and in real time.

BIBLIOGRAPHY

[1] Edison Foundation, *Utility-Scale Smart Meter Development, Plans, and Proposal*, IEE Report, May 2012, www.edisonfoundation.net/iee/documents/iee_smartmeterrollouts_0512.pdf.

[2] A. Allen, S. Santoso, and E. Muljadi, *Algorithm for Screening Phasor Measurement Unit Data for Power Systems Events and Categories and Common Characteristics for Events Seen in Phasor Measurement Units Relative Phase-Angle Differences and Frequency Signals*, National Renewable Energy Laboratory Technical Report, August 2013.

[3] US Department of Energy, *What the Smart Grid Means to Americans*, www.doe.gov/sites/prod/files/oeprod/DocumentsandMedia/ConsumerAdvocates.pdf.

[4] Demand, Response and Advanced Metering Coalition, *Overview of Advanced Metering Technologies and Costs*, April 2004, www.drsgcoalition.org/resources/whitepapers/DRAM White Paper 2004-04.pdf.

[5] G. Heydt, "Electric Power Quality: A Tutorial Introduction," *IEEE Comput. Appl. Power* **11**, 1 (January 1998).

[6] S. Darby, *The Effectiveness of Feedback on Energy Consumption: A Review for DEFRA of the Literature on Metering, Billing and Direct Displays*, April 2006.

[7] A. Schatz, "Cellphone Users to Get Billing Alerts under New Voluntary Standards," *Wall Street Journal*, October 17, 2011, http://online.wsj.com/article/SB10001424052970203658804576635053172551850.html.

[8] National Energy Technology Laboratory, *Advanced Metering Infrastructure*, February 2008, www.netl.doe.gov/smartgrid/refshelf.html.

[9] Federal Energy Regulatory Commission, *Assessment of Demand Response and Advanced Metering*, Staff Report, February 2011, www.ferc.gov/legal/staff-reports/2017/DR-AM-Report2017.pdf.

[10] B. Heile, "Surveying the International AMI Landscape," *Electric Light & Power* **15**, 7 (July 1, 2010), http://www.elp.com/articles/powergrid_international/print/volume-15/issue-7/Features/surveying-the-international-ami-landscape.html.

[11] S. Renner, M. Albu, H. van Elburg, et al., *European Smart Metering Landscape Report*, Feburary 2011, www.sintef.no/globalassets/project/smartregions/d2.1_european-smart-metering-landscape-report_final.pdf.

[12] M. Bauer, W. Plappert, C. Wang, et al., "Packet-Oriented Communication Protocols for Smart Grid Services over Low-Speed PLC," IEEE International Symposium on Power Line Communications and Its Applications, March 2009.

[13] T. Tran, A. P. Auriol, and T. Tran-Qouc, "Distribution Network Modeling for Power Line Communication Applications," IEEE International Symposium on Power Line Communications and Its Applications, April 2005.

[14] S. E. Collier, "Delivering Broadband Internet over Power Lines: What You Should Know," Rural Electric Power Conference, May 2004.

[15] T. Khalifa, K. Naik, and A. Nayak, "A Survey of Communication Protocols for Automatic Meter Reading Applications," *IEEE Commun. Surv. Tut.* **13**, 2, (2011).

[16] G. N. Ericsson, "Communication Requirements: Basis for Investment in a Utility Wide-Area Network," *IEEE T. Power Deliver.* **19**, 1 (January 2002).

[17] H. Tan, C. Lee, and V. Mok, "Automatic Power Meter Reading System using GSM Network," IEEE Power Engineering Conference, December 2007.

[18] A. Abdollahi, M. Dehghani, and N. Zamanzadeh, "SMS-Based Reconfigurable Automatic Meter Reading System," IEEE International Conference on Control Applications, October 2007.

[19] B. Brown, "802.11: The Security Differences between B and I," *IEEE Potentials* **22**, 4 (Oct.–Nov. 2003).

[20] P. Kulkarni, S. Gormus, F. Zhong, et al. "A Self-Organising Mesh Networking Solution Based on Enhanced RPL for Smart Metering Communications," IEEE International Symposium on a World of Wireless, Mobile and Multimedia, June 2011.

[21] D. Bernaudo, et al., *SmartGrid/AEIC AMI Interoperability Standard Guidelines for ANSI C12.19 End Device Communications and Supporting Enterprise Devices, Networks and Related Accessories, Association of Edison Illuminating Companies, Meter and Service Technical Committee Report*, version 2, May 2010, www.smartgrid.gov/files/AEIC_Guidelines_v20_SmartGridAEIC_AMI_Interoperability_Stdar.pdf.

[22] E. Queen, *A Discussion of Smart Meters and RF Exposure Issues*, An EEI-AEIC-UTC White Paper, March 2011, www.occeweb.com/pu/SMARTGRID/smart%20meters%20and%20rf%20exposure_031511.pdf.

[23] V. Aravinthan, V. Namboodiri, S. Sunku, et al., "Wireless AMI Application and Security for Controlled Home Area Networks," IEEE Power and Energy Society General Meeting, October 2011.

[24] C. Bennett and D. Highfill, "Networking AMI Smart Meters," IEEE Energy 2030 Conference, November 2008.

[25] Open SG Users Group, "AMI-SEC Task Force Overview," http://osgug.ucaiug.org/utilisec/amisec/default.aspx.

[26] DLMS User Association, www.dlms.com.

[27] L. Atzori, A. Iera, and G. Morabito, "The Internet of Things: A Survey," *Comput. Netw.* **54**, 15 (October 2010).

[28] J. Northcote-Green, and R. Wilson, *Control and Automation of Electric Power Distribution Systems*, CRC Press, Boca Raton, FL, September 2006.

[29] K. Fehrenbacher, "AEP Calls for Dedicated Wireless Spectrum for Smart Grid," Gigaom, August 25, 2009, https://gigaom.com/2009/08/25/aep-calls-for-dedicated-wireless-spectrum-for-smart-grid.

[30] J. G. Kassakian, R. Schmalensee, W. W. Hogan, et al., *The Future of the Electric Grid: An Interdisciplinary MIT Study*, 2011, http://energy.mit.edu/wp-content/uploads/2011/12/MITEI-The-Future-of-the-Electric-Grid.pdf.

[31] Y. Hu, A. Kuh, and A. Kavcic, "A Belief Propagation Based Power Distribution System State Estimator," *IEEE Comput. Intell. M.* **6**, 3 (2011).

[32] S. Gong, H. Li, L. Lai, et al., "Decoding the 'Nature Encoded' Messages for Distributed Energy Generation Control in Microgrid," IEEE International Conference on Communications, June 2011.

[33] P. Chiradeja, "Benefit of Distributed Generation: A Line Loss Reduction Analysis," IEEE PES Transmission and Distribution Conference and Exhibition, August 2005.

[34] C. H. Lo and N. Ansari, "Decentralized Controls and Communications for Autonomous Distribution Networks in Smart Grid," *IEEE T. Smart Grid* **4**, 1 (2013).

[35] Y. Wang, P. Yemula, and A. Boss, "Decentralized Communication and Control Systems for Power System Operation," *IEEE T. Smart Grid* **6**, 2 (2014).

[36] R. Ma, H. H. Chen, Y. R. Huang, et al., "Smart Grid Communication: Its Challenges and Opportunities," *IEEE T. Smart Grid* **4**, 1 (2013).

[37] C. H. Lo and N. Ansari, "The Progressive Smart Grid System from Both Power and Communications Aspects," *IEEE Commun. Surv. Tut.* **14**, 3 (2012).

[38] V. Giordano, I. Onyeji, G. Fulli, et al., *Guidelines for Conducting a Cost-Benefit Analysis of Smart Grid Projects*, Reference Report by the Joint Research Centre of the European Commission, 2012, pp 10–72.

[39] J. Reilly, "Microgrid Controllers: Standards for Specifications and Testing," Advanced Grid Technologies Workshop Series, National Renewable Energy Laboratory, July 9, 2015, www.nrel.gov/esif/assets/pdfs/agct_day3_reilly.pdf.

[40] C. Natesan, S. K. Ajithan, P. Palani, et al., "Survey on Microgrid: Power Quality Improvement Techniques," *ISRN Renewable Energy* 2014 (2014).

[41] M. Vandenbergh, C. Garza, and A. Notholt-Vergara, "Overview of Supervisory Control Strategies including a MATLAB Simulink Simulation Tool," International Energy Agency Photovoltaic Power Systems Programme, August 2012.
[42] Software Engineering Institute, Cyber-Physical Systems, www.sei.cmu.edu/cyber-physical.
[43] G. R. Kunkolienkar and J. G. Priolkar, "Phasor Measurement Units for Power Systems," *Electrical India*, November 5, 2015, www.electricalindia.in/blog/post/id/8522/phasor-measurement-units-for-power-systems.

CHAPTER 17

ENERGY PROCESSING AND SMART-GRID TEST BEDS

17.1 INTRODUCTION

This chapter discusses test-bed design for energy processing and the smart grid. The energy-processing concepts are developed based on the knowledge of classical energy-conversion machines such as generators, induction machines, and transformers. The integration of renewable-energy resources (RER) as a stand-alone and grid-connected system is also studied with simple experiments to determine performance measures and characteristics.

Energy-processing components based on power electronics devices for inverters and converters are introduced. Hands-on experiments to validate real-time measurements such as smart meter PMUs are determined. Application of real-time data simulations, as hardware-in-the-loop, is included.

The integration of these devices is used to develop two working test beds, one as a smart-grid platform for system design, simulation, and optimization, and control, and the second as a classical machine test bed, which includes power electronics and real-time measurements tools.

The application of the smart grid needs to be scaled down to the university-laboratory level to test possible improvements on a small scale; results should be evaluated through simulations and random tests.

This chapter aims to provide guidelines for the structure of a smart-grid test bed and the requirements at each level, e.g., generation, transmission, distribution, and user. It also discusses features, components, and architectures for a smart-grid test bed and details various simulation software, hardware, and other equipment essential for the test bed.

17.2 STUDY OF AVAILABLE TEST BEDS FOR THE SMART GRID

The smart-grid concept is an evolutionary technology of interest, so it has attracted the focus of many researchers, utilities, power industries, and others. It is discussed

Energy Processing and Smart Grid, First Edition. James A. Momoh

by researchers working in power arenas. Many universities throughout the world are working on smart-grid technologies, with many definitions and attributes and a variety of names: intelligent grid, modernized grid, and smart grid, based on the selected functions and applications within the bulk electric grid. By any name, it includes the latest devices—from advanced controls and sensors to advanced communication. So the smart-grid concept includes electrical, electronics, and communications knowledge. Moreover, full-fledged working smart grids will involve customers as well so they even catch the attention of individual human beings.

In this chapter, we discuss some worldwide activities that pertain to design and development of a smart-grid test bed and improvements needed in current research.

Features and Architecture

The smart grid has special features that are lacking in the traditional grid, such as two-way communication, advanced sensors, advanced power electronics, and renewable-energy penetration as a distribution source or as centrally connected. These features require the latest devices and components to be added into the system whose performance still needs to be evaluated under different contingencies and improvements. These components should be tested in a laboratory environment and before being considered for practical use. Moreover, their interoperability in new grid environment should also be verified and required changes should be made. Different components and architectures are discussed below.

Wide-area measurement systems (WAMS) are required to be studied in following areas:

- ideas for wide-area monitoring and control for smart grids
- increase in overall system reliability via significant dependence on WAMS
- depth of observability using a PMU
- optimal location for placement and number of PMUs within a wide-area power network
- communication requirements for a higher-bandwidth data network

Plug-in hybrid electric vehicles (PHEVs) are an essential part of the future smart grid. To optimize these assets, their interoperability with the smart grid should be studied in detail for:

- researching and developing smart energy management
- power sharing and utilization of the energy stored in batteries
- utilization of power available in charging stations for voltage stability assessment on the utility grid
- real-time smart load-management algorithms to improve smart-grid performance with high penetration of EVs and PHEVs

Advanced power electronics will be implemented at each stage of the smart grid and their interoperability in the new grid needs to be examined considering the complexity and disturbance they add into the system. Such studies should include:

- developing smart controllers for power electronic converters
- developing smart controllers to maximize the stable operating range of DC–DC converters in PV systems by dynamically tuning the controller parameters at various loading conditions
- developing smart controllers to enhance transient responses

Wide-area monitoring, control, and protection (WAMPAC) includes the use of systemwide information and communication infrastructure to send local information to a control center in order to counteract the propagation of large disturbances. New synchronized measurement technology (SMT), which enables WAMPAC, is being developed by researchers.

Synchrophasors have a wide area of applications and some of the possible benefits have not been evaluated, particularly their benefits for protection in practical systems. Synchrophasors should be given emphasis for study and verified under different applicable areas such as wide-area control system and wide-area protection systems. They should be evaluated for:

- real-time visualization and control of power systems
- detecting isolated networks and preventing large blackouts
- load management and other load-control techniques such as demand response mechanisms to maintain power system stability
- increasing reliability of the power system by fast detection of faults, isolation of healthy systems, and prevention of power outages
- increase power quality by precise analysis and automated correction of power sources of the system
- wide-area measurement, protection, and control, in broad areas of the network and local grids
- interoperability with advanced relaying systems
- direct transmission of phasors between protection devices

Inclusive of all these components and devices, smart-grid test-bed architecture should be similar to that provided on Chapter 15 in Figure 15.1.

17.3 SMART MICROGRID TEST-BED DESIGN

Smart grid testbed includes hardware and software subsystems. The Hardware consists of Distributed Generation (DGs) and a Central Generator (CG) to implement grid connected operations as a power supply to the microgrid which is basic block of smart grid. The microgrid operates either as standalone or grid connected mode. These operations are performed by power electronic devices serving in transfer switches and as inverter/converter. Storage devices such as batteries, super capacitors and flywheels are used in the system to stabilize the supply by storing surplus power generation. The RERs and storage devices are connected to the system through power electronics interface devices. The hardware subsystem also includes control system for different parts of the system.

Introducing communication technologies between components and between controls allows the system to operate in a smart way by adding smart automation functions. Automation functions are applied from grid side or distribution side depending on the connection of the smart grid to the main system. Distribution automation, when connected from distribution side, works with a goal of real time adjustment to changing loads, generation, and failure conditions of distribution system usually without an operator intervention. For example, it consists of power quality controls, protocols and application programs to check stability and vulnerability of entire system.

The software subsystem of smart grid testbed includes advanced real time simulator such as Real Time Digital simulator (RTDS) and OPAL RT. The software subsystem requires communication subsystem and optimization subsystem for automation functions such as Optimal Power Flow (OPF), loss management, fault analysis and others.

Various experiments can be performed on smart grid testbeds to maintain sustainable operation of the power system. These studies include: evaluation of RERs as distributed sources under different load conditions and contingencies, operation of power system by optimizing generations, locational marginal pricing (LMP) and related topics.

17.4 SMART-GRID TEST BEDS

With increasing proponents for more sustainable power systems, i.e., smart grid, numerous efforts have been embarked on by several institutions, bodies, and organizations to address the overall goals of the smart grid. To ensure that all features of the smart grid are rightly implemented, smart-grid proponents have come up with several test beds to address features and architecture and ensure that smart grids do not fall short when in operation in the real world of power systems.

Smart-grid test beds hope to achieve the following:

(a) To provide an open, unbiased opportunity for testing programs, technologies, business models, and other smart-grid-related activities.

(b) To provide on-grid locations for testing such potentially innovative smart-grid-related technologies and services, including, but not limited to, those funded at the federal or state levels.

(c) To facilitate testing of business models or services that help integrate smart-grid-related technologies into the electric grid, especially those business models that may help promote new products and services for retail customers, subject to appropriate customer protections and limitations required by law.

(d) To offer opportunities to test and showcase smart-grid technologies and services, especially those likely to support economic-development goals of a nation.

(e) To adequately protect the safety and security of the grid and its related operations, both through how applicants and applicant-sponsored technology are selected and how the test bed itself is operated.

17.5 SMART-GRID CASE STUDIES

17.5.1 Ireland

The smart-grid test bed development in Ireland is based on architecture that employs open-source and widely available software and hardware tools. The developers of Ireland's smart-grid test bed aimed to encourage free and open use of the smart-grid test bed for development of a live project. The scope of the test bed has not been feasible to perform a comprehensive analysis of the technology, nor to implement a complete test bed to support all high-level requirements. The project instead is intended as a starting point for a test bed that will be extended further in subsequent work with a more comprehensive solution. Currently, the test bed focuses on the emulation of the TCP/IP stack using open-source tools that communicates with SCADA protocols. As a result of a diverse range of SCADA protocols in use, a Modbus protocol was selected for use due to its simplicity and ease of implementation. The smart-grid test-bed implementation therefore attempted to produce a Modbus with RTU that communicates over a network to produce actual SCADA network traffic, which may be captured for analysis using common networking tools. This SCADA traffic can be used for security analysis and evaluation of power flows.

17.5.2 South Korea

The current test bed for the smart grid in South Korea is well implemented in Design of the Smart Power Grid Management System [19]. The architecture highlighted for the smart-grid test bed comprises:

- **Intelligent distribution automation system (DAS)** that uses a remotely operated monitoring and control system of breakers and switches on the distribution network in real time. The DAS encompasses distribution substations and supports optimal network operation such as fault processing, loss minimization, and load balancing. The DAS comprises the control center, automated switches, RTUs, a communications network, and a modem. For the DAS to be intelligent, additional functions such as monitoring and diagnosis of the equipment, power-quality monitoring, and interconnection of DG to the power grid are added.

- **Substation automation system (SAS)** is the automated system that utilizes substation operation with a management system under one network. Different functions such as measurement, control, monitoring, diagnosis, and protection of the power facilities in the substation are standardized, integrated, and interconnected. It is composed of intelligent electronic devices (IEDs), operating systems including human–machine interfaces (HMIs), gateways, network devices such as Ethernet switches, and optical cables.
- **Intelligent transmission system** is composed of an online monitoring system for transmission facilities, a crisis-management system, with a satellite and reactive management system in the power grid. In the online monitoring system for transmission facilities, different sensors are instilled on the transmission monitoring network and implemented in the HMI system of the main center. This HMI system consists of the terminal devices, HMI servers, and communication devices that connect with each other for data communication. The terminal devices monitor wind direction/speed, insulation, and temperature.

17.5.3 United States

The smart-grid lab at Georgia State University consists of four main components:

- **Intelligent power switch (IPS)**, a critically important component for achieving distributed and scalable structure, needs intelligence to efficiently control the interconnection of components. The purpose of the IPS is to switch power from one port to another. So, any component connected with the IPS should be connected with the others.
- **Energy supplier**—main supply and renewable-energy sources as solar panels and wind turbines.
- **Energy demander**, e.g., appliances and consumers. In this test bed either wall outlet power or renewable-energy sources can be used as a power supply. Energy demanders can be connected to the IPS to provide energy to the rest of power network and are small enough to be used in laboratory experiments to evaluate smart-grid protocols and algorithms.
- **Information network with a power meter**, an important component in this test bed. It can measure how much current is flowing in the test line.

Features of a Smart-Grid Components

- **Architecture of power network**: IPSs can be connected to energy sources including renewable energy and power grid, smart appliance, energy storage, power meter, and more IPSs. Therefore, an IPS is a critically important component for achieving smart-grid structure. Based on this feature, the grid that contains IPS can be configured into different topologies. This IPS can be connected as mesh, tree, or ring, and the configuration can be changed as desired. The distributed power suppliers and consumers are connected to the cloud of the IPS. By connecting to the IPS, a new component can easily be added into the power grid. In this power network, no centralized control is needed, rather it is like a peer-to-peer network. The IPS can be connected to the current power-grid system and can also act like a microgrid [8]. It can group the devices connected to it and can isolate from the main power grid if any disturbance is detected.
- **Architecture of information network**: In the smart-grid test bed, two-way communication allows information exchange. A variety of communication media could be used in achieving this in the test bed, including copper wiring, optical fiber, power-line carriers, and wireless. A wireless network can be configured as a wireless mesh network for a test-bed prototype to emulate the network [9]. Although Wi-Fi or wired media can also be used for communication, a specific wireless network is more advantageous for low power and flexibility for a smart-grid test bed. This can be configured to any topology. This information network can be configured into a centralized network or a distributed network. For the centralized configuration, a power meter can send data to an energy management center (EMC), which can compute the whole power-grid status and send out control signals. In distributed configuration, each microcontroller on an IPS will compute its own status based on the information it receives from other IPSs and power meters. Each IPS and power meter can communicate with each other and exchange their status.
- **IPS design**: IPS is a critically important component for achieving distributed and scalable structure, and it needs intelligence to efficiently control the interconnection of components. The purpose of the IPS is to switch power from one port to another. So, any component connected with the IPS should relate to the others. At the same time, multiple pairs of ports can be connected. That means if there are two supplies, A and B, and two consumers C and D, A can provide power to C and B can provide power to D in a parallel manner.
- **Power meter**: This is another important component in any smart-grid test bed. It could be used to measure how much current is flowing in the test line. A plug node is used as a current transformer for the current sensor.
- **Energy supply and energy demand**: In any good proposed smart-grid test bed, either wall outlet power or a renewable-energy source can be used as a

power supply. This can be connected to the IPS to provide energy to the rest of the power network. This power supply is small enough to be used in laboratory experiments to evaluate smart-grid protocols and algorithms. As energy demanders (i.e., consumers), lamps can be useful as well as computers and other appliances.

17.6 SIMULATION TOOLS, HARDWARE, AND EMBEDDED SYSTEMS

Design of embedded systems for the smart grid containing both hardware and software requires solving several unique and difficult problems. One of the interesting problems is that of debugging the embedded software in conjunction with the hardware.

The traditional co-design process, where the software is debugged after hardware is fabricated, produces large design delays due to late discovery of errors in the hardware and in the interface between hardware and software. Integration on a chip could make this problem worse because currently used tools cannot be used and signals on a chip cannot be easily observed. There is an obvious need for a change in the co-design methodology whereby software and hardware can be jointly debugged earlier in the design cycle. However, this change in methodology can only happen when appropriate design tools are available.

There are two approaches to debugging hardware and software for better simulation in smart grids [9] without building the actual hardware.

The first one is based on emulation of hardware using, for example, field-programmable gate arrays (FPGA) and using a separate board for the processor and memory. A designer can generate a prototype relatively quickly and debug the smart-grid software and interfaces on the prototype.

After these bugs are detected, the entire system can be recompiled within a relatively short time. In most cases, hardware emulators run only an order of magnitude slower than the actual system, allowing the designer to test the system with a large number of test cases. However, due to its high cost, this technique is economically feasible only in certain cases. It also cannot be used to model timing constraints accurately and in many cases designs must be modified to suit emulation. Recompilation of hardware takes more time than compiling software or a hardware description language (HDL) model for simulation. Finally, it is not always possible to observe the internal state of the circuit, both in the HDL and processor, making debugging complicated and slow.

A complementary approach is to build software models for all the components of the system and use simulation to analyze behavior. There are many advantages to this approach. First, software can be combined with behavioral-level hardware descriptions to detect bugs as early as possible in the design phase. Hardware, software, and interface routines can be designed and debugged in parallel. Second, timing constraints can be accurately modeled. Third, recompilation of either hardware or software is quick. Detailed debugging, where internal states of all components can be accessed and altered at all time points, can be easily supported. Finally, this approach is not as expensive as emulation.

Simulators have been mostly used for the design of hardware and there are few tools for co-simulation. We describe a hardware-software co-simulator that can be used in the design, debugging, and verification of embedded systems. This tool contains a simulation backplane that can be used to integrate processor, hardware (HDL), and dedicated simulators for peripherals, forming a co-simulator capable of efficiently simulating hardware, software, and their interaction.

Each simulator implements debugging functions like setting break points, examining and altering internal states, and single stepping. To feed stimulus to the system and to observe its response, a set of virtual instruments have been created. The co-simulator and the virtual instruments can be used to create a virtual laboratory that will provide users with a platform for rapid virtual prototyping. Performance metrics (like clock cycles needed to execute software) can be easily evaluated, allowing the user to explore different algorithms, hardware and software implementations, and hardware-software trade-offs. The main drawback of simulation is its speed. In many cases, simulation runs orders of magnitude slower than the actual system. Simulation time depends on the (timing) accuracy of the models, with time increasing with increased accuracy. Therefore, reducing simulation time without sacrificing timing accuracy becomes a very important problem. Our main contribution is a set of techniques to speed up simulation of processors and peripherals without significant loss in timing accuracy. Processor simulation speed is improved by accurately (in terms of timing) simulating only those cycles where there is interaction with peripherals and by caching results of instruction decoding. Suppression of periodic signals and other techniques speed up simulation of peripherals. Simulation overhead is kept low by managing time more efficiently.

This tool is expected to be used at any point after the initial architecture is determined. Software designers may use behavioral hardware models for initial debugging, evaluation, and exploration of algorithms and implementations. System architects may use the tool to determine hardware-software trade-offs to be implemented in smart grids. Hardware designers can use prototype software to evaluate, test, and debug hardware. When hardware and software are ready, designers can work on testing and debugging the entire smart-grid system.

17.6.1 Simulation of Smart-Grid Hardware

Custom smart-grid prototype hardware represented using software is simulated using a commercial simulator. Since a commercial simulator is designed to be a stand-alone tool and does not implement the interface functions required by the smart grid, its integration poses certain problems. The interface functions can only be implemented using different programming languages for the simulator. This approach may only allow user-defined functions during hardware simulations such as with Real-Time Digital Simulator (RTDS) and Opal-RT. It also allows these functions to call certain functions for simulation control in RTDSs. In hardware implementation, the user is required to call some functions using custom circuit descriptions. There are some requirements on the manner of input, output, and bidirectional lines.

GridLink and GridIQ are under-development hardware for the smart grid that can be represented at any level of the power system—generation, transmission, or distribution. These are currently the only accurate timing hardware simulators for the smart grid.

Simulators like GridSim are only phase accurate and their interface functions may introduce errors during rounding of event times if proper care is not exerted in describing the hardware.

To properly verify the features of the smart grid, hardware-software co-simulators for embedded system design and debugging are necessary. This would provide a natural environment for joint debugging of software and hardware. They will also be useful for evaluating system performance, selection of algorithms, implementations, and exploring hardware-software for smart-grid architectures.

The improved speed of the co-simulator comes from various sources. First, the co-simulator is targeted toward phase-accurate simulation. Switching between transmission and distribution during simulation, caching of decoded instructions, and not simulating instruction fetch cycles all contribute to the increased speed of simulation processes. Speed is increased through the use of:

- dedicated simulators
- suppression of periodic signals and associated events
- specific short-cuts that reduce the time required for simulation of whole architectures

Making time discrete and using a statically allocated timing wheel helps keep coordination overhead low. We have demonstrated the use of the tool in three design examples and have shown that the simulation speed is adequate.

The expected result from a co-simulator for the smart grid is first to cut down cost, which may be wasted for a smart grid that lacks basic features. Hence, co-simulators when properly implemented is aimed at promoting:

- data and resource sharing
- allows to study different features of the smart grid (e.g., power and communication)
- identification of errors and mismatch between different tools
- improving efficiency due to data exchange between tools

17.6.2 Smart Grid Simulators in the Marketplace

Real Time Digital Simulator (RTDS)

RTDS is a real time simulator used for studying power system operation and perform testing control and protection equipment. It is a super computer constructed from workstation interface cards, input/output cards, processors, synchronization cards and others. it comes with a software package called RSCAD on which most modeling and sizing of the system is performed. In RSCAD, various libraries of power system, control and interface libraries are included for modeling and setting up requirements. The compiled model is then run on main simulator (RTDS) interfaced through runtime window of the software.

Component models run on the simulator will be used to study automation functions such as frequency load control and V/Var optimization. Hence, the simulator is used as a good add on for the design of future deployment of microgrid operation in grid connected and standalone mode.

OPAL RT Simulator

The OPAL RT simulator is also another family of real time simulator. It consists of analog and digital I/O signal modules and power electronic circuits. The Simulator works on computational software called eMEGAsim which is fully integrated with MATLAB. MathWorks enable the use of a wide variety of software toolsets including Control System Design and Analysis toolboxes, Code Generation toolboxes, and Physical Modeling toolboxes. Power System models are built on MATAB platform before being loaded to the simulator. It is also use as hard-ware in the loop (HIL) to solve various control option for performance analysis of micro grids and smart grids.

17.7 LIMITATIONS OF EXISTING SMART-GRID TEST BEDS

In any smart-grid test bed, the main components are IPS (for power routing), different sources of power supply, energy demanders, and power meters. IPS enables the power grid to be configured as any kind of topology for power distribution, and power meters sense the energy on each line and report the data. The information network is co-designed with the power network. Therefore, this approach can significantly help researchers analyze and compare various algorithms and protocols. Several experiments are performed to show the applicability of this test bed for the smart-grid research community.

However, currently the IPS has to be plugged into another power supply [91]. Even though this could be replaced with a rechargeable battery, the power meter cannot measure active and reactive energy consumption. This limits the scope of application of the smart-grid test bed until further research approaches are proposed to add this feature.

The test-bed designs and implementations only partially fulfill the requirements outlined in the IEC and NIST guidelines for general test-bed design. A major concern is that there are no single test-bed designs that address both the traditional SCADA system as well as the modern smart-grid approach to electrical infrastructure management.

17.8 CHAPTER SUMMARY

In this chapter, we presented discussions on test-bed design and motivation for it, test-bed components, architecture, and functions. Case studies of test beds around the world were presented for the benefit of the student. Though it may be accepted that smart-grid infrastructure will eventually replace traditional SCADA technology, it would be unreasonable to assume that this would occur at a rate such that traditional SCADA systems would be rendered obsolete overnight. As such, it is important to test smart-grid designs and implementation to accommodate different architectures and allow for a transition from the traditional SCADA approach to the modern smart-grid approach of management. Work being done in the research community in the design of the micro/smart grid was cited. This chapter provided technology, concepts, and schematics for future development of smart-grid systems using the knowledge and concepts from an energy-processing scheme.

BIBLIOGRAPHY

[1] J. A. Momoh, *Smart Grid-Fundamentals of Design and Analysis*, Wiley-IEEE Press, 2012.

[2] US Department of Energy, *User Guide for the US Department of Energy Smart Grid Computational Tool (SGCT)*, version 2.0, August 2011, www.smartgrid.gov/files/US_DOE_Smart_Grid_Computational_Tool_User_Guide_Version_2.0.pdf.

[3] W.-Z. Song, D. De, and G. Lu, "A Smart Grid Testbed: Design and Validation," *in Smart Grid Lab*, World Scientific, Singapore, 2011.

[4] A. Narayan, "GridSpice: A Virtual Test Bed for Smart Grid," https://web.standford.edu/group/tomcat/cgi-bin/doc/amit-narayan.

[5] M. Ilie, *Dynamic Monitoring and Decision Systems (DYMONDS) and Smart Grids: One and the Same*, EESG Working Paper R-WP-21, 2009.

[6] J. Ilić, *Interactive Power Systems Simulator User's Manual*, December 16, 2007, www.ece.cmu.edu/~nsf-education/UsersManual.pdf.

[7] SMB Smart Grid Strategic Group, "IEC Smart Grid Standardization Roadmap," June 2010, www.iec.ch/smartgrid/downloads/sg3_roadmap.pdf.

[8] M. Pipattanasomporn, H. Feroze, and S. Rahman, "Multi-Agent Systems in a Distributed Smart Grid: Design and Implementation," IEEE/PES Power Systems Conference and Exposition, March 2009.

[9] S. Karnouskos and T. N. de Holanda, "Simulation of a Smart Grid City with Software Agents," Third UKSim European Symposium on Computer Modeling and Simulation, November 2009.

[10] H. Farhangi, "The Path of the Smart Grid," *IEEE Power Energy M.* 8, 1 (January–February 2010).

[11] M. Hommelberg, C. Warmer, I. Kamphuis, et al., "Distributed Control Concepts Using Multi-Agent Technology and Automatic Markets: An Indispensable Feature of Smart Power Grids," IEEE Power Engineering Society General Meeting, June 2007.

[12] K. I. Geisler, "A Smarter Greener Power Grid," IEEE Power Systems Conference and Exposition, March 2009.

[13] C. Potter, A. Archambault, and K. Westrick, "Building a Smarter Smart Grid through Better Renewable Energy Information," IEEE Power Systems Conference and Exposition, March 2009.

[14] US Department of Energy, "Grid Modernization Initiative," https://energy.gov/under-secretary-science-and-energy/grid-modernization-initiative.

[15] M. Albadi and E. El-Saadany, "Demand Response in Electricity Markets: An Overview," IEEE Power Engineering Society General Meeting, June 2007.

[16] D. Hart, "Using AMI to Realize the Smart Grid," IEEE Power and Energy Society General Meeting, July 2008.
[17] S. Collier, "Ten Steps to a Smarter Grid," IEEE Rural Electric Power Conference, April 2009.
[18] G. Lu, D. De, and W.-Z. Song, "SmartGridLab: A Laboratory-Based Smart Grid Testbed," First IEEE International Conference on Smart Grid Communications, 2010.
[19] J. Min, "Jeju, Test-Bed for the World's Smart Grid Industry," February 2011, www.jujeweekly.com/news/articleview.html?idxno=1222

INDEX

Energy Processing and Smart Grid, First Edition. James A. Momoh

IEEE Press Series on Power Engineering

Series Editor: M. E. El-Hawary, Dalhousie University, Halifax, Nova Scotia, Canada

The mission of IEEE Press Series on Power Engineering is to publish leading-edge books that cover the broad spectrum of current and forward-looking technologies in this fast-moving area. The series attracts highly acclaimed authors from industry/academia to provide accessible coverage of current and emerging topics in power engineering and allied fields. Our target audience includes the power engineering professional who is interested in enhancing their knowledge and perspective in their areas of interest.

1. *Electric Power Systems: Design and Analysis, Revised Printing*
Mohamed E. El-Hawary

2. *Power System Stability*
Edward W. Kimbark

3. *Analysis of Faulted Power Systems*
Paul M. Anderson

4. *Inspection of Large Synchronous Machines: Checklists, Failure Identification, and Troubleshooting*
Isidor Kerszenbaum

5. *Electric Power Applications of Fuzzy Systems*
Mohamed E. El-Hawary

6. *Power System Protection*
Paul M. Anderson

7. *Subsynchronous Resonance in Power Systems*
Paul M. Anderson, B.L. Agrawal, and J.E. Van Ness

8. *Understanding Power Quality Problems: Voltage Sags and Interruptions*
Math H. Bollen

9. *Analysis of Electric Machinery*
Paul C. Krause, Oleg Wasynczuk, and S.D. Sudhoff

10. *Power System Control and Stability, Revised Printing*
Paul M. Anderson and A.A. Fouad

11. *Principles of Electric Machines with Power Electronic Applications, Second Edition*
Mohamed E. El-Hawary

12. *Pulse Width Modulation for Power Converters: Principles and Practice*
D. Grahame Holmes and Thomas A. Lipo

13. *Analysis of Electric Machinery and Drive Systems, Second Edition*
Paul C. Krause, Oleg Wasynczuk, and Scott D. Sudhoff

14. *Risk Assessment for Power Systems: Models, Methods, and Applications*
Wenyuan Li

15. *Optimization Principles: Practical Applications to the Operation of Markets of the Electric Power Industry*
Narayan S. Rau

16. *Electric Economics: Regulation and Deregulation*
Geoffrey Rothwell and Tomas Gomez

17. *Electric Power Systems: Analysis and Control*
Fabio Saccomanno

18. *Electrical Insulation for Rotating Machines: Design, Evaluation, Aging, Testing, and Repair, Second Edition*
Greg C. Stone, Ian Culbert, Edward A. Boulter, and Hussein Dhirani

19. *Signal Processing of Power Quality Disturbances*
Math H. J. Bollen and Irene Y. H. Gu

20. *Instantaneous Power Theory and Applications to Power Conditioning*
Hirofumi Akagi, Edson H. Watanabe, and Mauricio Aredes

21. *Maintaining Mission Critical Systems in a 24/7 Environment*
Peter M. Curtis

22. *Elements of Tidal-Electric Engineering*
Robert H. Clark

23. *Handbook of Large Turbo-Generator Operation and Maintenance, Second Edition*
Geoff Klempner and Isidor Kerszenbaum

24. *Introduction to Electrical Power Systems*
Mohamed E. El-Hawary

25. *Modeling and Control of Fuel Cells: Distributed Generation Applications*
M. Hashem Nehrir and Caisheng Wang

26. *Power Distribution System Reliability: Practical Methods and Applications*
Ali A. Chowdhury and Don O. Koval

27. *Introduction to FACTS Controllers: Theory, Modeling, and Applications*
Kalyan K. Sen and Mey Ling Sen

28. *Economic Market Design and Planning for Electric Power Systems*
James Momoh and Lamine Mili

29. *Operation and Control of Electric Energy Processing Systems*
James Momoh and Lamine Mili

30. *Restructured Electric Power Systems: Analysis of Electricity Markets with Equilibrium Models*
Xiao-Ping Zhang

31. *An Introduction to Wavelet Modulated Inverters*
S. A. Saleh and M. Azizur Rahman

32. *Control of Electric Machine Drive Systems*
Seung-Ki Sul

33. *Probabilistic Transmission System Planning*
Wenyuan Li

34. *Electricity Power Generation: The Changing Dimensions*
Digambar M. Tagare

35. *Electric Distribution Systems*
Abdelhay A. Sallam and Om P. Malik

36. *Practical Lighting Design with LEDs*
Ron Lenk and Carol Lenk

37. *High Voltage and Electrical Insulation Engineering*
Ravindra Arora and Wolfgang Mosch

38. *Maintaining Mission Critical Systems in a 24/7 Environment, Second Edition*
Peter M. Curtis

39. *Power Conversion and Control of Wind Energy Systems*
Bin Wu, Yongqiang Lang, Navid Zargari, and Samir Kouro

40. *Integration of Distributed Generation in the Power System*
Math H. J. Bollen and Fainan Hassan

41. *Doubly Fed Induction Machine: Modeling and Control for Wind Energy Generation*
Gonzalo Abad, Jesús López, Miguel Rodrigues, Luis Marroyo, and Grzegorz Iwanski

42. *High Voltage Protection for Telecommunications*
Steven W. Blume

43. *Smart Grid: Fundamentals of Design and Analysis*
James Momoh

44. *Electromechanical Motion Devices, Second Edition*
Paul Krause, Oleg Wasynczuk, and Steven Pekarek

45. *Electrical Energy Conversion and Transport: An Interactive Computer-Based Approach, Second Edition*
George G. Karady and Keith E. Holbert

46. *ARC Flash Hazard and Analysis and Mitigation*
J. C. Das

47. *Handbook of Electrical Power System Dynamics: Modeling, Stability, and Control*
Mircea Eremia and Mohammad Shahidehpour

48. *Analysis of Electric Machinery and Drive Systems, Third Edition*
Paul Krause, Oleg Wasynczuk, Scott Sudhoff, and Steven Pekarek

49. *Extruded Cables for High-Voltage Direct-Current Transmission: Advances in Research and Development*
Giovanni Mazzanti and Massimo Marzinotto

50. *Power Magnetic Devices: A Multi-Objective Design Approach*
S. D. Sudhoff

51. *Risk Assessment of Power Systems: Models, Methods, and Applications, Second Edition*
Wenyuan Li

52. *Practical Power System Operation*
Ebrahim Vaahedi

53. *The Selection Process of Biomass Materials for the Production of Bio-Fuels and Co-Firing*
Najib Altawell

54. *Electrical Insulation for Rotating Machines: Design, Evaluation, Aging, Testing, and Repair, Second Edition*
Greg C. Stone, Ian Culbert, Edward A. Boulter, and Hussein Dhirani

55. *Principles of Electrical Safety*
Peter E. Sutherland

56. *Advanced Power Electronics Converters: PWM Converters Processing AC Voltages*
Euzeli Cipriano dos Santos Jr. and Edison Roberto Cabral da Silva

57. *Optimization of Power System Operation, Second Edition*
Jizhong Zhu

58. *Power System Harmonics and Passive Filter Designs*
J. C. Das

59. *Digital Control of High-Frequency Switched-Mode Power Converters*
Luca Corradini, Dragan Maksimovic, Paolo Mattavelli, and Regan Zane

60. *Industrial Power Distribution, Second Edition*
Ralph E. Fehr, III

61. *HVDC Grids: For Offshore and Supergrid of the Future*
Dirk Van Hertem, Oriol Gomis-Bellmunt, and Jun Liang

62. *Advanced Solutions in Power Systems: HVDC, FACTS, and Artificial Intelligence*
Mircea Eremia, Chen-Ching Liu, and Abdel-Aty Edris

63. *Operation and Maintenance of Large Turbo-Generators*
Geoff Klempner and Isidor Kerszenbaum

64. *Electrical Energy Conversion and Transport: An Interactive Computer-Based Approach*
George G. Karady and Keith E. Holbert

65. *Modeling and High-Performance Control of Electric Machines*
John Chiasson

66. *Rating of Electric Power Cables in Unfavorable Thermal Environment*
George J. Anders

67. *Electric Power System Basics for the Nonelectrical Professional*
Steven W. Blume

68. *Modern Heuristic Optimization Techniques: Theory and Applications to Power Systems*
Kwang Y. Lee and Mohamed A. El-Sharkawi

69. *Real-Time Stability Assessment in Modern Power System Control Centers*
Savu C. Savulescu

70. *Optimization of Power System Operation*
Jizhong Zhu

71. *Insulators for Icing and Polluted Environments*
Masoud Farzaneh and William A. Chisholm

72. *PID and Predictive Control of Electric Devices and Power Converters Using MATLAB®/Simulink®*
Liuping Wang, Shan Chai, Dae Yoo, Lu Gan, and Ki Ng

73. *Power Grid Operation in a Market Environment: Economic Efficiency and Risk Mitigation*
Hong Chen

74. *Electric Power System Basics for the Nonelectrical Professional, Second Edition*
Steven W. Blume

75. *Energy Production Systems Engineering*
Thomas Howard Blair

76. *Model Predictive Control of Wind Energy Conversion Systems*
Venkata Yaramasu and Bin Wu

77. *Understanding Symmetrical Components for Power System Modeling*
J. C. Das

78. *High-Power Converters and AC Drives, Second Edition*
Bin Wu and Mehdi Narimani

79. *Current Signature Analysis for Condition Monitoring of Cage Induction Motors: Industrial Application and Case Histories*
William T. Thomson and Ian Culburt

80. *Introduction to Electric Power and Drive Systems*
Paul Krause, Oleg Wasynczuk, Timothy O'Connell, and Maher Hasan

81. *Instantaneous Power Theory and Applications to Power Conditioning, Second Edition*
Hirofumi Akagi, Edson Hirokazu Watanabe, and Mauricio Aredes

82. *Practical Lighting Design with LEDs, Second Edition*
Ron Lenk and Carol Lenk

83. *Introduction to AC Machine Design*
Thomas A. Lipo

84. *Advances in Electric Power and Energy Systems: Load and Price Forecasting*
Mohamed E. El-Hawary

85. *Electricity Markets: Theories and Applications*
Jeremy Lin and Fernando H. Magnago

86. *Multiphysics Simulation by Design for Electrical Machines, Power Electronics, and Drives*
Marius Rosu, Ping Zhou, Dingsheng Lin, Dan Ionel, Mircea Popescu, Frede Blaabjerg, Vandana Rallabandi, and David Staton

87. *Modular Multilevel Converters: Analysis, Control, and Applications*
Sixing Du, Apparao Dekka, Bin Wu, and Navid Zargari

88. *Electrical Railway Transportation Systems*
Morris Brenna, Federica Foiadelli, and Dario Zaninelli

89. *Energy Processing and Smart Grid*
James A. Momoh

Printed and bound by CPI Group (UK) Ltd, Croydon, CR0 4YY
17/09/2026
14964317-0007